Forestry: Planning and Managing Forests

Forestry: Planning and Managing Forests

Editor: Roland Winters

STATES
ACADEMIC PRESS
www.statesacademicpress.com

States Academic Press,
109 South 5th Street,
Brooklyn, NY 11249, USA

Visit us on the World Wide Web at:
www.statesacademicpress.com

ISBN: 978-1-63989-212-9 (Hardback)

Cataloging-in-Publication Data

Forestry : planning and managing forests / edited by Roland Winters.
p. cm.
Includes bibliographical references and index.
ISBN 978-1-63989-212-9
1. Forests and forestry. 2. Forests and forestry--Planning. 3. Forest management. I. Winters, Roland.
SD131 .F67 2022
634.9--dc23

Table of Contents

Preface

The science of creating, using, conserving, repairing and managing forests, and associated resources for environmental and human benefits is known as forestry. It is majorly practiced in natural stands and plantations. Forestry is concerned with the management of forest ecosystems and genetic improvement of various tree species. It also deals with the challenges of protecting, thinning, extracting, controlled burning and processing of timber. The field uses elements from biological, social, physical, political and managerial sciences. Some of the primary concerns of forestry are wildlife habitat protection, landscape and community protection, erosion control, watershed management, biodiversity management, etc. This book provides significant information about this discipline to help develop a better understanding of forestry. It includes some of the vital pieces of work being conducted across the world, on various topics related to this field. Students, researchers, experts and all associated with forestry will benefit alike from this book.

The information shared in this book is based on empirical researches made by veterans in this field of study. The elaborative information provided in this book will help the readers further their scope of knowledge leading to advancements in this field.

Finally, I would like to thank my fellow researchers who gave constructive feedback and my family members who supported me at every step of my research.

Editor

1

Bound by debt: Nutmeg trees and changing relations between farmers and agents in a Moluccan agroforestry systems

Messalina Lovenia Salampessy[1,*], **Indra Gumay Febryano**[2], **Dini Zulfiani**[3]

[1] Nusa Bangsa University Bogor. Email: meis_forester@yahoo.com
[2] University of Lampung. Email: yanogumay@gmail.com
[3] University of Mulawarman. Email: dini.qt19@gmail.com
* Correspondence author

Abstract: Agroforestry is a land management system long practiced by communities in the Moluccas. The practice is commonly known as "Dusung", where one cash crop in particular, nutmeg, is interspersed throughout farmer groves. Farmers have faced a number of challenges in recent years, especially concerning a system of debt bondage inflicting undue losses upon them. This study aims to explain the involvement of farmers within the debt bondage system, otherwise known as the *tree mortgage system*. We utilize a case study methodology, whereby data were collected through interviews and participant observation and results analyzed using principal agent theory. Findings highlight that nutmeg farmers, what we call the Principal, incur high risks when debt bondage is applied by an Agent that has the increasing ability to deny and change the terms of an agreement. This occurs when Agents exploit information unavailable to farmers about nutmeg marketing prospects, which weakens farmer negotiating positions. Improving institutional support for contracts in the tree mortgage system could help to ensure a more equitable arrangement, improving the terms for nutmeg farmers, meanwhile also encouraging the continued preservation of the dusung system, which has long helped to manage forest resources sustainably.

Keywords: Agroforestry; nutmeg dusung; tree mortgage system; principal agent; debt bondage

---

1. Introduction

Agroforestry has long been practiced for generations by people in the Moluccas Islands, a system known regionally as "Dusung". This land use patterns combines crop farming (agriculture) and forest trees (forestry) in the same area and is managed over time. One of the most important perennial plant species introduced by communities in the Dusung system is nutmeg (Myristica sp). According to (Silaya.Th, Tjoa M, Lelloltery H, Siahaya L, Bone I, 2012) nutmeg is native to the Moluccas Islands and has long played an important role in the global economy since the Dutch colonial era. Even today the world demand for nutmeg remains very high.

The Moluccas Islands, as a regional hub of nutmeg production with the highest quality, has an important role in the global economy of nutmeg. Unfortunately, most nutmeg farmers in the Moluccas Islands remain below the poverty line. According to data from the (Dinas Pertanian Prov Maluku, 2012) community plantation on nutmeg commodity in Moluccas province has a total area of 22,271 ha with a total production of 2,347 tons and productivity of 337.56 kg / ha. The number of nutmeg farmers also increased from 17,173 people in 2008 to 20,199 people in 2010.

National and regional efforts to increase community forestry schemes has presented a fresh new focus for potential government support. Such changes present opportunities for new mechanisms to support access provision on community-based sustainable forest management to increase participation of local communities in forest management, as well as the potential for increasing community incomes in the forestry sector. This juncture of policy developments therefore require a closer look at the infrastructure and mechanisms for various market arrangements and the ways that forest commodities travel across the commodity chain.

unfortunately, the formal banking institutions are still unwilling to fund small-scale community ventures, especially for businesses with small-scale plantations like the Nutmeg Dusung. This paper explores ways in which market exchanges affect rural agroforestry farmers, and thus affect various landscapes.

As a form of land management considered community forestry, nutmeg farmers have developed the Dusung system with little formal education and knowledge of markets. The weak financial management services and marketing systems for nutmeg in the Moluccas is one reason farmers have become involved in a system of debt bondage known locally as the (ijon) system. Research about the involvement of farmers in the ijon system is not well documented and has important development implications in an era that prioritizes social forestry policies. For that reason, this study aims to explain nutmeg farmer involvement in a form of bonded labor, which can be considered a "tree mortgage system." Collecting information and building knowledge around the reasons and processes for farmer involvement in the Dusung nutmeg tree mortgage system can support in development efforts to empower communities and provide more equitable terms to those living around the forest.

2. Materials and Methods

This research was conducted in the village of Hutumuri, Salahutu District, Moluccas province. Hutumuri village was selected as the research site because it is a village that has a high potential to produce Nutmeg within the Dusung system. It is also the largest nutmeg producer in the District of South Leitimur, Ambon.

This research uses a case study method from the Hutumuri experience, where data collection was conducted through in-depth interviews and participant observation. Interviews involved 15 owners of Dusung nutmeg systems. Data analysis was performed using the Principal Agent Relationship Theory to help explain relationships in which one or more persons give their trust (Principal) to influence others as partners to receive trust (of the Agent) to perform some tasks the Principal requires by delegating decision-making authority to the partner in question (Agent). This relationship implies either: (1) problems related to the transfer of ownership rights (transfer of rights), (2) the problem of a mismatch of information (asymmetric information) in a relationship of principal and agent, and (3) the emergence of transaction costs (information costs, coordination costs and strategy cost). I chose this theory because it helps to illustrate the relationship of interests involved in the lending process of capital that exist between the owner of Dusung and those that apply the Ijon. This theory helps explain the rules of the game and the risks that both parties experience through this process. Principal agent theory herein especially highlights the unfair terms experienced by the Principal (Dusung).

3. Results and Discussion

3.1. Transfer of Rights

Communities usually harvest nutmeg three times a year: March-April; July-August; December-January. Nutmeg harvests have also developed a tree mortgage system, which was established in the region a long time ago. The tree mortgage system is one form of an agreement established between nutmeg farmers as principal (P) and debt bondage as agent (A). The system shows the temporary transfer of rights between P and A. The agreement consists of a certain number of nutmeg trees used as collateral for borrowing money by farmers and in turn bonds labor to that venture. The system works due to the economic disadvantages and households needs among farmers (P), whereby during certain circumstances, farmers need capital to finance various family obligations, such as school or college fees for children, home improvements, and ceremonial events such as weddings.

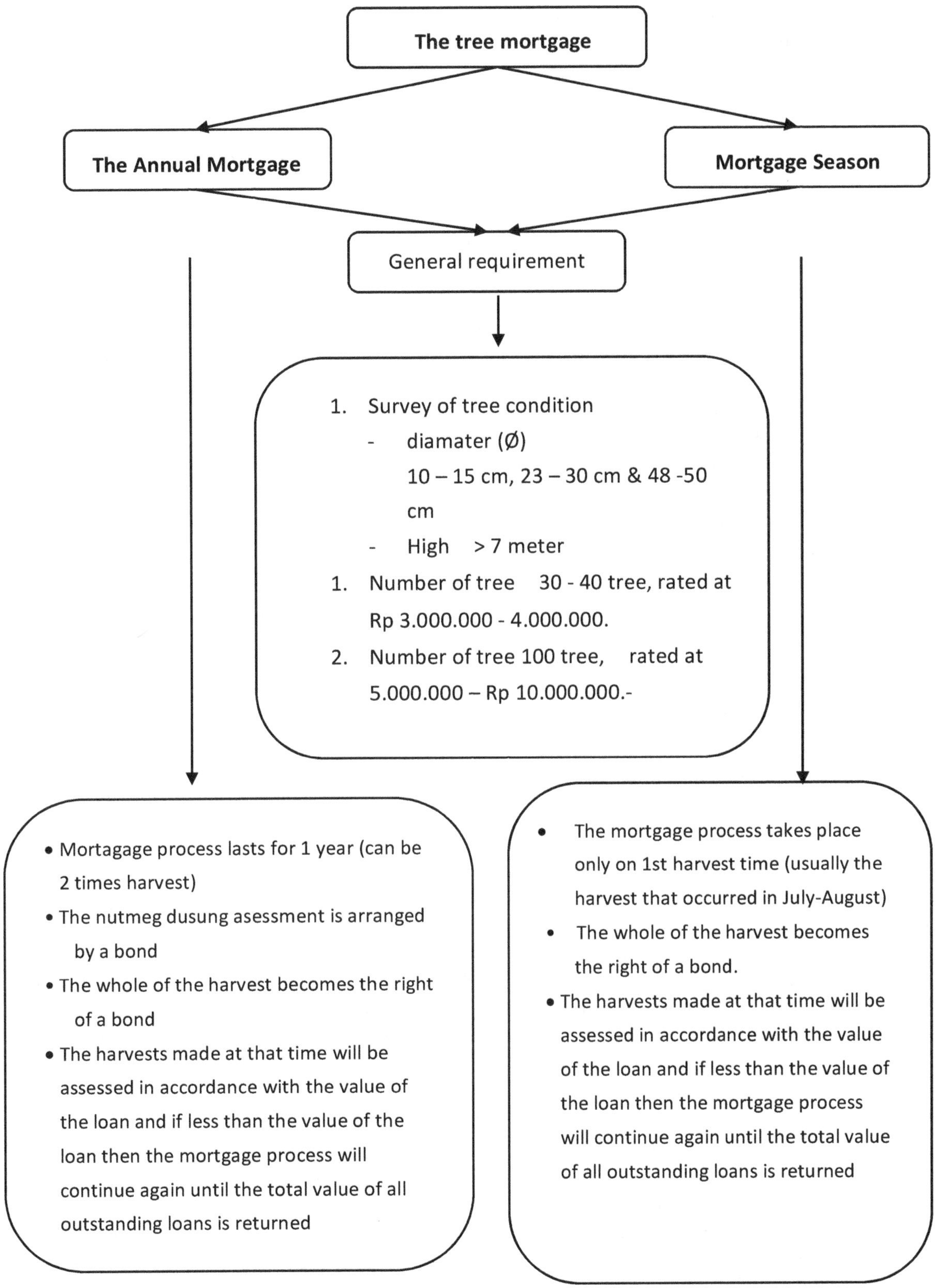

Figure 1. Tree mortgage system diagram

The tree mortgage system is described as follows:

1) The two sides, both farmers and ijon know each other well and sometimes have family ties so it requires no physical evidence (letter loan) which binds both parties. Mutual trust between the two sides forms a major asset in this pledge system.
2) There are 2 other forms of the tree mortgage system, namely:
   a. The annual mortgage, which is an agreement where the trees are mortgaged as collateral for ≥ 1 year according to the agreement that was built. The annual pledge requirements as follows:
      - Ijon will survey the physical condition of trees (diameter and height) as well as the number of trees to forecast its value as collateral for the loan .
      - During the mortgage time, the process of maintenance and harvesting is fully arranged by ijon. Maintenance tasks can be performed together or not, depending on the agreement. Sometimes ijon hires special labor to nurture the trees or work alongside the tree owner.
      - During harvest, when yields exceed the value of the loan, then the ijon receive windfall advantages, and when yields are less than the value of the loan ijon will incur a loss.
      - Farmers entitled to participate in the harvesting process fully pay the ijon during the harvest period.
   b. Mortgage season, is a tree mortgage agreement applicable at the time of great harvest . Terms of the mortgage seasons are as follows:
      - In general, the main requirement is not much different from an annual mortgage. The difference is that a tree used as collateral mortgage is only valid during the great harvest. For Nutmeg Dusung, large harvests with plentiful good quality fruits occur during the rainy season (July-August).
      - During the great harvest season, all yields become fully ijon rights and if the result is less than the value of loans that have been received by the farmer, the mortgaged trees remain valid for the next harvest in order to cover the value of the loan. However, when the result exceeds the value of the loan, the deal is only valid at that time and the advantage is for the ijon.
      - Farmers entitled to participate in the process of harvesting will be fully paid by the debt to ijon during the harvest period.

3.2. *Asymmetric Information*

Relationships that develop in the tree mortgage system between nutmeg farmers as P and ijon as A has led to asymmetric information. In other words, a mismatch in available information takes place because A exploits their advantages about nutmeg marketing prospects. This can happen because the marketing and determination of nutmeg selling price is fully controlled by A. P trusts, or rather has no option to trust, the nutmeg market distribution process conducted by A. Almost all information on the management and marketing of the Dusung is controlled by A. This also creates a moral hazard by A because although they strongly control the outcomes, they are also able to neglect maintenance and security of Dusung. Therefore, Dusung can seem poorly maintained during the contract period. If A enforces these coercive terms, the final harvest also suffers, and in turn P is likely to receive unfair shares from selling prices of the prevailing harvest. In accordance with (Petrie, 2002), agency theory has focused on issues of asymmetric information, where the agent has more information about the actual performance, motivation, and goal, which potentially creates moral hazards and adverse selection. The principal themselves should have opportunities to better determne the costs of monitoring the performance of A and establishing incentives structure and efficient monitoring.

One of the rules in the tree mortgage system agreed upon between P and A is the maintenance process, which until the time of harvest is organized completely by A. However, this form of rights transfer may put farmers in a weak position, especially in the event that trees are attacked by pests or incur diseases. In such a scenario, the responsibility for treatment falls on P, while A does not have to bear the longer term consequences in this matter. This further indicates that P incurs higher risks compared to A, which could simply ignore the investment, break promises, or find other ways to get out of the agreement. In the tree mortgage system there is rarely any contract letter signed between P and A. Rather, it is based on trust and a sense of dependency between the two parties. This can cause opportunistic attitudes among one of the parties, who can change the terms and violate an agreement that has already been negotiated.

3.3. *Transaction Cost*

Transaction costs are incurred when nutmeg farmers as P put their trust in ijon as A, where P and A held the exchange of their rights and mutual desires to enforce the exclusive rights. The Relationship between P and A would be more efficient if there were also a greater balance in the level of reward, as well as a more honest discussion about the ways to minimize transaction costs related to the tree mortgage system.

In this tree mortgage system, P takes on a much larger percentage of the risk than A. The risk is an accumulation of the work performance conducted by A and the supervision is carried out by P. When A commits unfair practices toward their agreement, which is then followed by the lack of informed supervision by P, then P takes on risks that are difficult to foresee. Therefore, P cannot fulfill production results as may have been anticipated. Therefore, the working performance of A and supervision of P greatly affects the final result in the management of Dusung.

The emergence of the transaction costs for any coordination process cannot be separated from a marketing system that is fully controlled by A. When a tree mortgage system is put in place, generally every month P will meet to coordinate with A. Coordination is conducted to discuss Dusung conditions and allows the opportunity to determine appropriate harvest time. This implies P incurring coordination costs every month.

Information costs will be incurred by both parties, P and A. However, P has more control upon the information, especially about the marketing products of Nutmeg Dusung. On the other hand, A does not have the ability to access this information. The presence of a local cooperative would greatly improve the situation for P because they could better negotiate the terms and give greater weight to P's role in the transaction. This would also result in better Dusung management and likely imorive harvests. The absence of such institutions however, results in power and resources better controlled by A. Therefore A is able to expend strategic costs to secure maximize short term gains in their business ventures.

3.4. *Improving Contracts in the Tree mortgage system*

To avoid the risks associated with agency theory, or the unfair weighting of power to A, the nutmeg farmers as P and ijon as A could try to improve conditions by reducing overall risks to mortgage system implementation. Risks by P and A are jointly incurred because the contractual mortgage system is not designed to maximize harvest opportunity. The tree mortgage system is skewed in favor of A, and A has lesser stake and knowledge in protecting the Dusung. Case in point that shows the favorable terms to A is that if the yield is less than the value of P's loan, then the tree mortgage remains valid for the upcoming harvest to cover the P's loan shortage. However, if the yield exceeds P's loan value, then the deal is valid only at the time of the harvest, thus providing full advantages to A. Any reformulation of the mortgage contract system should pay special attention to these arrangements in order to highlight the unequal terms under conditions of payment shortages or excess harvest. According to (Zubayr, Darusman, Nugroho, & Nurrochmat, 2014) improving process contracts should be performed if P wants to reduce their risks.

Determining the optimal written contract could be one way to increase clarity on the terms, rights and obligations between P and A. The contractual relationship between P and A could guarantee that P gains greater incentives from Dusung management activities, especially as such efforts take place over many years. P could also highlight contractual arrangements and other strategies to modify the behavior of A. To design an efficient P-A relationship, a contract should be self-enforcing, whereby it would take action against or suppress overly opportunistic behavior common among A in the current situation. Such measures would help to ensure compliance over expectations by P towards A.

We therefore suggest three ways that P could design an efficient and self-enforcing contract to limit unequal power sharing between P and A:

a. P can work to carefully filter out the different A known to enforce excessive or unfair dealings. In this case, P would identify who will work with them, and P could work together to only work with eligible A.
b. P can monitor A activities, such as assessing the terms and the different benefits between parties. When the mortgage system is established, all access to dusung is fully controlled by A and in turn places P in a more equal position.
c. P can draw up contractual terms that can increase the credibility of the relationship between P and A. Articulating these terms at the outset allows them to be written down, signed, and thus self-enforcing. Furthermore, it can also include penalties for wrongdoing, and reward agreements in the event of windfall profits.

In addition to improving the mortgage tree system, complementary strengthening of local institutions would provide a much needed facilitation role. The lack of nutmeg farmers bargaining position in the Hutumuri village is inseparable from the lack of cooperatives that can help with product collection and marketing. This is one of the central reasons why farmers are increasingly dependent on the ijon, because they do not have the option to sell their products and access markets elsewhere. Breaks in this supply chain and lack of market information also disadvantages the farmer position. Moreover, the absence of organizational support for farmers and other parties related to the bargaining position of farmers is very low. According to (Febryano, Suharjito, Darusman, Kusmana, & Hidayat, 2014) collaboration between local institutions, local NGOs, national and international universities, research institutions, private companies through CSR, and others can improve the bargaining position of local farmers and their institutions, and could advocate to encourage local government policies that favor local farmer institutions.

Increasing the bargaining position of nutmeg farmers will also encourage the sustainability of forest resources. Dusung management systems have historically performed well to protect resources. Weakening their stakes in the management of the Dusung as indicative in the ijon relationship described above could further weaken the overall ecological system. Studies conducted by (Salampessy, Bone, & Febryano, 2012) in the villages of Soya (Ambon Island), Alang (Ceram), and Paperu Booi (Saparua) shows how dusung with nutmeg have various other benefits in terms of economic development and environmental management. Dusung with nutmeg and their performance at all these locations showed that: (1) productivity and cost structure of nutmeg farmer income correlates with the number of cultivated plants; (2) cultivation and security efforts undertaken by the owner with the help of Kewang (traditional forest police) indicates the orientation toward productivity and sustainability; (3) community priorities in managing Dusung are for commercial purposes, especially for nutmeg and clove, while interspersing vegetables for subsistence purposes; and (4) cultivation of Dusung are relatively efficient in their ratio of input and output production. Furthermore, from an institutional perspective, rights relating to ownership, control, management and the existence of co-management arrangements between Dusung proved beneficial to owners, managers, harvesters and their communities.

4. Conclusion

Relations between nutmeg farmers and ijon in the tree mortgage system shows the implications of the current conditions of changing relations, especially in the temporary transfer of rights from the farmer to ijon. Asymmetric information has perpetuated debt bondage to exploit opportunities for profit accumulation among the ijon. Their additional access to information and marketing advantage privileges them in the profits from nutmeg over the farmers. Improving system contracts for tree mortgage, as shown in this paper, could help to emphasize the cultivation issues in the context of marketing concerns. Strengthening local institutions is also one way to improve the bargaining position of nutmeg farmers and encourage sustainability of forest resources through Dusung nutmeg agroforesty systems, as has been proven the case in various other areas.

Acknowledgments

I wish to thank Mr Polly Pessy, one of the co-owners of a dusung nutmeg agroforestry grove who accompanied the researcher in Hutumuri Village. I also acknowledge the support I have received from Ina Hutubessy who also assisted in this research.

References

Dinas Pertanian Prov Maluku. (2012). Prosiding : Agroforestry berbasis Pala untuk Kesejahteraan Masyarakat Maluku. In *Presentasi :Strategi Pengembangan Komoditi Pala di Maluku* (pp. 98–104). Ambon: Program studi Manajemen Hutan PPs Universitas Pattimura Ambon, KPM & Cordaid.

Febryano, I. G., Suharjito, D., Darusman, D., Kusmana, C., & Hidayat, A. (2014). The Roles and Sustainability of Local Institutions of Mangrove Management in Pahawang Island. *Jurnal Manajemen Hutan Tropika*, *20*(2), 69. https://doi.org/10.7226/jtfm.20.2.69

Petrie, M. (2002). A Framework for Public Sector Performance Contracting. *A Framework for Public Sector Performance Contracting*, 117–153.

Salampessy, M. L., Bone, I., & Febryano, I. G. (2012). Performansi Dusung Pala Sebagai Salah Satu Agroforestri Tradisional di Maluku. *Tengkawang*, *2*(2), 55–60. Retrieved from http://download.portalgaruda.org/article.php?article=111696&val=2344

Silaya.Th., Tjoa, M., Lelloltery, H., Siahaya, L., & Bone, I. L. M. (2012). Prosiding : Agroforestry berbasis Pala untuk kesejahteraan Masyarakat Maluku. In *Agroforestry berbasis Pala di kepulauan Maluku* (pp. 5–15). Ambon: Program studi Manajemen Hutan PPs Universitas Pattimura Ambon.

Zubayr, M., Darusman, D., Nugroho, B., & Nurrochmat, D. R. (2014). Principal-agent relationship in policy implementation of the use of forest area for mining activity , Indonesia, *3*(January), 181–188. https://doi.org/10.11648/j.aff.20140303.17

2

Mysterious ginger: Enclaves of a boom crop in Thailand

Sukanlaya Choenkwan[1,*]

[1] Department of Agricultural Extension, Faculty of Agriculture, Khon Kaen Univesity, Khon Kaen, Thailand. Email: sukanl@kku.ac.th

* Corresponding author

Abstract: Ginger (*Zingiber officinale*) is a perennial herb revered as one of the most popular and valued spices of the world. Ginger is the main cash crop supporting the livelihood and improving the economic level of many ginger growers of India, Indonesia, Nigeria and Bangladesh. Although Thailand was the fifth highest ginger producing country in the world, documentation and general recorded information on ginger cultivation, both for domestic production and export value trade, is very limited. The ginger production system is also confounding. Moreover, ginger in Thailand has received little attention from researchers. There is no research or publications about Thai production of ginger and ginger farmers in both Thai and English publications. This paper is an initial attempt to establish a ginger story in Thailand by using the Plaba sub-district in Loei province as a case study to document this overlooked story and bring attention from researchers or government officials who are interested in the possibility of identifying new agriculture possibilities and recognize potential crops for further development to support rural livelihoods.

Keywords: high value crop; land use; mountain agriculture

1. Introduction

Ginger (*Zingiber officinale)* is a perennial herb revered as one of the most popular and valued spices of the world. It is native to the tropical rainforests of Asia. The most useful part of this plant is the underground rhizome. It is widely used for culinary and medicinal purpose, including food seasoning, candy, and ginger wine (Sawa, 2017). Ginger is also used to treat nausea, vomiting, indigestion, fever and infection and to promote vitality and longevity (Singh and Singh-Dhillon 2015).

Ginger is cultivated in more than 30 countries over the world. In 2014, the global planted area of ginger was 396,724 ha with the production of 2,156,453 tons (FAOSTAT, 2017). India was the top producing country accounting for 17% of total global production and China was second, accounting for 12%. Ginger is generally sold in its raw form in local markets but there are several other products, including dry ginger, ginger powder, ginger oil and oleoresin for the international market. The five leading non-crushed and non-ground ginger exporting countries in 2016 were China with US$337.3 million (58.7% of total non-crushed and non-ground ginger exports), Netherlands (12%), Nigeria (4.9%) and Thailand (4.7%) (Workman, 2017).

Thailand was fifth highest ginger producing country in the world amounting to 135,514 tons, and accounting for about 6.28% of the world's total ginger production in 2014. Thailand was also the fourth leading ginger exporting countries in the world with a value of US$ 27.2 million and accounting for about 4.7% of non-crushed and non-ground ginger exporting countries in 2016. In Thailand, ginger is mainly grown in some mountainous districts in Petchabun, Loei, Chiang Rai and Pha Yao provinces (Lakchaikoon, nd). The most recent available data, found in 2011, shows that ginger covered about 15,241 rai[1] and there were 8,145 growers. These figures amount to a very small proportion compared to total agricultural land (114.6 million rai) and total number of

[1] 1 ha = 6.25 rai

farmers in the country (5.9 million farmers) (Agricultural Statistics Group, 2013). Productivity of ginger in Thailand is around 2,000 - 3,000 kg/rai which is much more than the average productivity in the world (400 kg/rai). The United States has the highest productivity (8300 kg/rai) of ginger in the world (Singh and Singh Dhillon, 2015). Records of Loei Provincial Agricultural Extension Office during 2007 to 2013 show that farm gate ginger prices fluctuated dramatically. For example, prices were 6 Baht[2]/kg in 2009, but in the next year (2010) had increased to 25.48 Baht/kg, and in 2012 dropped to 4 Baht/kg.

This paper began as a result of a short initial visit to a village in Phu Ruea district, Loei province of Thailand, where I discovered a village had continuously grown ginger for more than 30 years. Everyone in Plaba village grows ginger, and yet the surrounding areas do not. My conversations with ginger growers in this village reveals that ginger is the most valuable crop in the area. In late 1980s, the ginger was first planted in the village by Hmong people who migrated 160 km from Khao Kho district in Petchabun province to Phu Ruea district. About 30 years ago, Hmong began to rent the land in the village to grow ginger. They did this for a few years and then migrated to other areas to acquire new lands for growing ginger. After the Hmong left, the villagers also began to farm ginger that had been left behind by the Hmong. Transplanting the ginger to their lands, they discovered that ginger could be 100 times more profitable than maize. Maize in this area had been a low value field crop continuously grown for more than 50 years. These changes began to transform the landscape.

Ginger prices have always been unstable and cultivation poses difficulties as it is sensitive to disease. Once ginger is infected, the farmers will not get any yield for that year. However, for those still willing to take the risk, in a very good year they can earn windfall profits. Such high profits are used to purchase a car and tractor, generally bought in cash within a given year; while in other cases farmers boast being able to buy both a car and build a new house in just one year. All farmers proudly expressed that their wealth is because of ginger (large homes, cars, and tractors). They all revere ginger and its power to boost incomes. It is a king crop for them. In contrast with other agricultural extension programs however, this romantic boom story about ginger in this village has nothing to do with the government. There was no project or plan to develop ginger farming from government. A farmer told me that the government does not have any policy or development plan to include ginger or support its local farmers. In fact, the government actively discourages ginger cultivation because they believe it destroys the forest.

Therefore, I have tried to examine this unique village and their successful ginger ventures in this paper. I try to understand how Phu Ruea became an enclave of ginger cultivation. Ginger has long been vastly cultivated in Plaba sub-district of Phu Ruea, but why has it not been introduced in any of the neighboring sub-districts? In short, I try to explain how and why this mystery crop emerged. In order to answer these questions, I began by reviewing literature on ginger production in Thailand. Although Thailand was the fifth highest ginger producing country in the world, documentation and general recorded information on ginger cultivation, both for domestic production and export value trade, is very limited. The ginger production system is also confounding. Moreover, ginger has received little attention from researchers. Most of research about ginger were focused on the study of taxonomy, molecular, and diversity (e.g. Triboun, 2006; Sertwasana, 2008; Wongsuwan, 2010). There is no research or publications about production of ginger and ginger farmers in both Thai and English publications. Thus, this paper is an initial attempt to establish a ginger story in Thailand by using the Plaba sub-district case study to document this overlooked story and bring attention from researchers or government officials who are interested in the possibility or responsible for agricultural development, to recognize this potential crop and help to seek ways for further development. Information in this paper are based on a reconnaissance survey and semi-structured interviews with ginger growers and agricultural extension officers undertaken in ginger growing areas of Plaba subdistrict, Phu Ruea distrct, Loei province in December 2016. Secondary information comes from government databases, previous

[2] 1 US$ = 34 Baht in June 2017

reports, research, published articles, and some online news (as relevant) were also reviewed for supplementary information.

This paper is divided into four parts. First I provide some literature reviews related to ginger cultivation. Secondly, I examine the physical, social and economic characteristics of Plaba sub-district in order to explain how ginger took root in this particular area. Third, I provide a history of ginger cultivation in the study area related to the Hmong growers that initially brought the crop to the area. Finally, I provide a description of the ginger production system in Phu Ruea. Indonesia.

2. Literature reviews related to ginger cultivation

Available research related to relatively similar ginger production systems have been conducted in India (Rahman et al., 2009; Muenster, 2015), Ethiopia (Geta and Kific, 2011), Nepal (Micro-Enterprise Development Program, 2013) and Nigeria (Nandi et al., 2011; Makarau et al. 2013; Ezra et al., 2017), where ginger is an important commercial crop. Singh and Singh Dhillon (2015) analyzed economic returns of ginger crops in Sirmour distrct of Himachal Pradesh, India, and found that income measured were found to be positive. Cost and benefit ratio of ginger crop was about 2 which means ginger crop was very profitable for farmers. In the South Indian district of Wayanad, Some farmers blamed ginger as a killer because it causes some cultivators to lose their money and in some cases, it drives farmers to suicide. However, others also labeled the crop "ginger king" because some farmers and rural entrepreneurs have become millionaires in just a few years of cultivating ginger. Ginger is also known as a gamble because returns are unpredictable. The cultivators will be rich or will be ruined (Muenster, 2015).

Ginger prefers well-drained soils like sandy or clay loam, red loam or laterite loam. Muenster (2015) indicates that ginger is most suitable for recently cleared forest soil. Yadav et al., (2004: citing Gosh, 1984) also indicates that ginger is an exhaustive crop by nature and, therefore does not advise cultivation in the same field year after year. In the Northeast region of India, ginger is rotated with French bean or soybean, which not only improve the physical condition of the soil but also provides additional income to farmers (Yadav et al., 2004). The traditional systems of ginger cultivation in the Northeastern Region of India is called the *jhum* system, incorporated as part of shifting cultivation. The system involves clearing a patch of forest land which is usually done by burning, but retaining useful trees and plant varieties, cultivating gingers for two or three years and then abandoning it for 10-20 years to allow the natural forest to grow back and the soil to regain its fertility. However, with increasing populations, pressure on land has increased and the time-period for this cycle has shortened to around 3-5 years, which does not allow the same recovery time for the soil to regain its fertility. This has taken place as part of large scale deforestation taking place in the region, accompanied by soil degradation and depletion of the resource base (Yadav et al., 2004; Rahman et al., 2009).

Constraints of the production and marketing of ginger in India, Nigeria and Ethiopia are similar. Ginger is mainly constrained by input such as supply of fertilizer and varieties which is both insufficient and low quality, lack of source of credit, high cost of labor and high cost of transportation for marketing (Geta and Kifle, 2011; Singh and Singh Dhillon, 2015; Ezra et al., 2017).

3. Physical, social and economic characteristics of Plaba sub-district

Plaba is a sub-district in Phu Ruea district, Loei province in the Northeastern region of Thailand. The sub-district covers about 9,499 hectares located in the mountainous area with elevation about 800-900 meters above mean sea level. These are the highest place that people live in the Northeast. Climatically, the rainy season occurs from May through October and there is a prolonged dry season during the remainder of the year. Average annual rainfall is about 1,428 mm. Average minimum temperature is 11.5 °C in December and average maximum temperature is 34.5 °C in April (Agricultural Technology Transfer and Service Center, 2013) There were 3,466 inhabitants with a population density of 36 people/km^2 in 2016 (Department of Provincial

Administration, 2016). Ethnically, they are called Tai Loei. The Tai Loei speak the Lao Isan dialect of the Thai Language, which is similar to other Northeastern lowlanders. Moreover, many cultural patterns and agricultural practices are also similar with other nearby lowland areas, which contrasts with mountain communities in the North whom are ethnically and culturally much different from the northern lowland (Choenkwan et al, 2014).

More than 90% of Plaba people are engaging in agriculture with average landholdings at about 15 rai, but range from 3 to 50 rai. Average labor is about 2-3 people per household. This sub-district is surrounded by forests. About 90% of total sub-district's area is under protected area (Office of Soil Resources Survey and Research, 2008). Land titles of those agricultural land in the sub-district is Por.Bor.Tor 5 which is evidence showing that the occupier of a plot of land has been issued a tax number and has paid taxes for using the benefit of the land. This confers no ownership rights but was formerly used to establish that the holder was occupying a plot of land (Thailand law online, 2016).

Because of suitable environments for agriculture, there are various crops grown in this sub-district including rice (both upland rice and wet rice), maize, cassava, black-eyed peas, ginger, dragon fruit, pineapple, longan, lychee, banana, pineapple and rubber. However, the majority of land is used for growing field crops, especially maize. Generally, the farmers have more than one plot of land. They usually diversify crops into different land holdings. Rice are mostly grown for household consumption whereas other crops are grown for cash income. They also diversify cash crops to avoid risk of crop failure or low prices. In cases that one crop fails, they still have other reserves. Another reason is that each crop has different timing for production and harvest. In such systems they can earn income throughout the year from different crops and one crop can be used to invest in another crop.

Table 1. Calendar of agricultural activities in Plaba sub-district

Activities	Jan	Feb	Mar	Apr	May	Jun	Jul	Aug	Sep	Oct	Nov	Dec	Yield* (Kg/rai)	Price* (B/kg)	Value* (B/rai)
Wet rice					■	■	■	■	■	■	■	■	545	15.5	8447
Upland rice				■	■	■	■	■	■	■	■		415	17.7	7,345
Maize					■	■	■	■	■				679	5.5	3,734
Black eyed bean	■									■	■	■	287	21.5	6,170
Ginger			■	■	■	■	■	■	■	■	■	■	2,580	6.7	17,285
Banana	■	■	■	■	■	■	■	■	■	■	■	■	1,850	9.7	17,945
Dragon fruit	■	■	■	■	■	■	■	■	■	■	■	■	1,790	9.9	17,721
Longan	■	■	■	■	■	■	■	■	■	■	■	■	686	15	10,290

*Source: Loei Provincial Agricultural Extension Office (2013)

4. History of ginger cultivation in Praba sub-district and the Hmong

Ginger cultivation in Plaba sub-district began 30 years ago. Initially, it was grown by Hmong, a hill tribe living in Khoa Kho, a mountainous district in Petchabun province about 160 kilometers from Phu Ruea district. The Hmong came to the sub-district to rent lands for growing ginger. However, given that ginger cannot grow over repeated cultivation on the same land, the Hmong stay in the area for a few years and then move to other areas to find new land. Plaba farmers noticed the Hmong received high income for ginger and learned how to grow ginger from them. One farmer noted that the most attractive aspects of ginger was that it only uses a small area but can produced several tons. Although the price of ginger was low at the time, the value was still much higher than maize or other field crops. After the Hmong left the area, ginger has been widely grown in Plaba sub-district ever since.

The Hmong are a hill tribe living in the high mountainous lands, mostly in the northern part of Thailand. Hmong ginger growers are in Kheknoi sub-district, Khao Kho district, Petchabun province located in the central part of Thailand with an elevation of about 800 meters above mean sea level. Kheknoi has been establish as a community by the Thai army since 1971 for political reason (Sanyakul, 2004). However, the Hmong have been living there since at least 1922 (Xiong, 2015). Historically, this group of Hmong originally lived in the northern part of Thailand. They migrated from Nan province through the high mountain ranges, continuously settling new areas but remaining linked to the northern regions. They settled in this area because the climate is cooler, suitable for their lifestyle, and advantageous for growing opium (Xiong, 2015).

Sanyakul (2004) studied the economic and cultural development of Hmong in Kheknoi and reveals that during the 1960s, the Thai government was fighting the Communist Party of Thailand (CPT). Members of CPT escaped from the Thai government and fled into forest areas all over the country including Khao Kho district where the Hmong live. At that time, Khoa kho district was still contained by dense forests. The Thai army came to the area to fight the CPT. A large battle ensued between the army and CPT. As a result, the Hmong community were overcome and driven out. Most of the Hmong had supported the CPT and fought against the army. Some Hmong members however, supported the army and others migrated to other areas. Four years later, fighting between CPT and the Thai army had not ended. The Thai army enlisted male Hmong between the ages 18-25 years old to be a part of the army. These men were trained to be solders to fight the CPT, and they were called the Khoa Kho volunteer solders. Finally, the army seized control of the area. The army allowed the volunteer soldiers to bring their families who migrated to other areas back to live with them. Many groups of Hmong migrated to the area and built a large Hmong community there. However, the area was still controlled by the army and the Hmong still lived in fear because the fighting had not completely ended. The CPT hid in the surrounding forest. In 1971, the Thai army asked the Royal Forestry Department for 20,000 rai of degraded forest land for these Hmong to live which it was a reason for security of the country. Most of the trees in these degraded forest had been cleared during the fighting with CPT in order to destroy places where CPT hidden in the forest (Xiong, 2015). Kheknoi was officially established as a village in 1975.

Hmong livelihoods were mainly driven by agricultural production systems. The traditional agricultural systems of the Hmong were a complex process of shifting cultivation driven by two main crops, namely rice for household consumption and opium for cash income. However, due to the restrictions of agricultural land and forest areas – some were under protected area status and other areas were seized and controlled by the CPT – the Hmong were restricted from practicing their traditional agricultural system of shifting cultivation. Moreover, they could not grow opium for cash income because it was against the law. As a result, maize became an important cash crop among the Hmong, which was also supported and encouraged by the Thai government. Maize was grown after the CPT surrendered to the Thai government because the Hmong began to expand the area of maize cultivation into the forest areas that were seized back from the CPT. Maize had been a popular crop for almost 10 years. In 1985 however, maize area cultivation began to decrease because certain areas were reclaimed and announced under protected area status, thus limiting the area to grow maize. In turn, a new cash crop became popular: ginger.

Ginger came to the area because a pickled ginger factory was established in Lom Sak district, Petchabun province. The factory is about 40 kilometer far from Khao Kho. Ginger had become popular because it yielded 5-8 times the profits in comparison with maize, and required less areas for cultivation. This was particularly advantageous because of the loss of cultivable areas re-designated as forest and protected lands. Hmong thereafter received tremendous early success and income from ginger and the Hmong continue to cultivate ginger until today. However, as ginger prefers virgin soil, land also became a concern, especially for the need to incorporate new cultivable lands. For that reason, Hmong have moved further afield into other mountainous districts such as in Phayao, Nan, Chiang Rai and several in Loei province to rent land to grow ginger. As Hmong dispersed into new regions to cultivate ginger, their reputation as forest destroyers became increasingly common. In news reports Hmong from Kheknoi sub-district, Kho Kho district were arrested by the authorities of the Royal Forestry department for encroaching on forest and protected areas to grow ginger in Nan province (ASTV Manager Online, 2015).

Figure 1. Maize at Khao Kho in 1980 (Fukui Hayao, 2009)

5. Ginger production system in Plaba sub-district

5.1. Land management

In 2015, Ginger was planted in Plaba sub-district and amounted to 3,507 rai with 458 farmers. Productivity is about 2-5 tons per rai. Gingers in Plaba sub-district are produced by small-scale farmers. They grow ginger in about 10-15 rai per household, but they split it into many plots, about 2-3 rai per plot. Farmers noted that if one plot gets infected by fungus and bacteria they still get something from other plots. This is a protection measure so that they will not lose everything.

Due to these land considerations however, ginger farmers every year search further afield to find new land to avoid rot problems. Thus, in order to find new land to grow ginger, farmers rent land in other districts or sub-districts. Farmers express that available land for rent to grow ginger is getting rarer and the price of the land is steadily getting higher. Presently, the price of the land that never grown ginger is about 3,000 – 4,000 baht per rai whereas the land that is used to plant ginger is about 1,000 baht per rai. Currently, some farmers rent land as far as 40-50 kilometers away from their homes to cultivate ginger. One farmer said that although ginger cultivation requires these distances and can be very costly, it is still worth a try because of the potential windfall profits. Some are willing to take the risk.

5.2. Cropping system

Suitable land for ginger cultivation are located on terrain with a slight slope, high altitudes, and clay loam soil. Nevertheless farmers express they cannot be picky about land selection. As it stands, farmers will try to find land that was not previously planted with ginger. However, because of land limitations, farmers also cultivate lands that were previously plant with ginger. The general criteria however, is that the land must have been free from ginger for at least three years. Thus, the ginger growing system in Plaba could be considered as a rotation system, which contrasts with the Hmong systems who do not return to grow ginger on the same land. Farmers also express that after growing ginger (which usually includes upland rice or maize) other crops will get a good yield because of the fertilizer left behind from growing ginger.

Ginger is propagated by planting pieces or portions of rhizomes, which have at least one good bud. Plaba farmers keep the rhizome in the ground (they do not harvest the ginger until they are ready to grow the following crop) from their previous crops to grow for the next crop. Rhizomes are cut into pieces of the required sizes (usually 2.5 to 5 cm length) and are immediately planted.

The ginger farmers start plowing the land allocated for ginger production and planting the rhizomes during March and April depending on rainfall. The rhizomes are planted in one row in beds and use chicken manure as fertilizer on the bottom layer, which is then covered with soil. The spacing between rhizomes is around 15-20 cm. Immediately after planting, beds are covered with mulch, such as grasses of up to 8-10 cm thickness. Mulching protects seedlings from the sun, prevents weed growth, and keeps soil soft and moist. Chemical fertilizers are applied one month after planting the ginger and again before harvesting to accelerate the growth of rhizomes. The farmers are very careful with the ginger during the 2-3 months after planting because these are the highest risk period for disease. If they can identify the rot in the ginger early on, they will destroy the rotted plants before it infects the entire plot. Ginger can be harvested and sold in three stages, including baby ginger (3-4 months age), mild ginger (4-6 months age) and mature gingers (more than 9 months age), the latter of which is kept to plant for the subsequent crop. Usually baby gingers are harvested in July.

5.3. Marketing channels

The middlemen are local people who live in the area, or come from Lom Sak district, Petchabun province – a distance of about 100 kilometers from Plaba. These middlemen come to farmer homes and buy the product directly. There are 2 types of selling systems, either to sell as the rhizome or directly as land. Selling rhizomes means that farmers harvest the ginger and sell it according to the weight and market price. Selling as land means the farmer sells all the production process to the middlemen.

Figure 2. A ginger farmer in Plaba sub-district, Phu Ruea district, Loei

The middlemen then pay the farmer according to the size of the land and considers the production costs that the middle men predict. This requires the middlemen to organize and conduct the harvest. The farmers usually like to sell their products as land because they can save time and reduce hiring costs for harvesting. The farmers sell the rhizome into 3 types. Firstly, the highest price, the farmers grade the larger rhizomes and clear the skin (no damage to the peel). The second grade consists of the rhizome that is large but which consist of some damages the the peel. The third grade is those of the lowest quality in size and appearance. This type gets the lowest price.

Farmers said they will evaluate their products and then decide what type of rhizome they will sell. If production quality is good, they will sell as the first or second grade. If the quality is poor, they will combine the stock and sell it as the third grade for a lower price. The ginger price fluctuates every day. Farmers will evaluate the price of ginger from the buyer. If the price is acceptable, they will harvest. If the price is unacceptable, they will not harvest the ginger until they can justify an acceptable price. Ginger requires high investment costs, especially among new growers. It requires several thousand baht for initial investment including the cost of the rhizomes, chicken manure, land preparation, and land rent. Moreover, growing the ginger also risks price fluctuations and the potential for rot. Therefore, according to one farmer, it would be difficult for new growers to enter the market. Ginger also requires high labor inputs but Plaba does not have this problem because migrant labor is readily available from neighboring Laos.

6. Conclusion

The aim of this paper was to begin to reveal some of the broader agricultural dynamics taking place among lesser known rural areas. I have examined ginger as the main economic crop for an ethnic hill tribe, the Hmong, who have long been associated with cultivating ginger and destroying the forest. One community described herein however, has been continuously growing ginger for almost 30 years in the Plaba sub-district by ethnic Tai-Loei, at times with high income potential and without the commonly associated issues of ginger and deforestation. Ginger is a king crop for some of the farmers there, although I have identified challenges associated with unstable ginger prices and challenges in cultivation. That said, although I have begun to expose some of the more perplexing dynamics of the lesser known cultivation histories, markets, and potential for ginger, much of the complexity remains a mystery.

From an agricultural systems perspective, ginger could provide a potential for expansion to other areas that can help to lift rural incomes and improve quality of life. One main challenge however are the risks associated with disease sensitivity. Once a ginger crop is infected, farmers could lose entire yields for that year. Understandably, especially among more vulnerable communities, farmers are unwilling to take this risk. If the disease aspects could be addressed however, there is great potential for expanding this cash crop. One barrier also identified in my research is the perception that ginger continues to act as a forest destroyer. Without a better articulation of the potential for ginger and assurances that it will not impinge on forests there is likely to be little formal interest through government support programs.

Therefore, although this research has scratched the surface of the more mysterious elements of ginger in a particular region, findings indicate that more focused attention and research should be given to the potential diversification of valuable crops such as ginger. In particular, research is needed to explain the potential for ginger production and the practices for implementing such a program. Such research requires various expertise and backgrounds, such as social scientists that can evaluate the broader effects on these upland communities. Furthermore, agricultural systems analysts can help to respond to some of the more volatile market aspects that need to be addressed between producers and middlemen. One approach could be to generate greater understanding about why Plaba farmers have been able to continuously cultivate ginger, especially why some farmers succeeded in growing ginger and why others do not. Plant pathologists or soil scientists could help to find ways to overcome disease threats and reduce risks for farmers. Environmentalists and ecologists could also help to examine the broader environmental interactions between ginger cultivation and the forest in ways that could provide a more nuanced understanding on the agro-forest landscape. If for example, successful ginger cultivation could occur without constant relocation to new landscapes as has occurred in Plaba, there are opportunities to work with government agencies to make cultivation practices more sustainable elsewhere.

References

Agricultural Statistics Group. (2013). Preliminary Report 2013 Agricultural Census. http://popcensus.nso.go.th/web/kaset/file/Preliminary_Report_2013.pdf; accessed on 15 June 2017

Agricultural Technology Transfer and Service Center. 2013. *Agricultural Development Plan of Plaba sub-district, Phu Ruea district, Loei Province*. Department of Agricultural Extension, Bangkok.

ASTV Manager Online. (2015). Hmong Kheknoi were arrested for encroaching forest. http://www.manager.co.th/local/viewnews.aspx?NewsID=9580000034047; accessed on 20 June 2017 (in Thai)

Choenkwan, S., Fox, J.M., & Rambo, A.T. (2014). Agriculture in the mountains of Northeastern Thailand: Current situation and prospects for development. *Mountain and Research Development* 34(2), 95-106.

Department of Provincial Administration. (2016). Number of population in Plaba in 2016. http://stat.dopa.go.th/stat/statnew/upstat_age.php; accessed on 20 June 2017 (in Thai)

Ezra, D., Akinola, M.O., Banta, A.L., Makarau, S.B. & Hussaini, A. (2017). Socio-ecomomic Assessment of Ginger Production in Jaba Local Government Area of Kaduna State, Nigeria. *Asian Journal of Agricultural Extension, Economics & Sociology* 15(2), 1-11.

FAOSTAT. (2017). Ginger. http://www.fao.org/faostat/en/#data/QC/visualize; accessed on 15 June 2017.

Fukui, H. 2009. *Soil color and land use in the topics. Teaching material for program on system approach in agricultural*, Khon Kaen University Thailand. (unpublished)

Geta, E., & Kifle, A. (2011). Production, processing and marketing of ginger in Southern Ethiopia. *Journal of Horticulture and Forestry* 3(7), 207-213.

Lakchaikoon, nd. (2017). Ginger. http://www.agriman.doae.go.th/home/news/Year%202013/017_Gingr.pdf; accessed on 8 December 2107

Loei Provincial Agricultural Extension Office. 2013. Annual report in 2012. Department of Agricultural Extension, Bangkok. (in Thai)

Makarau, S.B., Damina, A., Daneji, M.I., & Garba, O.A. (2013). Socio-Economic Factors Influencing the Adoptionof Ginger (Zingiber Fficinale) Farming Technologies in Samaru Zone of the Kaduna State Agricultural Development Project (Kadp). *International Journal of humanities and Social Science Invention* 2 (7), 39-44.

Micro-Enterprise Development Program. (2013). *Mirco-Enterprises Development for Poverty Alleviation Volume II.* Ministry of Industry, Nepal.

Muenster, D. (2015). "Ginger is a gamble" Crop booms, rural uncertainty, and the neoliberalization of agriculture in South India. *Journal of Global and Historical Anthropology* 71(2015), 100-113.

Nandi, J.A., Yurkushi, E.N., & Ashiko, T.G. (2011). Resource use efficiency among ginger farmers in Kaduna state, Nigeria. *Journal of Agriculture, Forestry and the Social Sciences* 9(1), 110-121.

Office of Soil Resources Survey and Research. (2008). Land Use Zone. Department of Land Development, Bangkok. (in Thai)

Rahman, H., Karuppaiyan, R., Kishore, K., & Denzongpa, R. (2009). Traditional practices of ginger cultivation in Northeast India. *Indian Journal of Traditional Knowledge* 8 (1), 23-28

Sanyakul, W. (2004). Factors Affecting the Development of Economic Culture of the Hmong in Kheknoi Communities, Amphoe Khaokho, Chagwat Petchaboon. Master degree dissertation, Kasetsart University, Thailand (in Thai)

Sawa, B.E. (2017). The Leading Ginger Producing Countries in the World. http://www.worldatlas.com/articles/the-leading-ginger-producing-countries-in-the-world.html; accessed on 17 June 2017.

Sertwasana, A. (2008). The development of ginger drink powder and ginger powder processing. Thesis for the Degree of Master of Science in Food Technology, Khon Kaen University. Khon Kaen, Thailand. (in Thai)

Singh, S., & Singh-Dhillon, S. (2015). Socio-Economic Analysis of Ginger Crop in Himachal Pradesh. *Indian Journal of Hill Farming* 28(1), 35-42

Thailand law online. 2016. Land laws, land title documents. http://www.thailandlawonline.com/article-older-archive/land-title-deeds-land-documents-in-thailand; accessed on 18 June 2017

Triboun, P. (2006). Biogeography and Biodiversity of the Genus *Zingiber* in Thailand. Thesis for the Degree of Doctor of Philosophy in Biology, Khon Kaen University. Khon Kaen, Thailand.

Workman, D. Top Ginger Exporters. (2017) http://www.worldstopexports.com/ginger-exporters/; accessed on 15 June 2017.

Wongsuwan, P. (2010). *Taxonomy and Molecular Study of Genus Hedychium (Family Zingiberaceae) in Thailand.* Thesis for the Degree of Master of Science in Pharmaceutical Chemistry and Natural Products, Khon Kaen University. Khon Kaen, Thailand.

Xiong, D. (2015). Hmong Khek Noi. http://hmongthaiculture.blogspot.com/2015/04/blog-post_30.html; accessed on 20 June 2017 (in Thai)

Yadav, R.K., Yadav, D.S., Rai, N., Sanwal, S.K., & Sarma, P. (2004). Commercial prospects of ginger cultivation in Northeast-Eastern Region. *ENVIS Bulletin: Himalayan Ecology* 12(2)

3

Forestry, illegibility and illegality in Omkoi, Northwest Thailand[1]

Bobby Anderson[1, 2,*] **and Patamawadee Jongruck**

1 Bobby Anderson is a Myanmar-based Research Associate at the School of Oriental and African Studies, University of London. Email : rubashov@yahoo.com

2 Professor Patamawadee "Jan" Jongruck teaches at Chiang Mai University's School of Political Science and Public Administration. Email : patamawadee.j@cmu.ac.th

* Corresponding author

Abstract: Opium poppy cultivation in Thailand fell from 12,112 hectares in 1961 to 281 ha in 2015. One outlier exists: Chiang Mai province's remote southwestern district, Omkoi. 90% of the district is a national forest reserve where human habitation is illegal. However, an ethnic Karen population has lived there since long before the law that outlawed them was created, unconnected to the state by road, with limited or no access to health, education and other services: they cultivate the majority of Thailand's known opium poppy, because they have little other choice. They increasingly rely on cash-based markets, their lack of citizenship precludes them from land tenure which might incentivize them to grow alternate crops, and their statelessness precludes them from services and protections. Nor is the Thai state the singular Leviathan that states are often assumed to be; it is a collection of networks with divergent interests, of whom one of the most powerful, the Royal Forestry Department, has purposely made Omkoi's population illegible to the state, and has consistently blocked the attempts of other state actors to complexify this state space beyond the simplicity of its forest. These factors make short-term, high-yield, high value, imperishable opium the most logical economic choice for poor Karen farmers residing in this "non-state" space.

Keywords: Alternative Development, Drug Control, Forestry, Hill Tribes, Karen, Swiddening, Zomia

---

1. Introduction

Thailand's opium poppy cultivation fell from 12,112 hectares (ha) in 1961 to 281 ha in 2015. By most standards of measurement, the country's "war" on drug production succeeded, especially in comparison to other drug-producing countries such as Afghanistan, Myanmar[2], Laos, Colombia, Peru, Bolivia and Mexico. The United Nations Office on Drugs and Crime (UNODC) declared Thailand "opium free" a decade ago, and has not included Thailand in its World Drugs Report since 2008. Thailand is also widely heralded as a successful example of "alternative development" programming, which seeks to replace illicit crops with licit ones, and which is conducted in tandem with coercive policies to eliminate the opium economy, comprised of poppy cultivation, refinement into opium, conversion to heroin, and export.

One outlier exists: Chiang Mai province's remote southwestern district, Omkoi, where the majority of known opium cultivation in Thailand occurs. 90% of Omkoi is classified as a national forest C-Zone reserve area: people are prohibited from inhabiting, cultivating or otherwise utilizing it. But a large ethnic Karen population of swidden cultivators has lived there since long before the law that outlawed them was created. They reside in areas unconnected to the state by road, with

---

[1] This article is the second of a two-part series for Forestry and Society; the first, "People, Land and Poppy: the Political Ecology of Opium and the Historical Impact of Alternative Development in Northwest Thailand", was published in Forest and Society 1/1, April 2017.

[2] For the sake of continuity this article uses the country names Thailand and Myanmar throughout, including in pre-1939 Instances when Thailand's official name was Siam, and in pre-1989 instances when Myanmar's official name was Burma.

limited or no access to schools, health centers, and other services: select among them are the grassroots cultivators of Omkoi's opium supply, sold onward to opaque networks the state is struggling to identify. While most Karen no longer engage in cultivation or consumption—indeed, most have been incorporated into the Thai state as well as Thai culture—in remoter parts of Omkoi, Karen cultivation rates are the highest in Thailand. Thai and US drug control efforts historically overlooked this district, and possibly others hosting drug production not yet identified by the state, because it did not host an interrelated communist insurgency, nor were cultivation rates as high as other areas of NW Thailand.

Opium's profit hardly accrues to the farmer. Those in Omkoi who grow opium do so because they have little other choice. The encroachment of "development" upon Omkoi's Karen has already occurred through their growing reliance on lowland markets where currency is the only form of exchange, and opium is the highest-value cash crop around. A lack of citizenship precludes much of the rural population from land tenure which might incentivize the growing of estate crops with a lower rate of return and a longer cultivation period. Their statelessness also precludes them from services, credit and other protections. Even if they had citizenship, their presence in a C-Zone reserve area still renders them "illegal". And if these issues are all overcome, a stark truth remains: there is no alternative crop that can equal the price a farmer earns from opium. Previous programs did not succeed because of agriculture, but because of the increased presence of the state.

Nor, in the case of Omkoi's Karen, is the state the Leviathan it is assumed to be; it is a set of agencies and networks of divergent and contradictory interests, and this hobbles the state's alternate development approach. Omkoi's complexity has been simplified by state actors around environmental concerns, but the attempts of other state actors to make Omkoi's inhabitants legible leads to contradictions. For example, forestry officials refuse to allow the General Administration Department (GAD) to gazette settlements and provide services, and state extension agents cannot provide substitute crops which require land clearance or supporting infrastructure; in a reversal of "classic" statebuilding practice as described by Anderson (2006), Scott (1998), Tilly (1985, 1990) and others, this government department purposely makes a population illegible. These factors serve to make short-term, high-yield, high value, easily transportable and imperishable opium the most logical choice for poor farmers, especially given the lack of law enforcement presence.

That presence, however, is growing, most palpably in eradication and law enforcement. Omkoi has already been bureaucratically constructed—the state certainly doesn't see it as stateless—but the complex realities on the ground do not match the deliberately simplified prisms through which Omkoi is viewed. The district hosts the historical extension of lowland Padi state power into an ungoverned, untallied, ephemeral highland (Scott 2009)—one of the last areas in Thailand to undergo this process. As the government destroys opium poppy but does not replace it with alternates, the potential for future unrest cannot be discounted. However, a middle ground between the state's contradictory positions can be discerned.

This analysis seeks to discover how the state can reduce opium cultivation and preserve the watershed whilst legitimating both Omkoi's people and the Thai state, as well as the recent historical circumstances in which Omkoi was "forgotten" by the state in the first place. It is based on field research and interviews with Thai government officials, Karen leaders, security actors, academics, civil society representatives, local businesspersons, ex-cultivators, and recovering opium addicts in Omkoi, Chiang Mai, and other areas, from December 2015 to June 2017. These findings emerged from an earlier project at the National University of Singapore's Lee Kuan Yew School of Public Policy, in collaboration with Chiang Mai University and the Government of Thailand's Office of the Narcotics Control Board.

2. Contemporary Opium Poppy Cultivation in Northwest Thailand

Omkoi lies 170 kilometers (a four hour drive) from Chiang Mai town, in the far southwest of the province near the Myanmar border, which is as porous as it is under-patrolled[3]. Of the 281 Ha

[3] While select interviewees state that the border is well-patrolled, the majority of them state otherwise.

of opium poppy uncovered by Thai authorities in 2015, 142 Ha occurred there. This was a significant decline from the 246.37 Ha identified in the previous year; in 2016, however, The Thai Office of the Narcotics Control Board (ONCB)'s surveillance revealed the highest cultivation identified since 2009. The price of a Joi (1.6 Kilos) of opium in Omkoi fluctuates between 120,000 and 170,000 Baht (US$3,650 – US$5,170); despite government assertions that Omkoi's opium is only for internal markets, these fluctuations causally match international demand fluctuations. Opium usage in Omkoi is also prevalent. The Chiang Mai University Medical Faculty's substance abuse research center estimates that Omkoi hosts 5,000 addicts, but their estimate is based on an arbitrarily assigned average of 5-10 addicts per village. No one knows the level of addiction in Omkoi, although anecdotally it is high. Omkoi's six sub-districts contain roughly 100 villages, although no agreed-upon figure exists: many of the "villages" are only a set of homes populated by a single extended family. Omkoi's population is estimated at 55,000; each government agency has a different population figure[4]. Most of Omkoi is inaccessible during the rainy season; beyond select main roads, the district lacks schools, medical facilities, roads and electrification, due to its national forest status (see below). Beyond the sub-district capitals, the area lacks government presence, and according to government interviewees, Omkoi serves as a place of exile for incompetent civil servants who are often absent from their duty stations[5]. The lack of educational opportunities in Omkoi over generations results in a waiver of the civil servant prerequisite of a secondary school diploma; in Omkoi, civil servants only need to speak Thai.

Approximately 80% of Omkoi residents are non-Thai hill tribes that still practice swiddening—mostly highland Karen with family links to Myanmar. Anecdotally, most of the indigenous population living in the remotest areas of Omkoi lack Thai citizenship (UNESCO, 2010); the further one travels from a road in the district, the less Thai citizens one encounters, even though nearly all Omkoi's Karen have lived in Thailand for generations.

2.1. The Karen in Southeast Asia

Contemporary issues of cultivation in Omkoi involve Karen in particular. In 1983 Peter Hinton wrote "Why the Karen do not Grow Opium: Competition and Contradiction in the Highlands of North Thailand", and this assertion continues to be made. But interviews with ONCB, academics, and Karen informants indicates that in particular areas of the highlands, they do grow opium, and they did in 1983. In Chiang Mai's Nong Tao, a village in the Mae Wang River basin, for example, Karen began growing opium a century before, after it was introduced by Hmong migrants from China[6] who planted poppy on hillsides cleared of teak. In Omkoi Karen have been growing opium for at least 50 years.

Southeast Asia's Karen predominantly reside in Myanmar; roughly half a million live in Thailand (Delang, 2003), although Karen interviewees state a population of close to 1 million. Thailand's Karen are generally concentrated along the Western border with Myanmar, from Mae Hong Son to Ratchaburi and Petchaburi. Chiang Mai Province contains one-third of Thailand's Karen. Sgaw are the largest Karen language sub-group, followed by Pwo (alt: Po). Two much smaller groups exist: Dtawng Soo/ Pa'o and Kaya/ Baway. Karen historically reside at elevations ranging from 600 to 1,000 meters[7], in small villages often comprised of extended family units of between 10 and 200 households. These village units were historically ephemeral, with villages subdividing constantly; entire villages would move as fields are planted, harvested for a few seasons, and then left fallow. The latter half of the 20th century has seen ever-larger numbers of Thailand's Karen move from the hills to valleys and cities, especially youth.

---

4 Interview, CMU Professors/ ONCB, Chiang Mai, December 2015.

5 Interviews, ONCB, Chiang Mai, December 2015.

6 Interview, Karen leader Phaw Luang Jorni Odochao, June 23, 2016.

7 In Myanmar, lowland Karen populations are found in delta areas that were once malarial forests and swamps prior to the British relocation of the capital from Mandalay to Yangon. Delta Karen allied with the British and against the Bamar who were their oppressors; against the historical norm, they did not flee, and Karen communities remain in the Ayeyarwady, although many no longer speak Karen.

The Karen identity appears to signify the external peoples and modes of governance which Karen wish to distinguish themselves *from*, rather than what Karen *are* in the absence of the other. The identity was promoted in early 20th century Myanmar by British officials, in order to simplify their census (Cheesman, 2002), and by missionaries who midwifed Karen nationalism. Cheesman, Scott, and other scholars note that the Karen, when viewed as a whole, share no common language, culture, history, or religious belief: their languages are not mutually intelligible. Their religions include Buddhism, Catholicism, and various Protestant sects; some are purely animist, while elements of animism are also found in Karen practices of other religions. The delineation between Karen sub-groups is unclear, as is their delineation from other non-Karen groups such as Kachin and Shan. Although all identities, including Han and Thai, are constructed, the Karen and numerous other highland tribes are stark recent examples of the ephemeral nature of ethnicity; Leach (1954), Scott (2009) and others note that highlanders have switched identities from Kachin to Shan within two generations, and multiple ethnicities can simultaneously occur in individuals. The same person can be a Kachin in one town, a Shan in another. In contemporary Chiang Mai province, a Karen in Omkoi becomes a Thai in Chiang Mai town, and these mimicries solidify into identities over time.

Thailand's Karen are, by and large, integrated into the Thai state to a greater degree than Hmong, Lisu and others; Hanks et al (2001) notes that they were not even considered to be a "hill tribe" by select authorities, although they were still considered to be non-Thai. Forsyth and Walker (2008) note the generalization of Karen by Thai authorities as "protectors" of the forest, in stark contrast to "destroyers" such as Hmong[8]. Karen traditionally cultivated opium for their own use, rather than for wider markets. The depth of this tradition amongst Karen is disputed by scholars as well as interviewees: nearly all agree that at in the last 50 to 100 years, opium was introduced by Hmong and integrated into Karen animist and other cultural practices, or simply used as a way to pass the time[9].

While most Karen no longer engage in cultivation or consumption, in remoter parts of Omkoi their cultivation rates are the highest in Thailand. Omkoi's Karen, then, are not only peripheral to Thais; they are generally peripheral to other, more integrated Karen.

While James C. Scott's research on Zomia (2009) is pivotal to analysis of Omkoi's recent history and in explaining the different strategies/ practices Omkoi Karen and other highlanders historically adopted to evade state control, the model has its limits when it is used as a prism through which to view contemporary Omkoi. Scott himself states that Zomia has not really existed since the end of the Second World War; it is anachronistic as per its intention. The Framework developed by Forsyth and Walker in *Forest Guardians, Forest Destroyers: The Politics of Environmental Knowledge in Northern Thailand* (2008) provides more clarity to Omkoi: their work demonstrates how the state uses environmental narratives to consolidate control over the land, resources, and people (Forsyth & Walker, 2008: 231), and of how environmental and security concerns are conflated, in border areas in particular; these prisms are being used in order to cast Omkoi as symptomatic of a crisis that must be solved to legitimate the state. Scott's earlier work, in *Seeing Like a State* (1998), is also broadly supportive of Forsyth and Walker in that it describes the narrowing of the vision of state officials as an aid to control (Scott, 1998, 11); something particularly obvious in the simplification in which state actors view Omkoi, where the area's complexity is reduced to manageable dimensions by state actors (ibid. 22).

2.2. Omkoi's Forest

"Utilitarian discourse replaces the term "nature" with the term "natural resources"... plants that are valued become "crops", the species that compete with them are stigmatized as "weeds", and the insects that ingest them are stigmatized as pests. Thus, trees that are valued become "timber",

[8] This dichotomy was most obvious in the Chom Thong upland – lowland dispute (Puginier 2003).

[9] Interviews, villagers, CRSPO, ONCB, Mae Tuen/ Omkoi town, Omkoi, March 2016.

whilst species that compete with them become "trash trees" or "underbrush". The same logic applies to fauna. Highly valued animals become "game" or "livestock," while those animals that compete with or prey upon them become "predators" or "varmints.""

- James C. Scott

Seeing Like a State, p. 13

"...In national forests that the authorities declare to be reserved or restricted, people have long been there. It is strange to enforce the law against people who live in such areas, which have always been non-reserved, but only lately declared reserved because of some lines drawn on paper. The problem occurs when those boundaries are drawn, causing the people inhabiting those areas to become "lawbreakers". In terms of legislation, they may be seen to violate the law, because the law is legally passed. However, if we consider the issue naturally as to who is actually breaking the law, it becomes clear that the lawmakers are, because the people lived in the area long before the law was enacted."

- King Bhumibol Adulyadej

June 27, 1973

Thailand's rapid population increase and explosive rates of economic growth and accompanying development between 1960 and 2006 resulted in a reduction of forest from 53.33% to 30.92% of overall land area (Ongprasert, 2011). The Royal Forestry Department (RFD) is tasked to enforce C-Zone regulations in Thailand. It was initially created to protect and regulate Thailand's teak supply in the late 1800s, after highland Karen began "illegally" selling teak to the British in Shan and the French in Laos. Ironically, RFD can be historically implicated in the introduction of opium to Karen: they banned Karen from growing rice, which would have required large-scale forest clearing, but allowed them to grow opium, which required much less clearing. The Karen would trade the opium for the rice they were prohibited from growing directly; the government taxed the trade as well.

Catastrophic flash floods in Thailand in 1989, widely linked to deforestation, led to enhancements in RFD's mandate, and soon subcontracted Thai demand for forest products into Myanmar (Smith, 1994), Karen National Union areas in particular[10], while Thailand's own watershed areas were preserved. Since its inception, the RFD has been at loggerheads with hill tribes, denouncing and seeking to end swiddening, which it and other agencies uniformly oppose under the mistaken impression that it is environmentally harmful, while many available studies demonstrate that it is, traditionally, a rejuvenating practice (Asia Indigenous People's Alliance, 2012; Bruun et al., 2009; Erni, 2009; Laungaramsri, 2005). The Department of National Parks, Wildlife and Plant Conversation (DNP) has been hived off from the RFD but they have the same policies and the same regard for highlanders as their parent department. Both agencies are reportedly corrupt (Phromlah, 2013).

RFD agents continually encroach upon rights recognized by Thai authorities, such as gathering fallen wood to use in cooking fires. Whilst ad-hoc actions to defend such rights occur regularly, according to interviewees, a broader movement to guarantee hill tribe rights to engage in traditional practices has not yet emerged. In the mid-1990s the Northern Farmer's Alliance, comprised of both highlanders and Thais, had promise, and the alliance organized a march from Nong Tao to Bangkok to protest the expulsion of Karen from Gampong Kit, state their rights over the forest and decry RFD abuses; in Bangkok, they were able to meet the RFD minister "for five minutes". According to interviewees in Nong Tao, this was the pinnacle of Karen participation in

[10] Interview, KNU representative U Mam Char, Taung Galay, Kayin State, Myanmar, December 2016. The KNU has recently declared a moratorium on forestry in its areas, and signed an MoU with the World Wildlife Foundation (Irrawaddy 2016).

civil society, and was followed by a fall in alliance members[11]. The last Karen community rights activist of note, Porlajee "Billy" Rakchongcharoen, was illegally detained by RFD officers in Kaeng Krachan National Park, Phetchaburi, on April 17, 2014, on the orders of then- park superintendent Chaiwat Limlikitaksorn, who had ordered a Karen village burned in 2011. Billy was never seen again[12]. The RFD continues to blame highlanders for lowland disasters, including the 2011 floods and a 2015 epidemic of lowland haze.

RFD has been instrumental in the configuration of Omkoi within the state as a C-zone forest reserve[13]: their prism through which they view the reserve prevents much of the population from state protection in the form of citizenship. While in most cases states seek legibility of populations, in Omkoi's case, the RFD purposely keeps populations illegible. And while Karen have traditionally found themselves regarded by the state as protectors, especially vis a vis Hmong, in contemporary Omkoi they are labeled the destroyers. Omkoi's watershed status can also be viewed as political in that it intends only to protect downstream (Thai) users; the designation is "a function of wider value judgements about the appropriate use of the uplands, the purpose of environmental management, and the status of different speakers" (Forsyth & Walker, 2008, 231).

2.3. A Topography of "Criminality"

In 2009, ONCB realized that much of the opium seized by authorities was originating in Omkoi. This occurred after roughly a decade where no cultivation surveys were undertaken in the district, theoretically because cultivation was thought to have ended[14]. Whether cultivation had ended at that time only to re-start later is, in hindsight, questionable: ONCB data reveal significant cultivation from at least 1995 to 2000, including a 485% increase in hectares under cultivation between 1997 and 1998. This increase in cultivation demonstrates Omkoi's connection to international markets; in the same year in Afghanistan, Taliban leader Mullah Omar issued a fatwa declaring that opium poppy cultivation and heroin manufacture was un-Islamic (Rashid, 2000). The price of heroin soon doubled in New York City, and nearly two decades ago unknown persons in Omkoi responded to opportunity.

In order to conquer a place, one must first map its people (Anderson, 2006; Scott, 1998; Scott, 2009; Winichakul, 1994). This is doubly true when a stateless area's resources are utilized by a stateless people in a manner different from what the state intends; contestations over resources are implicit challenges to legitimacy[15]. In Omkoi this is the continuing process of conversion of highland economies and highland cultures—which cannot be disentangled from one another—to serve lowland needs. Thailand's priority is security, and so it seeks the contours of a topography of crime. The Thai state does not know the identities of stakeholders in Omkoi's opium economy. But this is changing. ONCB has identified a network constituted of financial sponsors (some of them non-resident Thai businesspersons operating in partnership with Karen leaders) and local government officials, both Thai and Karen. The former encourages cultivation and purchases the crop, while the latter is either directly involved in the trade or accepts bribes to let it continue.

---

[11] Protests, interviewees note, are not easy to sustain over time, especially when "people have families and need to work" (interview with Jowalu "Oshi" Chindanai, Nong Tao, June 23, 2016.

[12] In May 2015 Chaiwat was appointed by the DNP to head the Tiger Corps Operation Unit, a forest and wildlife protection unit. He remains the only suspect in Billy's disappearance.

[13] Various forest management initiatives also occur in C-zone designated areas, including the establishment of protected areas such as watershed areas, national parks, wildlife sanctuaries (Omkoi contains Thailand's last populations of wild elephant and mountain goat), non-hunting areas, forest parks, biosphere reserves, botanical gardens, and arboreta.

[14] The sudden end of surveillance in Omkoi in 2000/ 2001 makes little sense when one considers that 185.92 Ha of opium poppy was identified there in the 1999/2000 season.

[15] The inverse can also be found. Indonesia's annexation of Dutch New Guinea (later renamed Irian Jaya, followed by Papua) was initially contested by an insurgency that fragmented across decades. Vast tracts of the territory are not contested and in those areas the state concentrates on resource extraction; when people pose no threat, they are provided with no services, and in many areas, aren't even counted (Anderson 2015).

Many sponsors are anecdotally local politicians who finance electoral campaigns with opium earnings[16]. Of 305 local leaders and civil servants in Omkoi, ONCB has initially identified 50 involved in the trade: this number will likely rise.

Anecdotally, in the vacuum of the state found in Omkoi much crime exists besides opium poppy cultivation: stolen-to-order motorcycles and cars flow toward Myanmar, and trafficked people emerge from there, destined for Thai farms, fishing boats, and brothels. Ivory trading and illegal logging also occur. In this Omkoi's isolation is convenient, even purposeful: according to one ONCB director, the government "intentionally makes Omkoi a forgotten place." And although interviewed government officials claim that Omkoi is a place of exile for incompetent civil servants, it is more plausible that particularly corrupt civil servants seek postings there, and likely pay a commission to the relevant human resource office to secure a placement in such a profitable district.

3. A Network Governance Approach

The Royal Thai Government created the multi-agency Centre for the Resolution of Security Problems in Omkoi (CRSPO) in 2012, mandated to suppress opium cultivation, human trafficking, and illegal logging[17]. As the CRSPO has more clearly discerned the complexity of the area, it has expanded beyond its initial security focus, moving beyond eradication and into alternative development and harm reduction. The ministries and departments represented in the CRSPO increased to 22 in December 2015, and now include education and health. It includes ONCB, which plays an overall coordination and eradication targeting role, and the Highland Research and Development Centre (HRDI), which introduces replacement crops and provides extension services to farmers in eradication areas. It also includes the Royal Forestry Department, the presence of which lays bare core paradoxes in the approaches of different state actors. The CRSPO has no hill tribe representation, civil society or otherwise: Karen and other voices are not contributing to the policies which the paternalistic state which regards them as *nak anurak* is using to shape their futures.

3.1. Alternative Livelihoods

HRDI began providing extension services in Omkoi in 2009. Their contemporary alternative livelihoods model, antithetical to highland traditions of swiddening and mobility, provides Omkoi cultivators with a geographically fixed short-term alternate livelihood while preparing them for a longer term one, generally an estate crop with a 3-5 year cultivation period before first harvest. This can include non-agricultural alternatives such as vocational and technical trainings. But this "Royal Project Extension" is less successful than its predecessors elsewhere in the Northwest. Firstly, while the project is implemented by HRDI, it is not officially the Royal Project, and therefore, it does not carry royal authority. The Royal Project, by law, supersedes other laws, most importantly RFD regulations (see below); the extension does not. Secondly, the Royal Project does not serve as a guaranteed, fixed-price purchaser of new substitute crops in Omkoi, as it does in other areas (Anderson 2017). Short-term crops such as cabbage, decorative flowers and lettuce provide quick returns on investment, but these crops are generally not travel-worthy. Good roads were an important success factor in historical alternative development (Renard, 2001); they ensured that produce arrived in markets in a sellable condition, and they keep costs down, but good roads in Omkoi are scarce. The state-resistant areas where substitute crops are most needed are those same areas where ruined (or no) roads would most damage crops transported to market.

---

[16] ONCB interviews, Chiang Mai, December 2015.

[17] Interview, 7rd Infantry Colonel and CRSPO Secretary, Chiang Mai, June 21, 2016. While Mae Sot is the primary entry point into Thailand for smuggled persons and illicit/untaxed/ illegal goods including gemstones, logs and drugs (BBC 2001), due to its less rugged geography, state pressures on the illicit traffic crossing the Salween River result in places like Omkoi acting as secondary conduits.

Omkoi's licit crop cultivators are open to other forms of abuse: farmers lack vehicles, and are often taken advantage of by middlemen with the means of transport who fix prices and reap the benefit at lowland markets[18]. Less-duplicitous middlemen would need to pay low costs to the farmer in order to compensate for the percentage of the crop which will be lost in transport. Other Omkoi farmers are caught in an exploitative contract-farming model, with fixed prices far below the costs accrued through cultivation and maintenance. In the remotest areas, crop substitution is diversified subsistence. Even without exploitation, the remoteness of the area means that the costs of business will be high. Alternative development remains constrained by a lack of land tenure: homes and plots in the C-Zone forest reserve will not be legalized. Cultivators have little reason to switch to legal crops such as Arabica that they will still cultivate illegally for less income over a longer time frame. A further impediment to land tenure is the inability of many Karen to prove Thai citizenship. Land tenure is believed by RFD to be antithetical to watershed preservation. These factors serve to make short-term, high-yield, high value, easily transportable and imperishable opium the most logical choice for those poor farmers rendered illegible by the state, especially given the lack of law enforcement presence.

The state's alternative development approaches are constrained by intra-state divisions which the state's Network Governance approach has yet to streamline (Jongruck, 2012, 2015): "State" is not a monolith. Personalities, networks, and interests shape the political ecology of opium in Omkoi. Omkoi's paradoxical designations amongst state actors results in varying degrees of legibility based on the priority of a given department. This leads to policies which serve to cancel one another out, and competing interests are most apparent in forestry policy. The Royal Forestry Department only grudgingly and occasionally acknowledge a human population, and they ignore the CRSPO, even though RFD is a member. RFD's mandate, captured in the 1964 National Forest Conservation Act, contradicts ONCB and HRDI cultivation reduction strategies, namely crop substitution, infrastructure, services, and the "legalization" of a long resident population in the watershed that RFD is tasked to keep free of human habitation. RFD has blocked HRDI attempts to build small infrastructure in support of livelihoods. RFD policy also makes the construction of roads in forest reserves difficult, if not impossible: the application process takes up to eight years[19]. Only recently the 3rd Army forced the refurbishment of the 107 KM Omkoi ring road, through the intersession of members of the military currently in power in Thailand and allies on the Privy Council, who forced RFD Bangkok to approve the request. RFD's prohibition of most road-building is not unfounded, however: road building enables illegal logging by facilitating access.

The Thai Department of Local Administration (DOLA), for its part, seeks to make Omkoi's population legible, and grant secure land titles: RFD prevents them from doing so. Despite the tenuous illegality of these settlements, it is unlikely that the Thai authorities will relocate Karen in the national forest to other areas. According to guidelines for relocation set by the Thai authorities in the 1970s, they must volunteer for relocation, and be relocated to a similar ecological zone at a similar elevation. Authorities have undertaken land exchanges where, in exchange for *not* swiddening a 50 ha upland area, they are provided with 5 ha of lowland rice padi land to cultivate. This approach has serious limitations: upland Karen areas are distinguished by a paucity of land and a scarcity of people, whilst lowlands are the inverse. There simply isn't enough lowland padi for upland cultivators. For those who do make the exchange, this is not simply a matter of livelihood; it is a matter of culture, and of incentivizing Omkoi Karen to be, in a sense, more Thai.

Despite these blockages, RFD are not actively seeking to evict Omkoi's Karen, and they allow the cultivation of crops which do not involve land clearance, namely coffee and avocados. They also haphazardly provide limited land use permits for Karen farmers. This is a foundation upon which to build, but whether or not this might lead to a modification of RFD policy, and a middle

[18] An interviewee noted that cabbages and tomatoes her village cultivated were purchased by Thai middlemen for as low as 1 baht per kilo; the introduction of Arabica caused a wholesale abandonment of other alternative crops, even though Arabica is also a buyer's market, with prices fixed at 20-25 Baht per kilo.

[19] Interview, ONCB and 3rd Army Representatives, Omkoi, February 2016.

ground arrived at through a Participatory Land Use Planning project, which "seeks to achieve a successful marriage of diverse land management perspectives in transparent and locally inclusive ways" (Forsyth & Walker 2008, 245), is anyone's guess.

3.2. Eradication

Despite five years of eradication, opium cultivation continues, with little discernable impact on price; CRSPO activities are not impacting supply. Early 2016 interviews with ONCB officials in Omkoi reveal record numbers of fields under cultivation in Omkoi's far east and northwest.

Eradication occurs in a non-systematic manner, partially due to the state's inadequate but improving surveillance capability in its remoter corners: ONCB only knows the number of plots it identifies and destroys, not the total number of plots under cultivation. They also only report on the number of plots destroyed *once*, whereas the evasion strategies Karen cultivators have developed result in staggered crops which will sprout on those same plots destroyed by the army, weeks after they depart. Each planting cycle is roughly 3 months, but the annual opium-planting season is 8 months; several overlapping planting cycles occur. Other evasion strategies create even more difficulties. For example, Omkoi's main opium poppy cultivation areas are at least 2 to 3 hours by road (20 to 30 km) from Omkoi's district center, followed by a hike. No roads lead directly to poppy fields, which lay far from settlements, and so no information connects cultivator to crop unless they are apprehended on site. Plots are purposefully kept small, averaging .25 Ha. Cultivation occurs on steep slopes, difficult to detect from satellite imagery. There remain areas that the state cannot see, even with satellites. The friction of terrain remains as important now as it was in colonial times.

However, cultivation will ultimately decrease as both the state continues gathering complex information, and cultivators gain an understanding of the technology being deployed against them. In Baan Mae Long Luang village[20] during a CRSPO meeting in February 2016, each village head denied opium was being cultivated in their area; the CRSPO representatives then demonstrated satellite technology, showing on a laptop computer the opium poppy plots in several areas surrounding the villages of the attendees. It was the first time they had seen such a demonstration; some appeared shaken.

4. Rethinking policy priorities:

Omkoi, as a "stateless" region with little available data, will likely experience new contestations around expanding administration and coercive authority. The following recommendations will hopefully inform future policy interventions:

a) *Alternative Development and Markets:* ONCB and HRDI should continue to concentrate on crops which do not require land clearance. Arabica coffee remains the ideal crop: it earns the highest value and travels the best on bad roads, unlike avocados or peaches. It requires some shade, so the watershed remains preserved, and RFD implicitly does not object to it. The prohibition on chemical fertilizer and pesticide in the watershed leaves open the possibility of internationally recognized organic certifications, a potentially lucrative opportunity. This is a process that HRDI could be instructed by the CRSPO to undertake on behalf of farmers. HRDI should robustly intercede with RFD to provide 20-40 year limited use permits for Omkoi farmers, specific to crops requiring shade. Citizenship should not be a pre-requisite for the issuance of such permits, but those seeking such permits should have to meet criteria established for long-term residency as part of a citizenship application process. HRDI should explore the possibility of banks recognizing these permits as collateral for credit to farmers. The CRSPO and ONCB should also concentrate on improving market linkages between farmers and larger coffee buyers that have reputational concerns over Fair Trade; existing middlemen need to be removed. Lastly, HRDI needs to establish a physical presence in the most severe opium-producing regions of Omkoi. This could involve contracting extension agents with the

[20] This area was so afflicted by opium use that "even the monks needed treatment", according to ONCB.

requisite knowledge of Arabica coffee, either from the CMU department of agriculture or from agriculture-focused NGOs.

b) *Citizenship:* The state should consider establishing a process by which residency and land tenure can be granted according to local testimony and committee investigation. Residency should include a citizenship pathway, subsidized by the state. In the case of DNA tests for children without birth certificates, the state should consider subsidizing these costly tests as well.

c) *Indigenous rights, legalization and representation:* the CRSPO should form a dedicated forestry sub-committee, co-chaired by RFD, ONCB and the army, and with Karen civil society representation and Karen testimony, to consider legalizing Karen villages known to pre-date the National Forest Conservation Act, with the stipulation that permanent land clearing be banned, that logging remain illegal except for the personal use of indigenous communities, and that no new construction occur (only refurbishment of existing structures) except for potential new health centers, schools, staff accommodation, and police stations. Road construction to existing settlements should be considered. The sub-committee will report to the CRSPO, which can then submit findings and recommendations to the national-level ONCB and army for them to intercede with relevant national-level actors in order to seek Omkoi-specific policy changes with the agreement of the national-level RFD, followed by instructions to provincial RFD.

5. Conclusion:

Under the auspices of a "war on drugs" and "alternative development", the highlands of Northwestern Thailand were, with a few exceptions, integrated into the Thai state starting in the 1960's, through the extension of roads, the positioning of civil servants, and the extension of citizenship and services to hill tribes. The conversion of highland economies to fit lowland markets began long before this, and opium was the catalyst for that economic conversion; it was the first monocrop imported by outsiders, to feed lowland demand, and highlanders continue to be blamed for that mutation of sustainable agricultural practices by non-highland elites, KMT and otherwise, who reaped the profits from that trade. However, population growth in the last 70 years has consigned Zomia's methods of evasion, agricultural and otherwise, to history. This was, and remains, traumatic to hill tribes who have seen a way of life end; younger generations gravitate to Thai and other cultures.

This process of statebuilding within non-state spaces continues in Omkoi, an area bureaucratically constructed in different ways by different government departments with differing, and often, contradictory, priorities. For some, Omkoi's people are an inconvenience, and so they are not seen in favor of resources useful to the Thai state; for others, Omkoi's illegibility may provide profitable opportunities. The current and varying bureaucratic constructions of Omkoi within the state, however, have been recognized by government actors as untenable. The CRSPO's network governance approach provides the best platform by which to mitigate these contradictions and create a cross-department government policy that, if created equitably, makes Omkoi's people as legible as its trees. The state wishes to reconcile a current overriding forestry-focused policy that does not fit with facts on the ground in Omkoi, and ultimately it may do so by creating citizens of Omkoi's stateless peoples, with the same rights and access to services and protections as other Thais.

Acknowledgements

The authors extend their sincerest thanks to Tian Yeow Tan and Samuel Francis Woodcock, who participated in an earlier NUS study and conducted research with the authors in Omkoi and further afield. Micah Fisher at the University of Hawaii also provided valuable commentary. The author wishes to thank the Office of the Narcotics Control Board (ONCB) Region 5, especially Jerrapan Mugura, former ONCB Director of Strategy and Administration, and Karn Thaiyapirom. ONCB Region 5 representatives Director Watin Damronglaohapan, Director Pipop Chamnivikaipong, Thippamet Sangawanna, Kraivudh Maneeratana, and Varisara Yasamuth; 3rd Army representatives including Colonel Suraput Numlong, 7th Infantry Division; Professor Ora-orn Poocharoen, Dean of the Department of Political Science and Public Administration at Chiang Mai University, and Professor Panom; Raks Thai Foundation Program Coordinator Sarinthip Promrit; Pattana, Royal Project extension director; Khun Tuen Noi village inhabitants including village leader Pamoti; Bi Po village Tambon Administrative Officer Som Chai; Musa Bat Tam Village inhabitants including the treatment camp interviewees; Mimi Saeju Win of the Chiang Mai Lisu Cultural Heritage Center; Nong Tao Village Leader Phaw Luang Jorni Odochao, and Jowalu Oshi Chindanai. We also thank the Faculty and Staff at the National University of Singapore's Lee Kuan Yew School of Public Policy, as well as the two anonymous reviewers Any errors of fact or interpretation are the fault of the authors.

Reference

Anderson, B.O.G. (2006). *Imagined Communities: Reflections on the Origin and Spread of Nationalism.* London: Verso.

Anderson, B. (2015). *Papua's Insecurity: State Failure in the Indonesian Periphery.* East-West Center Policy Studies 73. Honolulu: EWC.

Anderson, B. (2017). People, Land and Poppy: the Political Ecology of Opium and the Historical Impact of Alternative Development in Northwest Thailand. *Forest and Society, 1*(1), 48-59. doi:http://dx.doi.org/10.24259/fs.v1i1.1495

Anderson, B., Tan, T.Y., Woodcock, S., and Jongruck, P. (2016). *Thailand's Last Opium War: Governance and Illegality in a Highland Periphery*. National University of Singapore Governance Study Project.

Asia Indigenous People's Alliance. (2012). *Drivers of Deforestation? Facts to be considered regarding the impact of shifting cultivation in Asia.* AIPP. http://ccmin.aippnet.org/attachments/article/956/Driver_%20of_Deforestation.pdf, last accessed June 29 2016.

Bruun., ilde B., Andreas de N., Deborah, L., & Alan D. Z. (2009). *Environmental Consequences of the Demise in Swidden Cultivation in Southeast Asia: Carbon Storage and Soil Quality.* Human Ecology 37; ASB-Indonesia Report Number 4, Bogor.

Cheesman, N. (2002). Seeing "Karen" in the Union of Myanmar. *Asian Ethnicity* 3(2), 199-220.

Delang., C. O. (ed). (2003). *Living at the Edge of Thai Society: The Karen in the Highlands of Northern Thailand*. London: Routledge.

Erni, C. (2009). Shifting the Blame? Southeast Asia's Indigenous Peoples and Shifting Cultivation in the Age of Climate Change. Paper presented at Adivasi/ST Communities in India: Development and Change", Delhi, August 27-29.

Forsyth., Tim., & Andrew, W. (2008). Forest Guardians, Forest Destroyers: The Politics of Environmental Knowledge in Northern Thailand. Seattle: University of Washington Press.

Hanks., Jane, R., & Lucien, H. (2001). *Tribes of the North Thailand Frontier.* Monograph 51, Yale Southeast Asia Studies.

Hinton, P. (1983). Why the Karen do not Grow Opium: Competition and Contradiction in the Highlands of North Thailand. *Ethnology* 22(1): 1-16.

Irrawaddy. (2016). *KNU Signs Forestry Memorandum with WWF.* November 9. http://www.irrawaddy.com/news/burma/knu-signs-forestry-memorandum-with-wwf.html, last accessed January 4, 2017.

Jongruck, P. (2012). *Network Governance through Resource Dependence Theory: a Case Study of Illicit Drug Policy in Thailand*. PhD dissertation, University of Manchester (UK).

Jongruck, P. (2015). "From Bureaucracy to (Mandated) Network: A Changing Approach to Opium Eradication in Northern Thailand". 15th International Research Society for Public Management Conference, University of Birmingham, United Kingdom, March 30 – April 1.

Laungaramsri, P. (2005). Swidden agriculture in Thailand. Myths, realities and challenges. Indigenous A airs 2/05. Copenhagen: IWGIA.

Leach, E. (1954). *Political Systems of Highland Burma: a Study of Kachin Social Structure.* London: Anthlone Press.

Office of the Narcotics Control Board. 1995-2015. Opium Cultivation and Eradication Reports for Thailand.

Ongprasert, P. (2011). Forest Management in Thailand. *Participants Reports on Forest Resources Management*: 151.

Phromlah, W. (2013). A Systems Perspective on Forest Governance Failure in Thailand. GSTF International Journal of Law and Social Sciences 3/1, October.

Puginier, O. (2003). *The Karen in Transition from Shifting Cultivation to Permanent Farming-Testing Tools for Participatory Land Use Planning at Local Level,* in Delang, Claudio O. (ed). 2003. Living at the Edge of Thai Society: The Karen in the Highlands of Northern Thailand. London: Routledge, 183-210.

Rashid, A. (2000). Taliban: Militant Islam, Oil and Fundamentalism in Central Asia. New Haven: Yale University Press.

Renard, R. D. (2001). Opium Reduction in Thailand, 1970-2000. Chiang Mai: Silkworm Books.

Scott, J. C. (1998). Seeing Like a State: How Certain Schemes to Improve the Human Condition Have Failed. New Haven: Yale University Press.

Scott, J. C. (2009). *The Art of Not Being Governed: an Anarchist History of Upland Southeast Asia.* New York: Yale University Press.

Smith, M. (1991). *Burma: Insurgency and the Politics of Ethnicity.* London: Zed Press.

Tilly, C. (1985). *War Making and State Making as Organized Crime,* in Evans, Peter, Dietrich Rueschemeyer, and Theda Skocpol. Eds. Bringing the State Back In. Cambridge: Cambridge University Press.

Tilly, C. (1990). *Coercion, Capital and European States, AD 990 – 1992.* Hoboken: Wiley-Blackwell.

[UNESCO] United Nations Educational, Scientific and Cultural Organization. 2010. Highland Household Survey. Last assessed 7 Apr 2016. http://www.unescobkk.org/culture/diversity/livelihood/surveys/impacts-of-legal-status-on-access-to-education/

[UNODC] United Nations Office on Drugs and Crime. 2007-2014. Southeast Asia Opium Surveys.

[UNODC] United Nations Office on Drugs and Crime. 2014. Last accessed May 2016. https://www.unodc.org/documents/southeastasiaandpacific//cbtx/cbtx_brief_EN.pdf

Winichakul, S. (1994). *Siam Mapped: a History of the Geo-Body of a Nation.* Honolulu: University of Hawaii Press.

4

People, Land and Poppy: the Political Ecology of Opium and the Historical Impact of Alternative Development in Northwest Thailand

Bobby Anderson[1, *]

[1] Research Associate, School of Oriental and African Studies, University of London
* Correspondence author: rubashov@yahoo.com

Abstract: Thailand's near-total elimination of opium poppy cultivation is attributed to "alternative development" programming, which replaces illicit crops with licit ones. However, opium poppy cultivation was not drastically reduced because substitute crops earned the same income as opium: nothing can equal the price of opium to smallholder farmers, especially those without land tenure. Thailand's reduction in poppy cultivation was achieved by the increased presence and surveillance capability of state security actors, who, year by year, were able to locate and destroy fields, and arrest cultivators, with increasing accuracy. This coercion was also accompanied by benefits to cultivators, including the provision of health and education services and the extension of roads; both stick and carrot constituted the encroachment of the Thai state. The provision of citizenship to hill tribe members also gave them a vested interest in the state, through their ability to hold land, access health care, education and work opportunities, amongst others. These initiatives did not occur without costs to hill tribe cultures for whom a symbiotic relationship with the land was and remains disrupted. These findings indicate that alternative development programming unlinked to broader state-building initiatives in Afghanistan, Myanmar and other opium poppy-producing areas will fail, because short-term, high-yield, high value, imperishable opium will remain the most logical choice for poor farmers, especially given the lack of a farmer's vested interest in the state which compels them to reduce their income whilst offering them no other protections or services.

Keywords: Alternative Development; Crop Substitution; Drug Control; Forest Conservation; Hill Tribes; Opium Eradication; Thailand; Zomia

---

1. Introduction

Opium poppy cultivation in Thailand fell from 12,112 hectares in 1961 to 281 ha in 2015. By most standards of measurement, the country's "war" on drug production succeeded, especially in comparison to other drug-producing countries such as Afghanistan, Myanmar, Laos, Colombia, Peru, Bolivia and Mexico. The United Nations Office on Drugs and Crime (UNODC) declared Thailand "opium free" a decade ago, and has not included Thailand in its World Drugs Report since 2008. Thailand is also widely heralded as a successful example of "alternative development" programming, which seeks to replace illicit crops with licit ones, and which is conducted in tandem with coercive policies to eliminate the opium economy, comprised of poppy cultivation, refinement into opium, conversion to heroin, and export.

This paper analyses the historical success of alternative development in Northern Thailand and finds that opium poppy cultivation was not drastically reduced because substitute crops earned the same income as opium: nothing can equal the price of opium to smallholder farmers, especially those without land tenure and the consecutive inability to securely invest in longer-term crops. Without eradication, opium cultivation is determined by market forces: in 1984, the year eradication began, cultivation was again peaking. But alternate crops did provide income, aided by a market-distorting price floor the government used to make up the difference. That price floor remains. The near-total reduction in opium cultivation, however, was aided by much more than new crops and price guarantees. The era of eradication was heralded by the increased presence of state security actors, enhanced law enforcement and surveillance, the destruction of crops, and the arrest of cultivators. But this coercion was also accompanied by benefits to hill tribes, including the provision of health and education services, the extension of roads, the assignment of civil servants to administer areas they were previously absent from. Both stick and carrot constituted the encroachment of the Thai state. The provision of citizenship to hill tribe members also gave them a vested interest in the state, through their ability to hold land, access health care, education and work opportunities, and bank credit, amongst others. These initiatives did not occur without both economic and cultural costs to hill tribe cultures for whom a symbiotic relationship with the land was and remains disrupted. Far from being a standalone historical episode unlinked to others, alternative development and eradication has occurred as a part of the centuries-long expansion and consolidation of lowland Thai state power into egalitarian and state-resistant highland areas it was previously absent from. This paper therefore applies the theoretical frameworks of Van Schendel (2002), and Scott (2009) toward Thailand's contemporary expansion into remote areas of Northwest Thailand.

These findings indicate that alternative development programming unlinked to broader state-building initiatives in Afghanistan, Myanmar and other opium poppy-producing areas will fail, because short-term, high-yield, high value, imperishable opium will remain the most logical choice for poor farmers, especially given the lack of a farmer's vested interest in the state which compels them to reduce their income whilst offering them no other protections or services.

This analysis is based on field research and interviews with Thai government officials, Hill Tribe representatives, security actors, academics, civil society representatives, local businesspersons, ex-cultivators, and recovering opium addicts in Omkoi, Chiang Mai, and other areas, from December 2015 to June 2016. These findings emerged from an earlier project at the National University of Singapore's Lee Kuan Yew School of Public Policy, in collaboration with Chiang Mai University and the Government of Thailand's Office of the Narcotics Control Board.

2. Prologue: Lowland Intrusions in Southeast Asian Highlands

Put vegetables in the basket: put people in the Muang *(village).*

- Thai saying

Van Schendel (2002) coined the word "Zomia" to describe the upland southeast Asian land massif which stretches from Vietnam to Tibet, and which includes Northwestern Thailand: an area traditionally resistant to centralized rule due to its high elevation, extensive forest cover, rugged topography, and the friction which results. Unlike monocropping lowland Thais, Bamars and others who often cultivate Padi rice, highlanders, including Thailand's Karen, Akha, Hmong, Lahu, and Lisu, cultivate a wide variety of crops that are hard to seize and tax, such as tubers. Their swiddening keeps them mobile, and their societies do not allow for the emergence of centralized rule: egalitarianism predominates amongst highlanders, and Myanmar's Kachin would go so far as to kill emergent leaders who seemed too ambitious (Leach, 1954).

Scott (2009), Van Schendel (2002) and others postulate that much of the Zomia population, including Karen, migrated to the hills in order to escape from the centralizing rule of lowland and wet-rice cultivating states that sought to capture and transport populations near to wetland rice producing areas, putting them to work cultivating monocrops which could then be taken, taxed, and stored for long periods. In additional to monocropping, these lowland states were characterized by permanent settlements; uniformity of culture, language, and education; the propagation of dynastic myths to justify rule in cosmological terms; the record-keeping made possible by literacy; the levying of tax; and conscription in pursuit of all of the above. Such states thrived, to the benefit or emerging elites and dynasties, and to the chagrin of captured populations that eventually became the members of uniform ethnicities and religious groups. The Padi state which caused the most flight was Han China, which caused wave after wave of non-Han to flee, first from the river valleys of central China, and then from Sichuan and Yunnan, as changing dynasties seized more land and people. Those who did not flee eventually became Han. Tai people originally fled Han expansion, but they replicated Han practices in the lowlands they settled in, and eventually the most successful of the Tai statelets, Thailand, posed the same threat to other highlanders that Han once posed to them as they, and their lowland Bamar neighbors, encroached upon the hills. Highlanders fleeing an expanded state did not only constitute the constructed ethnicities of highland populations; they were also constituted by select Bamar, Han, Tai and other lowlanders who sought freedom in the forests and hills.

The lowland Thai state grew from struggles with Khmer and Bamar neighbors that warred against one another in order to capture populations rather than territories. Thailand's position between French and British colonies served as a buffer between the two and so the state consolidated internally, expanding rule into peripheral and highland areas not claimed by European powers.

King Chulalongkorn, arguably the father of the contemporary Thai state, viewed highland tribes paternally, seeing them as *nak anurak* or benign savages (Renard, 2001); he also regarded them as Thai subjects. Contemporary authorities may regard Karen and other hill tribes as a less civilized form of Thai, but they regard them as relatives nonetheless. This attitude of benevolent superiority has percolated into much of the civil service[1].

3. A Brief History of Opium in Southeast Asia

The opium poppy (*papaver somniferum*) is indigenous to the eastern Mediterranean, and was first described by Hippocrates. From Anatolia its cultivation spread to the Balkans, Iran, India and China, where it was first used as medicine a millennia ago.

Refugees entering highland southeast Asia in flight from Han Chinese expansion and successive failed rebellions likely brought opium poppy with them from Sichuan and Yunnan, where it was long cultivated and consumed, by Hmong and Lisu in particular[2], but. Opium usage became incorporated into the hill tribe cultures of Southwest China, Northern Thailand, Northern Laos, and Northwest Myanmar, as a social and a religious activity. Opium was used as currency; many products were until recently priced in *joi* of opium. Opium was also a means to store wealth: unlike other crops, it travels well, and can be stored for long periods without degrading. It became a representation of stability, continuity, and wealth, its seed given by parents as wedding gifts to children leaving the household. Opium was also broad-spectrum medicine for a variety of physical and mental ailments, its persistence enhanced by the remoteness of many highland communities to other treatment options. It reduces hunger pangs. Lastly—or perhaps firstly—people like to get high.

British India seized upon opium as their opportunity to "right" the balance of trade issue with China, and India's chronic shortage of silver, by exporting opium to China and other markets. Later waves of Chinese labor migration to Southeast Asia's British and French colonies as well as Siam in the 19th and 20th centuries brought their addiction with them. The Thai King Rama II first banned opium in 1811, but Siam ultimately bowed to British pressure; Rama IV legalized opium again (it would not become illegal again until 1958) instituted a government opium monopoly. The state benefitted: at the end of the 19th century, monopolies on opium, alcohol and gambling constituted at least 40% of Thai government revenue (Lintner, 2000). Whilst opium was wholly imported from British India, the possibility of cultivating opium outside state monopolies and selling at monopoly prices was soon acted upon by traders, corrupt officials, and holders of monopoly concessions. Highlanders were encouraged to plant more opium, which they would exchange for rice. This also occurred in China, with illicit opium cultivated in Yunnan finding its way into the Thai drug market. The British increased opium cultivation in Shan state, which borders NW Thailand, after they absorbed Myanmar into India in three stages of conquest.

Opium cultivation in Thailand did not begin in earnest until the 1940s. The fall of China's Guomindang (Kuomintang, hereafter KMT) to the communists in 1949 led to the remnants of the

---

[1] Interviews, ONCB, Chian Mai & Omkoi, December 2015/ March 2016. Many government interviewees regard their work as a civilizing mission.

[2] Interviews, villagers, CRSPO, ONCB, Mae Tuen/ Omkoi town, Omkoi, March 2016.

KMT fleeing to Myanmar, where they expanded cultivation of the only cash crop in northern Shan-opium, which would reach international markets via Bangkok. Systematic cultivation percolated into Thailand with highlanders fleeing KMT conscription and taxation (McCoy, 1973, Lintner, 1999); Thailand's hill tribes in remote areas became the primary cultivators of the crop, although the profit accrued elsewhere, with KMT and *Chiu Chau*[3] Chinese networks. Demand for Southeast Asian opium exponentially increased as a result of the success of eradication programs in the Balkans, Anatolia, and Iran; organized criminal syndicates searched for new sources of opium, and found it in the Golden Triangle (McCoy, 1973). In Thailand, this illegal economy operated with the implicit involvement of the Thai police and army that fought one another over control of the trade; despite the illegality of the substance, law enforcement did not begin until 1984 (see below). Hill Tribe swiddening (or pejoratively, "slash-and-burn") agriculture (Thai: *Rai Mun Wian;* Burmese: *Taungya*), historically a sustainable farming practice practiced by highlanders, mutated into the clearance of entire hillsides for poppy in Shan and northwest Thailand. Thai communist insurgents also profited from this trade; highlanders were their farmers and foot soldiers. Opium became central to the dominant lowland Thai discourse on hill tribes. It was in this era especially that, for many lowland Thai, highlanders—swiddeners, opium growers, traffickers, rebels—became associated with environmental destruction, crime, and threat. These stereotypes, which do not stand up to even casual scrutiny. have created an atmosphere of suspicion and distrust which continues to color the lowland – hill tribe relationship.

4. The "War on Drugs" and the advent of Alternative Development

Northern Thailand's opium economy began to wane in the early 1970's, when United States President Richard Nixon declared a "War on Drugs" in response to a rising urban heroin epidemic which was partly fueled by addicted American soldiers returning from Vietnam. The Thai government, with US pressure and funding, began treating opium as an illegal substance and poppy as an illegal crop. Elements of the Thai state once involved in cultivation and trafficking began their own disentanglement at this time. The government increased the scale of alternative livelihoods programs which had initially begun under the auspices of Thai King Bhumibol Adulyadej's *Royal Project* (est. 1969). The Royal Project's approach to alternative livelihoods included:

1. Research and development of geographically suitable alternative crops which would not compete with lowland products;
2. Agricultural extension services including provision of seeds, fertilizer, training, and construction of supporting infrastructure (dams, irrigation, farm feeder roads, etc.);
3. Post-harvest and value-added processes, such as transportation, grading and packing, food processing, market development, etc.

While the United Nations Development Program, working alongside the Royal Project, initially managed many alternative development programs on behalf of the state (Renard, 2001), the Thai Government soon created a specialized agency, the Office of the Narcotics Control Board (hereafter ONCB), to oversee Thailand's drug control policy, alternative livelihoods and (later) eradication in particular. In line with the implicit mission of alternative development as an exercise in the consolidation of state power and the weakening of threats to the state, the ONCB established its first five regional offices in Communist Party of Thailand (CPT) insurgent strongholds (Race, 1974), where they worked as extension agents[4]: earlier state prohibitions on swiddening and logging made many Highlanders eager CPT recruits.

The Royal Project introduced over 150 new crops to opium poppy farmers, including Arabica coffee and tea. Other less-valuable crops included apples, beans, cabbage, corn, decorative flowers, herbs (seasoning), lettuce, peaches, and so on. Padi rice was also encouraged. Issues arose: cabbage cultivation was surprisingly a better earner than opium but only when farmers over-used chemical fertilizers and pesticides (Renard, 2001), and only in the first year of cultivation; prices

---

[3] alt: Chao Chu / Teochew / Chaozhou

[4] Interviews, ONCB, Chiang Mai, December 2015.

soon crashed, and chemicals were soon banned due to the contamination of water sources in watershed areas. The same happened with tomatoes; these initial failures caused many former opium cultivators to abandon the project[5].

The government established price floors for alternative crops and became the guaranteed buyer for Royal Project produce, in order to match the price farmers once earned from opium poppy cultivation. The Royal Project initially focused on monocropping, which served to make highland participants more reliant on the cash which their monocrop could earn, which they could then use for further nutritional diversification. This was an aspect of lowland "best practices" transplanted to an area which was not appropriate for it, and the Royal Project de-emphasized monocropping after the first decade[6].

Poppy *eradication* did not begin until 1984, simultaneous to disarming and demobilizing (and, in the Pieng Luang area, killing) the last vestiges of KMT forces on Thai territory, as well as expelling Khun Sa. This was a full 15 years after alternative development began, when suitable alternative crops were in place; the ONCB coordinated eradication efforts which were primarily conducted by the 3rd Army. In addition to the destruction of crops, authorities began actively arresting cultivators. Between 1984 and 1985, Thailand's area under cultivation dropped from 8,290 to 2,428 Ha (Renard, 2001), and accordingly rose in Myanmar[7], where the opium harvest, from an average of 300-400,000 KG per year, would peak at 1,500,000 KG in the mid-1990s; this peak coincided with government ceasefires with numerous highland insurgents emerging from the collapsed Communist Party of Burma. The new groups, especially Wa and Kokang, were then able to concentrate on drug production. Cultivation also shifted to Laos, peaking at 126,654 KG in the 1990s.

Cultivation persisted, however, in remote areas which the state had difficulty accessing or maintaining a presence in. The evasion strategies cultivators developed resulted in staggered crops which sprouted on those same plots destroyed by the army, weeks after they departed. A poppy growth cycle is roughly 3 months, but the annual opium-planting season is 8 months; several overlapping planting cycles occurred. New poppy fields were established further from settlements; no information connected cultivator to crop unless they were apprehended on site. Plots grew smaller, ultimately averaging 0.25 Ha. Planting occurred on steep slopes, difficult to detect from spotter planes (and later, satellite imagery). Even today there remain areas that the state cannot see. The friction of terrain remains as important now as it was in colonial times.

Harm Reduction was also a late addition to Thailand's drug control policy. Thailand's initial response to its multifaceted drug problem was driven by foreign, primarily American, concerns. As the problem was considered international and not local, the Thai government's approach concentrated only on the elimination of supply. Only in the 1990s did Thai drug control policy begin to address issues of local addiction, firstly of heroin (which for a brief period in the 1980s and 1990s exploded in young highland populations, Hmong in particular, before AIDS just as rapidly killed the highland injecting population), and secondly of opium: authorities belatedly noted that grassroots farmers of opium poppy were also consumers of a culturally acceptable product, and therefore demand required addressing as well as supply (Anderson et al., 2016). Over the course of alternative livelihoods and eradication, rehabilitation and harm reduction services grew in demand

---

[5] Interviews with project participants, Nong Tao, June 24, 2016, and with ONCB, June 21, 2016.

[6] Interview, Phaw Luang Jorni Odochao, Nong Tao, Chiang Mai, June 24, 2016.

[7] By the late 1980s Myanmar was the world's largest supplier of heroin, which was also used as a currency because of a collapse in the value of the Kyat. Afghanistan would soon surpass Myanmar: US government pressure on Myanmar's State Law and Order Restoration Council to crack down on heroin-manufacturing insurgents, most notably Khun Sa's Mong Tai Army. The world epicenter of poppy cultivation and heroin manufacture soon shifted to Afghanistan, where the Mujahideen group Hizb e Islami began synthesizing heroin in Helmand after the Soviet withdrawal in 1989. The Taliban's declaration of heroin manufacture as haram or forbidden in the late 1990s caused cultivation to shift back to Myanmar.

due to the reduction in heroin and opium supply throughout the 1990s: qualitatively this is visible, but figures are unavailable[8].

Historical Experiences with Alternative Livelihoods: Doi Laan, Chiangrai

"Doesn't the state have a responsibility to protect us, to not let us be taken advantage of?"

- Mimi Saeju Win[9]

The experience of Doi Laan is emblematic of the wider disruption in cultures and economies caused by alternative development and eradication. Doi Lann is a Lisu village in Chiangrai province. The majority of Lisu there were involved in opium cultivation and sale. Alternative development projects began in the early 1970's but opium remained the mainstay of the village economy until 1985, when the military arrived. Mimi Saeju Win was a child at the time; she recounted the arrival of helicopters, followed by uniformed men in trucks. Doi Laan's people fled into the forest while their crops were torched. The Royal Project and a German bilateral development project introduced tomatoes and cabbages but the Lisu quickly fell under the sway of lowlanders who paid below-market prices and then cashed in on the guaranteed price floor paid by the Royal Project. Farmers switched entirely to Arabica, and grow it because it takes less effort, even though the price is still fixed by middlemen at 20 to 25 baht per kilo.

Eradication, for the Lisu, was a significant cultural and economic disruption; opium was the surest sign of wealth in Lisu society. Mimi notes that Lisu women lost much of their household authority at the onset of eradication: women served as opium business managers and traders, and the loss in income from alternative crops affected their status. Doi Laan's men became unhinged as household incomes plummeted. Alcoholism increased, as did use of heroin, amphetamine-like stimulants (ALS), and prostitution; injection and prostitution heralded an HIV epidemic that swept Doi Laan beginning in the early 1990s. Many Lisu previously involved in opium began smuggling ALS from Wa and Kokang areas of Shan state, Myanmar, into Thailand, and a number of these Lisu smugglers and dealers were executed during Thaksin Shinawatra's "war on drugs" in 2003.

Interviewees note that, if many Lisu and other hill tribes had a choice, they would cultivate opium poppy again.

5. Contemporary Impact of past policy

The success of the overall alternative livelihoods and crop substitution program in Thailand did *not* result from substitute crops increasing incomes; the threat posed by eradication and increased law enforcement/ state surveillance, and the potential loss of both income and freedom, were primary factors in the decisions of former cultivators to switch crops[10]. In order to conquer a place, one must first map its people (Anderson, 2006; Scott, 1998; Scott, 2009; Winichakul, 1994). This is doubly true when a stateless area's resources are utilized by a stateless people in a manner different from what the state intends; contestations over resources are implicit challenges to legitimacy. The state's bureaucracy and coercive authority led to changes more than any new crops—none of which provided the same level of income that opium poppy cultivation did. But the price floor established by the Royal Project partially bridged the gap between illicit and licit income. The Royal Project continues to pay inflated prices for substitute crops.

---

[8] Interview, public health official, Omkoi, March 2016.

[9] Interview, Mimi Saeju Win, Chiang Mai, June 22, 2016.

[10] Interviews with former cultivators, Mae Tuen, Feb 2016; with ONCB officials, Chiang Mai, Dec 2015- June 2016; with Mimi Saeju Win, Chiang Mai, June 22, 2016; with Phaw Luang Jorni Odochao, Nong Tao, Chiang Mai, June 24, 2016.

Citizenship: An important factor in the success of alternative livelihoods, according to Renard (2001) and others, was the awarding of Thai citizenship to hill tribe members: citizenship offered the possibility of land tenure. Encouraging crop substitution is implicitly an encouragement in long-term investment, particularly for estate crops such as coffee, tea, and orchard fruits. Arabica coffee is a good example; it grows at the same altitude as opium, travels well, offers better returns compared to short-term crops, and it requires the shade of larger trees, so deforestation is not an issue: it is an ideal alternate crop. But Arabica takes at least 3 years to reach maturity for harvesting. Providing this tenure to farmers encouraged them to shift away from an illegal crop with a short cultivation window and a high rate of return by giving them the security of knowing they would not be expelled from the land they cultivated. In areas designated as protected by the Royal Forestry Department (see below), limited tenure was also allowed, sometimes grudgingly (Renard, 2001). Citizenship gives those who hold it a vested interest in the state and its laws; it provides minimum levels of security and protection, even in areas where the rule of law is still in adaptive phases. In Thailand's evolving state-citizen compact, it provides free education and subsidized health care. It allows ownership of land and access to credit, and is necessary for longer-term investments. The further one enters remaining contemporary opium cultivation areas (such as Omkoi- see below), the larger the percentage of the population lacks it.

Roads and electricity: The impact of crop substitution cannot be disentangled from other forms of development, namely road building, followed by electrification. Roads served a dual purpose: to transport administrators and security actors in, and transport substitute crops out with as little damage to the product as possible before it arrived at market.

Crop substitution and the end of Swiddening: The alternative livelihoods model, focused on static agriculture and livestock, was antithetical to highland swiddening traditions: while alternative development was primarily concerned with opium poppy, it implicitly targeted swiddening as well. Alternative development projects largely ended the symbiotic relationship between hill tribe swiddeners and forests. And in the areas it was successful, it changed the very reason why highlanders cultivated crops: from household unit consumption to sale in markets.

The Royal Forestry Department was initially created to protect and regulate Thailand's teak supply in the late 1800s, after highland Karen began "illegally" selling teak to the British in Shan and the French in Laos. At that time all forests were declared the property of the state. Ironically, RFD can be historically implicated in the introduction of opium to Karen in Nong Tao and other areas of Chiang Mai: they banned Karen from growing rice, which would have required large-scale forest clearing, but allowed them to grow opium, which required much less clearing. The Karen would trade the opium for the rice they were prohibited from growing directly; the government taxed the trade as well. RFD's mandate expanded into watershed conservation as Thailand's forest cover declined from 53.33% to 30.92% of overall land area between 1960 and 2006 (Ongprasert, 2011). RFD's policing of forest subcontracted Thai demand for forest products into Myanmar (Smith, 1994), Karen National Union areas in particular[11]. Since its inception, the RFD has been at loggerheads with hill tribes, denouncing and seeking to end swiddening, which it and other agencies uniformly oppose under the mistaken impression that it is environmentally harmful and leads to large-scale deforestation, while all available studies demonstrate that it is a rejuvenating practice[12] (Asia Indigenous People's Alliance, 2012; Bruun et al., 2009; Erni, 2009; Laungaramsri, 2005). Karen leader Phaw Luang Jorni Odochao notes a particular irony: "The Thai government blames us for deforestation. But where Karen live, there are forests; where Thais live, there are none." Further, the wholesale land clearances by KMT, Hmong, Lisu and others in the heyday of opium cultivation was *not* swiddening, but for many lowlanders, the practices were one and the same. While opium cultivation declined, so did swiddening.

---

[11] Interview, KNU representative U Mam Char, Taung Galay, Kayin State, Myanmar, December 2016. The KNU has recently declared a moratorium on forestry in its areas, and signed an MoU with the World Wildlife Foundation (Irrawaddy, 2016).

[12] This would change if the populations practicing it increased, but anecdotally, the opposite occurred, with younger generations seeking livelihoods other than shifting agricultural practice.

However, the RFD was forced to adapt to the wishes of the Royal Project. Were they given a free hand they may have simply expelled highlanders from the areas they were tasked to protect, at the Conservation Law which defined RFD duties had declared vast swathes of highland territory "protected forest", in which human habitation was illegal, even though numerous settlements and peoples pre-dated the law. According to the late King Bhumibol Adulyadej "in national forests that the authorities declare to be reserved or restricted, people have long been there. It is strange to enforce the law against people who live in such areas, which have always been non-reserved, but only lately declared reserved because of some lines drawn on paper. The problem occurs when those boundaries are drawn, causing the people inhabiting those areas to become "lawbreakers". In terms of legislation, they may be seen to violate the law, because the law is legally passed. However, if we consider the issue naturally as to who is actually breaking the law, it becomes clear that the lawmakers are, because the people lived in the area long before the law was enacted[13]." The king forced a compromise: swiddening would end, but populations would remain.

The conflict between RFD and the hill tribes continues in 2017, with RFD/ DNP continuing to remove Karen, Hmong, Lisu and others from lands they have dwelt in for centuries. The current Thai junta, the National Council for Peace and Order (NCPO), issued order 64/2014 to protect and restore existing forests; shortly after, the NCPO issued order 66/2014, to shield the poor from t64/2014. However, DNP seems to have embraced the former and ignore the latter.

With the passing of King Bhumibol, a significant restraint on the RFD and DNP may have been lifted.

Nutrition: The nutritional diversification accompanying former swiddening techniques ended due to monocropping substitution, and rural food security declined as a consequence. This is the opposite of claims made by alternative development practitioners[14] who assert that the cash earned from monocropping allows for the purchase of varied foodstuffs which were once cultivated. The market price of a monocrop, however, is volatile, especially for high-value crops such as coffee, and so a collapse in its price signifies a collapse in household purchasing power, and therefore, a collapse in the collapse of a food supply which was previously a constant. This reduction in food security is not quantitatively measureable due to a lack of preserved data from past projects and a lack of previous monitoring and evaluation, but Karen and other hill tribe interviewees who experienced this attested to it.

Increasing dependence on markets and out-migration: The reduction in swiddening further integrated highland cultivators into Thailand's "cash" (non-opium) economy, thus increasing their dependence on the state and its markets, and necessitating non-traditional forms of work paid in currency. This integration also encouraged seasonal and sometimes permanent migration of highlanders to towns and cities, often for construction and other casual labor: in many highland areas, remittances became the primary income streams for cash-based needs. Select Chiangrai activists make a connection between declining opium cultivation and increasing trafficking in tribal women to brothels in the south, but this is anecdotal.

Gender imbalances: The end of opium cultivation also instigated a disruption in gender relations amongst tribes such as Lisu. Hutheesing (1989) reported that, among the Lisu she lived with in the early- to mid- 1980s, women's traditional roles as money managers for opium incomes were disrupted, with significant negative impacts on their status in the village. Opium was central to Lisu life, Hutheesing (ibid.) and others note[15], and the enforced absence of it resulted in high rates of infidelity, alcoholism, amphetamine and heroin abuse, with resultant increasing HIV transmission in Lisu villages.

Other Impacts: Hill tribes saw alternative development as an imposition from Bangkok that stripped them of their culture and sought to turn them into caricatures of lowland Thais (Jantakad and Carson, 1998; Renard, 2001): they were the passive recipients of such programs, and their

[13] From a speech given by King Bhumibol Adulyadej, June 27, 1973. Translation by Patamawadee Jongruck.

[14] Interview, HRDI, Chiang Mai, March 2016.

[15] Interview, Lisu activist, Chiang Mai, June 22, 2016.

voices played no role in the shaping of them; nor did they choose which new crops they might cultivate.

All of these impacts led Margaret Mead and other anthropologists to condemn alternative livelihood programming, erroneously alleging a cultural genocide (Washburn, 1998).

Overall, state surveillance and control implicitly increased in areas where it was previously not present, through roads, troops, and bureaucrats, and non-state areas were integrated into the Thai state over time by virtue of this coercion, as well as Thai in-migration to the highlands and the spreading usage of the Thai language and the increased embeddedness of its culture. This era was also marked by the elimination of non-state rivals such as former KMT and communist insurgents.

Opium, in this era, can be seen as less of a commodity and more of a proxy indicator for the state of hegemonic power relations (Siriphon, 2001) between lowland and highland, shifting inexorably in favor of the lowlands by the lever of alternative development.

This state imposition of rule on egalitarian highland areas, touched off decades before by flight of the KMT, is referred to as "development". But in its economic conversion of the highlands to make it serve lowland interests, it may be better described as "capitalist terraforming".

6. Remaining Opium Poppy Cultivation in Northwest Thailand

While opium poppy cultivation declined significantly, it did not end. The majority of known opium cultivation in contemporary Thailand occurs in Chiang Mai's Omkoi district: an area bypassed by a half century of alternative development and eradication, although residents note that opium was cultivated there since at least the 1960s. ONCB recorded significant cultivation increases in the 1990s, but inexplicably stopped monitoring the district around 1998, only beginning again in 2009; since then, cultivation rates have ebbed and flowed in a manner unconnected to eradication impacts (Jongruck, 2012; ONCB, 1995-2015). 90% of Omkoi is classified as a national forest C-Zone reserve area in order to protect a watershed so important that the Thai authorities consider it a national security issue: people are prohibited from inhabiting, cultivating or otherwise utilizing it. But a large ethnic Karen population of swiddening cultivators has lived there since long before the law that outlawed them was created. They reside in areas unconnected to the state by road, with limited or no access to schools, health centers, and other services: they are the grassroots cultivators of Omkoi's opium supply, sold onward to opaque networks the state is struggling to identify. While past studies (Hinton, 1983) allege that Karen don't cultivate opium, field research demonstrates that, not only are Karen growing opium now, but they have been growing in Omkoi and other areas of Chiang Mai for at least 50 years. While most Karen no longer engage in cultivation or consumption, in remoter parts of Omkoi their cultivation rates are the highest in Thailand. Thai and US drug control efforts historically overlooked this district, and possibly others hosting drug production not yet identified by the state, because it did not host an interrelated communist insurgency, nor were cultivation rates as high as other areas of NW Thailand. The government has recognized the scale of cultivation and since 2009 has taken a novel "network governance" approach to suppression (Jongruck, 2012), utilizing lessons from former cultivation areas (Anderson et al, 2-15)

Opium's profit hardly accrues to the farmer. Omkoi's Karen grow opium because they have little other choice. The encroachment of "development" upon Omkoi's Karen has already occurred through their growing reliance on lowland markets where currency is the only form of exchange, and opium is the highest- value cash crop around. Their lack of citizenship precludes them from land tenure which might incentive them to grow estate crops with a lower rate of return and a longer cultivation period, and their statelessness also precludes them from services, credit and other protections. Even if they had citizenship, their presence in a C-Zone reserve area still leaves them "illegal". And if these issues are all overcome, a stark truth remains: there is no alternative crop that can equal the price a farmer earns from opium. Previous programs did not succeed because of agriculture, but because of the increased presence of the state.

That presence, however, is growing, most palpably in eradication, law enforcement, and market encroachment. Omkoi is 1960s highland Thailand in microcosm; it hosts the historical extension of lowland Padi state power into an ungoverned, untallied, ephemeral highland—one of the last areas in Thailand to undergo this transformation.

7. Conclusion

Thomas Hobbes's Leviathan serves as the foundational text of contemporary Realist state philosophies: people voluntarily surrender themselves to the munificent state in order to benefit from its protections and escape from the state of nature- the unceasing war of 'all against all'. But the reality is that states, organized by violent entrepreneurs, came to dominate not because they brought peace to a State of Nature, but because they were logistically more effective and coercively more brutal than competing entities when it came to waging war, holding territory, mobilizing labor/ conscription, and levying 'tax' (Tilly, 1985): they held 'a decisive advantage in the power to kill" (Landes, 1998: 88). Over time, firstly in Europe, and later, in the colonies, these exploitative entities developed beyond their solely violent and coercive beginnings; they became 'legitimate' by providing protection not just from themselves, but also from others, and they began to provide services[16]: transportation and communications infrastructure, health and education, limited welfare and social protection mechanisms, a (usually) impartial police and judiciary protecting citizens from one another and from the state, and so on[17].

Northwestern Thailand's highland inhabitants once avoided expanding lowland empires, Thai as well as Bamar, by taking advantage of the friction of topography that the highlands offered. In a world devoid of cash, where land was rich and the people few, the state was an imposition. But the lowland coercive authority that these highland inhabitants avoided has changed.

The Hill Tribes have also changed: their lives would not continue in a vast outdoor museum. Ethnic identities have always been in flux; globalization increases this process. Highlanders are drawn to the state, and absorbed into its cash economy: they need the tools that will help them better navigate it. These tools did not prevent highlanders from becoming more Thai and less Akha, Karen, Hmong or Lisu; as traditional lifestyles change, painful decisions must be made, but education, citizenship, and market skills were all tools which highlanders could use to play more of a part in those decisions, rather than simply have decisions made on their behalf, by their own leaders or by the state. Development, in this case, was self-defense.

The Thai state eliminated its position as a source of opium for international illicit drug markets decades ago. That success occurred with a great cultural cost to the hill tribes: the symbiotic relationship between people and land ended. This was, and remains, traumatic. Eradication worked due not only to eradication, arrests, and crop substitution; it worked because the imposition of law occurred consecutive with benefits to hill tribes such as the provision of citizenship and the rights and responsibilities it embodies; the end of inaccessibility through the establishment of roads; increased access to schools and health care; access to markets and to credit, and all manner of other services and obligations that, together, constitute an ever-evolving social contract.

Conflicts of Interest: The author declares no conflict of interest.

---

[16] State theorists and propagandists built on Hobbes and developed mythologies to justify state primacy by expanding the notion of a pre-state war of perpetuity that the state served as an antidote for; the 'necessity' of states was expanded across the globe, with all 'primitive' people imagined to voluntarily surrender to nearby states in order to gain the peace and protection these entities offered. Whilst these claims may have some significant historical precedents (especially Hobbes, whose writings were entirely shaped by the 30 Year's War, when the war of 'all against all' killed one-third of the population of the German-speaking lands of Central Europe: the Treaty of Westphalia, which ended that war, was the beginning of the primacy of the state system), they are generally inaccurate outside of European contexts.

[17] Contemporary Thailand varies from a key attribute of the contemporary northern European state, namely a military that solely serves as the implementer of civilian-driven policy related to external defense.

Acknowledgments: The author extends his greatest thanks to Dr. Patamawadee Jongruck at the Chiang Mai University School of Political Science and Public Administration, as well as Tian Yeow Tan and Francis Woodcock; all participated in a previous research project at the National University of Singapore and conducted field research with the author in Omkoi and further afield. Micah Fisher at the University of Hawaii also provided valuable commentary. The author wishes to thank the Office of the Narcotics Control Board (ONCB) Region 5, especially Jerrapan Mugura, former ONCB Director of Strategy and Administration, and Karn Thaiyapirom. ONCB Region 5 representatives Director Watin Damronglaohapan, Director Pipop Chamnivikaipong, Thippamet Sangawanna, Kraivudh Maneeratana, and Varisara Yasamuth; 3rd Army representatives including Colonel Suraput Numlong, 7th Infantry Division; Professor Ora-orn Poocharoen, Dean of the Department of Political Science and Public Administration at Chiang Mai University, and Professor Panom; Raks Thai Foundation Program Coordinator Sarinthip Promrit; Pattana, Royal Project extension director; Khun Tuen Noi village inhabitants including village leader Pamoti; Bi Po village Tambon Administrative Officer Som Chai; Musa Bat Tam Village inhabitants including the treatment camp interviewees; Mimi Saeju Win of the Chiang Mai Lisu Cultural Heritage Center; Nong Tao Village Leader Phaw Luang Jorni Odochao; Jowalu Oshi Chindanai, and Yonatan Keller. The author also thanks Faculty and Staff at the National University of Singapore's Lee Kuan Yew School of Public Policy. Any errors of fact or interpretation are the fault of the author.

References

Anderson, B.O.G. (2006). *Imagined Communities: Reflections on the Origin and Spread of Nationalism*. London: Verso.

Anderson, B., Tan, T.Y., Woodcock, S., and Jongruck, P. (2016). *Thailand's Last Opium War: Governance and Illegality in a Highland Periphery*. National University of Singapore Governance Study Project.

Asia Indigenous People's Alliance. (2012). *Drivers of Deforestation? Facts to be considered regarding the impact of shifting cultivation in Asia*. AIPP. Retrieved from http://ccmin.aippnet.org/attachments/article/956/Driver_%20of_Deforestation.pdf in June 29 2016.

Bruun, T. B., De Neergaard, A., Lawrence, D., & Ziegler, A. D. (2009). "Environmental Consequences of the Demise in Swidden Cultivation in Southeast Asia: Carbon Storage and Soil Quality." *Human Ecology* 37(3), 375-388; ASB-Indonesia Report Number 4, Bogor.

Erni, C.. (2009). *Shifting the Blame? Southeast Asia's Indigenous Peoples and Shifting Cultivation in the Age of Climate Change*. Paper presented at Adivasi/ST Communities in India: Development and Change, Delhi, August 27-29.

Hinton, P. (1983). Why the Karen do not Grow Opium: Competition and Contradiction in the Highlands of North Thailand. *Ethnology* 22(1): 1-16.

Hutheesing, O. K. (1989). *Emerging Sexual Inequality among the Lisu in Northern Thailand: The Waning of Dog and Elephant Repute*. Leiden: Brill.

Irrawaddy. (2016). KNU Signs Forestry Memorandum with WWF. November 9. Retrieved from http://www.irrawaddy.com/news/burma/knu-signs-forestry-memorandum-with-wwf.html. in January 4, 2017.

Jantakad, P., & Carson, S. (1998) *Community Based Natural Resource Management from Villages to an Inter-Village Network: A Case Study in Pang Ma Pha District, Mae Hong Son Province, Northern Thailand*. International Workshop on Community-Based Natural Resource Management, World Bank, Washington, USA.

Jongruck, P. (2012). *Network Governance through Resource Dependence Theory: a Case Study of Illicit Drug Policy in Thailand*. PhD dissertation, University of Manchester (UK).

Landes, D. S. (1998). *The Wealth and Poverty of Nations*. New York: Norton.

Laungaramsri, P. (2005). *Swidden agriculture in Thailand. Myths, realities and challenges*. Indigenous A airs 2/05. Copenhagen: IWGIA.

Leach, E. (1954). *Political Systems of Highland Burma: a Study of Kachin Social Structure*. London: Anthlone Press.

Lintner, B. (1999). *Burma in Revolt: Opium and Insurgency since 1948*. Chiang Mai: Silkworm Books.

Lintner, B. (2000). *An Overview of the Golden Triangle Opium Trade*. Chiang Mai: Asia-Pacific Media Services.

McCoy, A.W. (1973). *The Politics of Heroin in Southeast Asia*. New York: Harper and Row.

ONCB. (1995-2015). Opium Cultivation and Eradication Reports for Thailand.

Ongprasert, P. (2011). "Forest Management in Thailand." *Participants Reports on Forest Resources Management*: 151.

Race, J. (1974). The War in Northern Thailand. *Modern Asian Studies* 8/1, pp. 85-112.

Renard, R.D. (2001). *Opium Reduction in Thailand, 1970-2000*. Chiang Mai: Silkworm Books.

Scott, J.C. (1998). *Seeing Like a State: How Certain Schemes to Improve the Human Condition Have Failed.* New Haven: Yale University Press.

Scott, J.C. (2009). *The Art of Not Being Governed: an Anarchist History of Upland Southeast Asia.* New York: Yale University Press.

Siriphon, A. (2001). *Opium and the Hmong: Dynamism, Diversity and Complexity of Identifies of a Marginal people*. Master's Dissertation, Chiang Mai University, Unpublished.

Smith, M. (1994). *Burma: Insurgency and the Politics of Ethnicity*. London: Zed Books.

Tilly, C. (1985). "War Making and State Making as Organized Crime", in Evans, Peter, Dietrich Rueschemeyer, and Theda Skocpol. Eds. *Bringing the State Back In*. Cambridge: Cambridge University Press.

Van Schendel, W. (2002). Geographies of Knowing, Geographies of Ignorance: Jumping Scale in Southeast Asia. *Environment and Planning D: Society and Space* 20: 647–668.

Washburn, W. E. (1998). Against the Anthropological Grain. New Jersey: Transaction Publishers.

Winichakul, S. (1994). *Siam Mapped: a History of the Geo-Body of a Nation.* Honolulu: University of Hawaii Press.

5

Litterfall, litter decomposition, soil macrofauna, and nutrient content in rubber monoculture and rubber-based agroforestry plantations

Uraiwan Tongkaemkaew [1]*, J. Sukkul [1], Narathorn Sumkhan [1], Phantip Panklang [2], Alain Brauman[3], and Roslan Ismail[4]

[1] Faculty of Technology and Community Development, Thaksin University, Phatthalung Campus, 93120 Thailand

[2] Faculty of Natural resources, Songklanakarin University, Hatyai Campus, 90112, Thailand

[3] Representative of Institut de recherche pour le développement (IRD), French

[4] Faculty of Agriculture, Universiti Putra Malaysia, 43400, Malaysia

*Corresponding author: t_uraiwan@hotmail.com, uraiwan@tsu.ac.th

Abstract: This study is a comparison of the litterfall, litter decomposition, soil macrofauna, and nutrient contents in rubber monoculture and rubber-based agroforestry plantations. The three intra-couple differences examined are rubber with pakliang (RP) compared with rubber monoculture (RMP), rubber with timber (RT) compared with rubber monoculture (RMT) and rubber with fruit (RF) compared with rubber monoculture (RMF). Rubber plantation systems were selected at 3 plantations located in nearest pairs at 18 plantations in total. Data collected included litterfall at monthly intervals from October 2016 to April 2017, as well as decomposition conditions for assessment at the end of the experimental trials. Soil samples examined the species and number of macrofaunal and decomposition measurements of mesofauna by using Lamina bait scale to analyze nutrient content. Results showed the litterfall of leaves, twinges, and fruits in rubber monoculture and rubber-based agroforestry plantations were not significant between pair comparisons. This showed leaves fell at a high incidence. However, RT experienced a higher trend in litterfall. Decomposing litterfall was also not significant between pair comparisons, but when compared by associated plant species found that RT trends were more likely to experience higher decomposition rates and the litter index was higher as well. Macrofauna in the topsoil (0-5 cm) and subsoil (5-10 cm) were not significant. The composition of mesofauna was found at high decomposition rates in RF, RP, and RT. Organic matter and nutrient contents were not significant in both soil layers. Our data emphasizes that rubber-based agroforestry plantations help regulate C and nutrient cycles, implying that external input fertilizer management requirements for rubber farmers decreased.

Keywords: litter index; mesofauna; nutrient content; organic matter; associated plant species

---

1. Introduction

Rubber (*Hevea brasiliensis*) first expanded cultivation in southern Thailand in 1899 by using natural plant species called jungle rubber plantations. In 1960, the Office of Rubber Replanting Aid Fund or ORRAF (now The Rubber Authority of Thailand: RAOT) promoted plantation rubber by using hybrid varieties instead of jungle rubber. This resulted in rubber monoculture expanding to other regions, especially in the northeastern region (ORRAF, 2006). This helped farmers to claim funds for rubber cultivation and integrating with other commercial tree crops as recommended by government agencies to maximize land use. Nevertheless, rubber agroforestry practices among farmers are not currently widespread. This may be due farmers lack of knowledge in managing land plots that include rubber, especially in relation to litterfall and litter decomposition effects to soil fertility and soil macrofauna.

Yang, et al. (2004) reported that litterfall and litter decomposition is a source of essential nutrients such as N, P, and K returned to plants. There were also incidence of Ca and Mg, depending on the species and the type of litterfall. Wang, et al (2008) found that litter production, leaf litter decomposition and nutrient returns (e.g. mean annual litter production) was significantly higher in mixed stands than monoculture ones. The returns of N, P and K via leaf litter were significantly higher in the mixed stands than pure stands, but the returns of C, Ca and Mg between both stands did not differ significantly. The introduction of the broadleaved tree into monoculture stands helped to increase litter production and nutrient returns, thus providing advantages in degraded soil conditions helping with sustainable land management. Chen, et al.)2017) found that rubber-based agroforestry treatments as *H. brasiliensis–C.arabica*, *H. brasiliensis–T. cacao*, *H. brasiliensis–F. macrophylla* and *H. brasiliensis–D. cochinchinensis* had significantly increased total soil organic carbon (SOC) and N contents, and furthermore enhanced the formation of macroaggregates compared to the rubber monoculture treatment. The soil surface cover with constant leaf litterfall and extensive root systems in the rubber agroforestry systems increased organic carbon and nitrogen in the soil, helping to improve soil aggregation, reducing soil erosion, decreasing carbon and nitrogen loss, and ultimately improving carbon and nitrogen accumulation rates.

In the plots of plants with a wide range of plant species, the amount of litterfall and the quantity of macrofauna in the soil increased nutrient returns in the soil. Therefore, the diversity of macrofauna in the soil is indicative of the richness of the ecosystem. This is especially evident in monoculture areas, where litter fall is significantly less, affecting species and quantity of macrofauna (Pabsta, et al., 2013). Macrofauna is the most important priority for decomposition, which works by eating residual plants and animal that release waste in the form of organic matter. This organic material gets incorporated into the soil. The organic matter is then decomposed by microorganisms and eventually becomes organic carbon compounds that help to improve soil structure and provide nutrients to plants (Austin, et al. 2014). Soil fauna important in soil includes termites, ants, nematodes. Their activities help move the soil from the upper layer to the lower layer or the lower layer to the upper layer. This incurs gaps in the soil, which then allows water and air to penetrate. The activity of these organisms influences soil processes that control the availability of plant nutrients such as nitrogen, and affects organic matter dynamics (Sarah, 2016). At the study sites, rubber-based agroforestry systems are considered the best way to improve soil properties.

In this study, we investigated litterfall, litter decomposition, soil macrofauna and nutrient content in rubber monoculture and rubber-based agroforestry systems in order to make a comparison. This is useful information for rubber farmers in the management of rubber farms and helps as a decision support tool for farmers who decide to grow rubber-based agroforestry systems.

2. Materials and Methods

2.1 Site study and plot pair trials

The study area was conducted through a purposive selection in Srinakarin and Si Banphot Districts, Phatthalung Province, Southern Thailand, whereby planting systems are conducted as both monoculture and single crop systems of rubber cultivation. We chose intra-couple differences models of rubber-based agroforestry systems, in which each of them were compared with the closest monoculture rubber plots at the same age (from 10 to 30 years old). These plot pairs include the following:

(1) Rubber monoculture (RM) + Rubber with Fruit trees (RMF) (Pair I);

(2) Rubber monoculture (RM) + Rubber with pakliang (RMP) (Pair II); and,

(3) Rubber monoculture + Rubber monoculture with Timber trees (RMT) (Pair III). In three

intra-couple differences, we investigated 3 replications (Figure 1).

Therefore, we investigated 3 cases* 2 systems* 3 replicates for a total of 18 distinct plots. Our choice in rubber plantations was based on a previous socio-economic study conducted by Stroesser et al. (2018). They found that Phatthalung province had the best agroforestry systems, both in terms of land and labor valorization, associated rubber trees with fruit and timber trees. Farmers also take on off-farm activities to complement their family income. We selected the 2 districts as representative due to around 50% of rubber-based agroforestry systems located in the study area. For terms and abbreviations, we used many abbreviations in this article including P: Pakliang (*Gnetum gnemon*), F: Fruit trees (Mangosteen: *Garcinia mangostana*), T: Timber trees (*Hopea odorata*) and RM: rubber monoculture (clone RRIM 600). The study was conducted from October 2016 to April 2017.

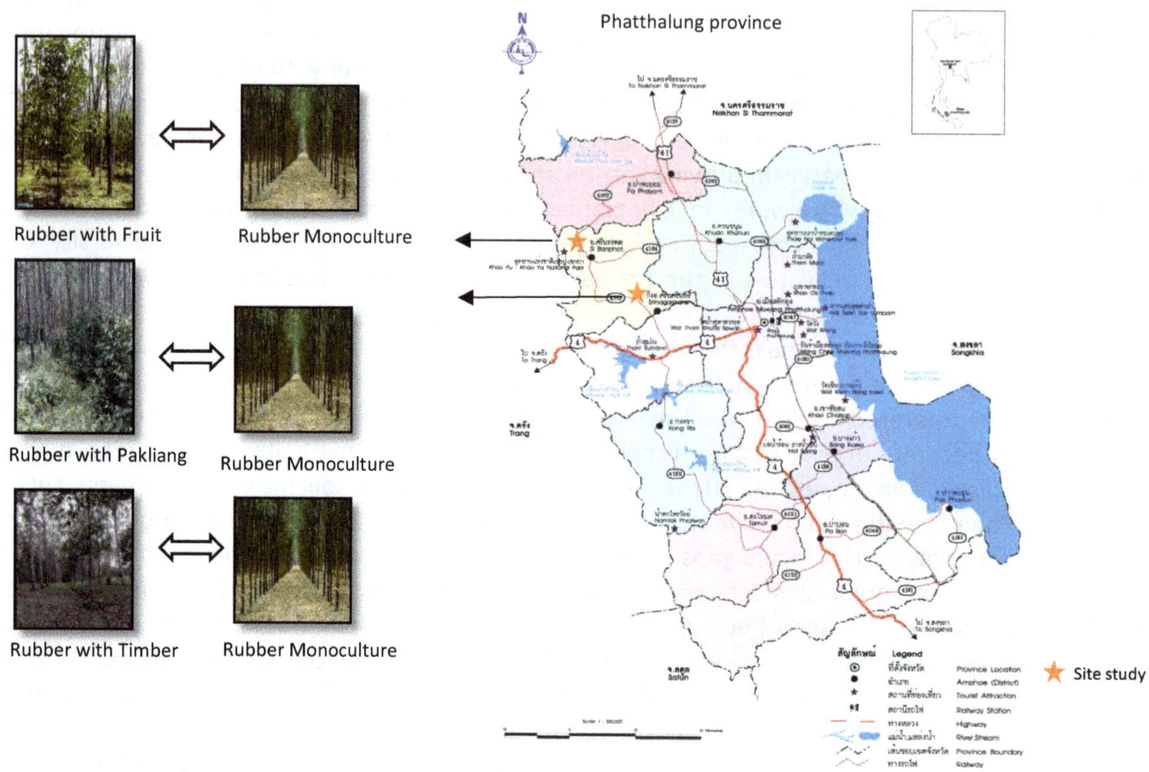

Figure 1. Site study and plot pair trials

2.2 Data collection and analysis

2.2.1 Litterfall

The selected locations of litterfall samples were collected at three points (replications) at the head, the middle, and the end of the plot trials by diagonal collection methods. The litter tray size sought to obtain litterfall samples harvested over a 10x10 cm^2 square plot area. Additional specifications included collections at 100 cm height from the ground by placing 2 litter trays in between the rubber rows and between rubber trees (1 point). Monthly litterfall samples were collected and dried over a period of 48 hours at 65°C and weighed. Thereafter, litter elements were separated, such as the leaves, twigs, and fruits to determine the dry weight of the various parts. Data obtained were analyzed for ANOVA on the intra-couple difference at 95%.

2.2.2 Litter decomposition:

Square clear plastic sheets of 20x20 cm^2 were placed at three points in the plot trials by the diagonal method, in which all the humus samples were collected from under the plastic sheet every month. Humus samples were collected and dried over a 48-hour period at 65°C. The entire skeleton, fragment, compact, fresh, cast, and wood were then weighed. Data obtained were calculated for the litter index and analyzed for ANOVA on the intra-couple difference at 95%.

2.2.3 Soil macrofauna

Three points in the plot trials were placed by the diagonal method between rubber rows that were marked by 25x25 cm^2 plots at a depth of 10 cm to identify soil macrofauna. Macrofauna was fixed in alcohol at 70%, then carried out to classification ordering at the laboratory under stereo microscope and specimen weight. The data obtained were analyzed for ANOVA on the intra-couple difference at 95%.

2.2.4 Decomposition of mesofauna

Lamina baits were used as an indicator directly associated with the feeding activity of soil invertebrates to provide an indicator of the activity and density of soil invertebrates. The method was to dry rubber litterfall over a 24 hour period at 80°C and mashed into powder. Mixtures of 7.5 g litter powder, 17.5 g of cellulose, and 50 ml of water were added then heated to 100 ° C, allowed to cool, and then poured into empty holes of strips. Eight strips per point were inserted vertically to a depth of 10 cm along the border of the 25x25 cm^2, for 1 month. The empty holes, partially degraded holes, and full holes were scored respectively by a 0, 1 and 2. The data obtained were analyzed for ANOVA on the intra-couple difference at 95%.

2.2.5 Physical-chemical soil properties

We collected soil from 2 layers at 0 to 5 cm and at 5 to 10 cm in three points in the plot trials by diagonal method between rubber rows. The dried soil was then taken to the Land Development Department laboratory for analysis of PH, EC, Organic matter, Phosphorus, Potassium, Calcium, and Magnesium content. The data obtained were analyzed for ANOVA on the intra-couple difference at 95%.

3. Results and Discussion

3.1 Litterfall

Litterfall in the three intra-couple differences found leaves, twigs, and fruits not statistically different from one another. However, the leaves and twigs were at high falling levels in pair II (RP and RMP) and pair III (RT and RMT), compared with the pair I (RF and RMF). Fruit deposited at the highest levels in pair I (RF and RMF), particularly in RF plots (Table 1). The average litterfall in rubber-based agroforestry systems were found in stand trees (RF and RT), which were higher than shrub trees (RP). Rubber monoculture (RM) still experienced high litterfall when compared with pakliang (RP) (Figure 1). Normally, litterfall in the forest areas of Phatthalung province is high from January to March because this is the period that the dry season affects trees in Southern Thailand. In these trials, RT had a tendency to have an increased fall of humus in the dry season, especially leaves, branches, and fruits respectively. Kaspari et al. (2008) reported that 90% of litter falling as leaves and twigs failed to increase with fertilization, but reproductive litter (fruits and flowers) resulted in increases of 43% with N. K enhanced cellulose decomposition. One or more micronutrients enhanced leaf-litter decomposition. P enhanced both. Apart from litterfall, Cizungu et al. (2014) reported that forests and mono-eucalypts had high fall of leaves followed by fruit and bark. However, this study focused on RT high falling rate of litterfall, but this does not necessarily mean the soil was higher in nutrients as well. It depended on the ability of the degradation of the humus in that area related to other physical, biological and chemical factors.

Table 1. Litterfall (kg/ha) in rubber monoculture and rubber-based agroforestry systems.

Litterfall	Pair I			Pair II			Pair III		
	RF	RMF	T-test	RP	RMP	T-test	RT	RMT	T-test
Leaves	8.78	8.34	ns	11.26	10.06	ns	11.26	10.06	ns
Twigs	1.98	1.92	ns	2.17	2.16	ns	2.17	2.16	ns
Fruits	0.93	0.23	ns	0.19	0.20	ns	0.19	0.20	ns
Mean	3.90	3.49		4.54	4.14		4.54	4.14	

Note: ns=not significant, RF=rubber with fruit, RMF=rubber monoculture compared with RF, RP =rubber with pakliang, RMP=rubber monoculture compared with RP, RT=rubber with timber and RMT =rubber monoculture compared with RT.

3.2 Litter decomposition

The litter index (entire, skeleton, fragment, compact, fresh, cast and wood) of the three intra-couple differences were not significant. On its entirety, the skeleton and fragment in the three pairs had less decomposition. RT and RMT in pair III had the residual highest results when compared in the pair I and pair II trials (Table 2). Comparing the average of litterfall and litter index found that RT had high levels both in fall and remains of plant parts from degradation. RP found less fall and high remains of plant parts from degradation. RM and RF had high falling and fewer remains of plant parts from degradation (Figure 2). This study is also consistent with Pausch et al. (2013) and Cizungu et al. (2014), which reported that plots with mixed-species were highly residual and it was found to have high degradation as well. Soil organic N in rubber-based agroforestry plots may be more significant than rubber monoculture plots. The litterfall of rubber-based agroforestry shows that carbon return system is higher than monoculture rubber. Constituent organic compounds and major elements involved in the fixation of energy by photosynthesis influences the carbon cycle and provides a valid means of examining similarities in ecosystem function.

Table 2. Litter index (kg/ha) in rubber monoculture and rubber-based agroforestry systems.

Litter index	Pair I			Pair II			Pair III		
	RF	RMF	T-test	RP	RMP	T-test	RT	RMT	T-test
Entire	0.64	0.60	ns	0.63	0.76	ns	0.82	0.64	ns
Skeleton	0.17	0.20	ns	0.32	0.30	ns	0.30	0.18	ns
Fragment	0.46	0.49	ns	0.46	0.64	ns	0.72	0.52	ns
Compact	0.00	0.00	ns	0.04	0.01	ns	0.10	0.04	ns
Fresh	0.48	0.48	ns	0.75	0.53	ns	0.77	0.51	ns
Cast	0.40	0.22	ns	0.49	0.37	ns	0.55	0.34	ns
Wood	0.57	0.20	ns	0.80	0.32	ns	0.71	0.42	ns
Mean	0.39	0.31		0.50	0.42		0.57	0.38	

Note: ns=not significant, RF=rubber with fruit, RMF=rubber monoculture compared with RF, RP =rubber with pakliang, RMP=rubber monoculture compared with RP, RT=rubber with timber and RMT =rubber monoculture compared with RT.

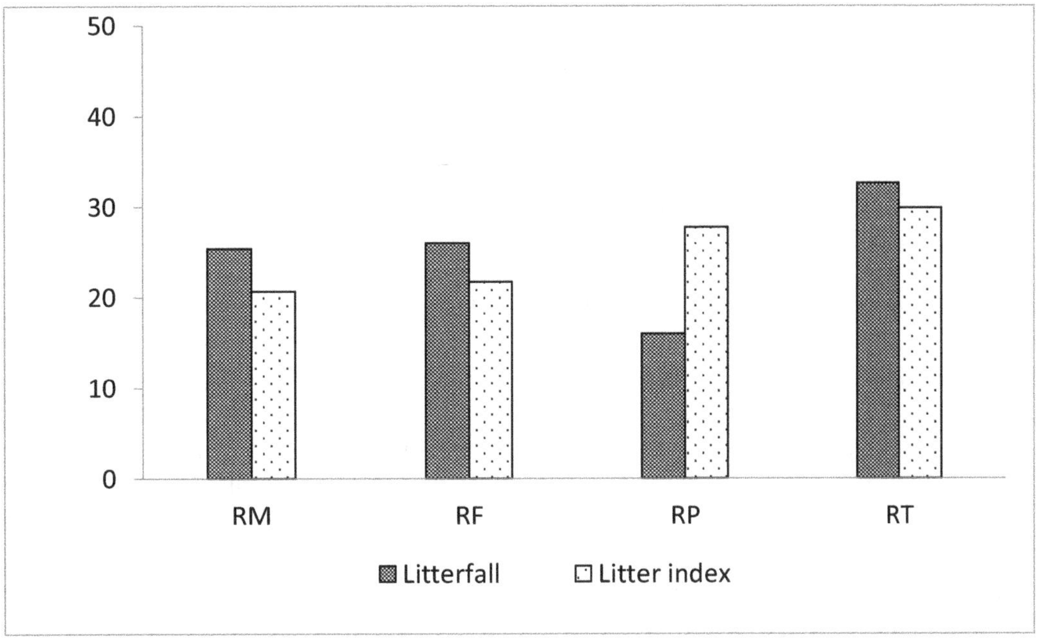

Figure 2. Percentage of litterfall and litter index in rubber monoculture and rubber-based agroforestry systems, rubber monoculture (RM), rubber with fruit (RF), rubber with pakliang (RP) and rubber with timber (RT).

3.3 Soil macrofauna

Macrofauna identified order-level classifications of Coleoptera, Hymenoptera, Isoptera, Isopoda, Diplura, Araneida, Chilopoda, Collembola, Stylommatophora and Earthworms at both rubber monoculture sites and rubber-based agroforestry plantations. These were not statistically different in the three intra-couple differences. Hymenoptera and Araneida were found in both systems. The macrofauna order communities were found in soil density at layers of 5-10 cm particularly in Pair III (RT and RMT). It was also noteworthy that earthworms were found in the topsoil, especially in rubber monoculture plantations, except in plots of RP (Pair II), which had a higher incidence of earthworms. Bruyn (1997) found that soil macrofauna, particularly earthworms, have been promoted as indicators of soil health. Earthworm function in soil processes is essential to understanding ecosystem function. Brown et al. (2001) reported that the diverse range of organisms and functions among macrofauna communities that perform in the soil combines to produce a diversity in biogenic soil structures that help regulate soil physical properties and processes, such as C and nutrient cycles. This further provides a whole host of ecosystem services that help increase heterogeneity in soils and improve soil ecosystem resilience and resistance to ecological disequilibria (e.g., pest outbreaks, degradation). Furthermore, several representatives of the soil fauna have also been proposed as bioindicators of soil health and sustainability. Many of the by-products of organisms are used as food resources by other soil fauna. Also included are the soil-feeding (geophagous) bioturbators, important in opening channels within the soil and on its surface, affecting hydrological processes and gaseous exchanges, as well as modifying soil structure, aggregate formation, and even soil formation rates. Finally, among the beneficiaries also counted are the predatory organisms that act at the top of the soil food chain, feeding on other soil and surface-dwelling or active organisms, controlling their populations and often helping to counteract pest outbreaks (thus acting as bio-control agents). Therefore, this study indicated that soil health in rubber-based agroforestry plantations had higher trends when compared with rubber monoculture plantations.

3.4 Decomposition of mesofauna

Decomposition factors in rubber monoculture and rubber-based agroforestry plantations had no statistical difference in the three intra-couple differences. Rubber-based agroforestry plantations tended towards higher decompose rates than rubber monoculture plantations (Table 4). However, the litter decomposition scale of mesofauna were found in RT plots, which decomposed at higher rates than other plots. This is probably due to factors resulted from high macrofauna communities (Figure 3). Bradford et al. (2002) reported that plant litter decomposition is a key process in C and nutrient cycling. The critical role of soil-faunal community composition in decomposition has been demonstrated using different mesh size litterbags to control the exposure of litter to different faunal size classes. Relative effects of manipulated faunal community composition on litter mass loss and C: N ratio was equivalent for green and senesced litter. The presence of meso- and macrofauna increased litter decomposition rates overall, despite inhibiting decomposition by microfauna, bacteria, and fungi through indirect effects. Ke et al. (2005) found that farming systems influenced decomposition by affecting both element turnover and mesofauna. In the soil at integrated farming sites, mesofauna reduced cumulative CO_2 production by 10% whereas in the soil from organic farming it increased CO_2 production by 20%. Villamor et al. (2014) reported that biodiversity in rubber agroforests indicate an eco-certification scheme opportunity by making the claim that such approaches help to reduce carbon emissions and species loss better than monoculture scenarios (monoculture rubber and oil palm). This study showed that mesofauna community dynamics implied that rubber-based agroforestry plantations have higher C and N turnover trends, which resulted in higher levels than in rubber monoculture plantations. However, a future study on the different management systems on soil mesofauna would be able to more accurately parse out these dynamics.

Table 3. Average of macrofauna orders found in soil layer at 0-5 cm and 5-10 cm in rubber monoculture and rubber-based agroforestry systems.

Macrofauna	Pair I						Pair II						Pair III					
	0-5 cm			5-10 cm			0-5 cm			5-10 cm			0-5 cm			5-10 cm		
	RF	MRF	T-test	RF	MRF	T-test	RP	MRP	T-test	RP	MRP	T-test	RT	MRT	T-test	RT	MRT	T-test
Coleoptera	0	0	-	0	0	-	0	0	-	0.67	0.57	ns	0.17	0.15	ns	0.67	0.77	ns
Hymenoptera	2	2.14	ns	0.33	0.33	ns	0.33	0.3	ns	1	0.85	ns	3.33	3.19	ns	5.67	5.22	ns
Isoptera	0	0	-	0	0	-	2.5	2.73	ns	0.33	0.38	ns	0.33	0.33	ns	0.5	0.43	ns
Isopoda	1.17	1.15	ns	0.5	0.48	ns	0	0	-	1.33	1.48	ns	0	0	-	2.17	2.39	ns
Diplura	0	0	-	0.33	0.3	ns	0.17	0.15	ns	1.67	1.62	ns	0	0	-	1.5	1.39	ns
Araneida	0.33	0.3	ns	0.17	0.15	ns	0.33	0.3	ns	0.5	0.52	ns	0.67	0.63	ns	0.33	0.33	ns
Chilopoda	0	0	-	0.33	0.33	ns	0.33	0.37	ns	0.33	0.38	ns	0	0	-	0.5	0.48	ns
Collembola	0.17	0.15	ns	0.33	0.3	ns	0.33	0.37	ns	0	0	-	0	0	-	0.33	0.28	ns
Stylommatophora	0	0	-	0	0	-	0.17	0.15	ns	0	0	-	0	0	-	0	0	-
Earthworm	0.11	0	ns	7.11	10	ns	0	0	-	14.33	12.11	ns	0.22	0.22	ns	8.89	12.33	ns

Note: ns=not significant, RF=rubber with fruit, RMF=rubber monoculture compared with RF, RP =rubber with pakliang, RMP=rubber monoculture compared with RP, RT=rubber with timber and RMT =rubber monoculture compared with RT.

Table 4. Comparison of litter decomposition in rubber monoculture and rubber-base agroforestry systems.

	Pair I			Pair II			Pair III		
	RF	RMF	T-test	RP	RMP	T-test	RT	RMT	T-test
Decomposition	0.99	0.82	ns	0.83	0.73	ns	0.86	0.78	ns

Note: ns=not significant, RF=rubber with fruit, RMF=rubber monoculture compared with RF, RP =rubber with pakliang, RMP=rubber monoculture compared with RP, RT=rubber with timber and RMT =rubber monoculture compared with RT.

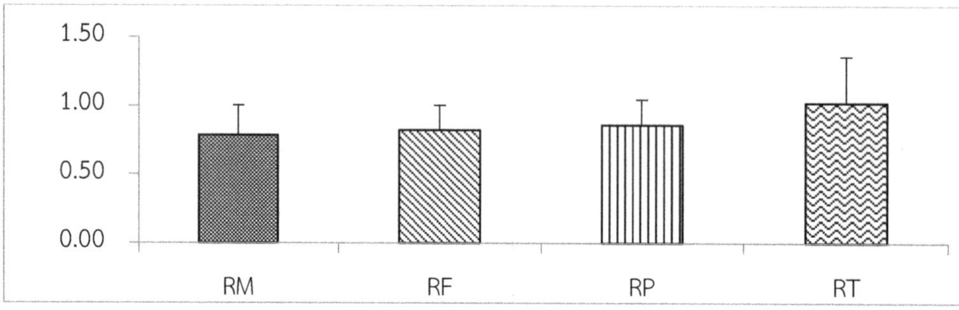

Figure 3. Litter decomposition scale of mesofauna fall in rubber monoculture and rubber-based agroforestry systems. Mean data calculated from a score of decomposition levels: 2 = complete decomposition, 1 = incomplete decomposition, and 0 = non decomposed between rubber monoculture (RM), rubber with fruit (RF), rubber with pakliang (RP) and rubber with timber (RT).

3.5 Physical-chemical soil properties

Three intra-couple differences were found by examining pH, EC, organic matter and nutrient contents, which were not statistically different in the two soil layers, except for calcium. Topsoil was more likely to have higher nutrient levels than subsoil, particularly in pair II (RP and RMP) and pair III (RT and MRT). Calcium, potassium, and magnesium in topsoil had a tendency to be high, especially in pair II (RP and RMP). The physical-chemical soil properties in intra-couples showed no differences due to the very similar areas. However, when compared between rubber monoculture with rubber-based agroforestry plantations, nutrient contents in the topsoil had higher values than the subsoil mainly evident from calcium content. RP and RT had soil nutrients in the topsoil higher than subsoil and also likely had higher nutrients than RF and RM. Calcium, potassium, and magnesium in topsoil were dominant among the intra-couples (Table 5). RP normally had less litterfall due to pakliang usage of young leaves for consumption like a vegeta ble, in which farmers needed to use more fertilizers for stimulation of the leaves during the juvenile phase. Nutrient content were found, probably a residual from fertilizer application. RT likely contained more nutrients due to the return levels from litterfall of rubber trees and timber trees. RF had fewer nutrient content due to the need for higher nutrients for fruit development and litter decomposition levels were lower. This study is consistent with Yang, et al. (2004), which reported that litterfall and litter decomposition acts as a source of essential nutrients. This includes N, P, and K that are returned to plants. There were also Ca and Mg, depending on the species and the type of litterfall. Pausch et al. (2013) mentioned that land uses also affect the amount of organic carbon produced by degradation of soil microorganisms. Farmland areas had organic carbon rather than natural and semi-natural areas. However, agricultural land use had differences in organic carbon and nitrogen content, such as the oak plots which had higher organic carbon and nitrogen content than the maize plots. Therefore, it can be seen the activities that affect land use also results in varying amounts of organic matter thus changing local ecosystem dynamics.

Table 5 Soil nutrients in topsoil (0-5 cm) and subsoil (0-5 cm) among rubber monoculture and rubber-based agroforestry systems.

Soil analysis	Pair I						Pair II						Pair III					
	topsoil (0-5 cm)			sub-soil (5-10 cm)			topsoil (0-5 cm)			subsoil (5-10 cm)			topsoil (0-5 cm)			subsoil (5-10 cm)		
	RF	RMF	T-test	RF	RMF	T-test	RP	RMP	T-test	RP	RMP	T-test	RT	MRT	T-test	RT	RMT	T-test
pH	5.07	5.07	ns	4.8	4.89	ns	5.01	4.92	ns	4.66	4.62	ns	5.14	4.78	ns	4.76	4.7	ns
EC (dS/m)	0.03	0.03	ns	0.02	0.02	ns	0.03	0.02	ns	0.03	0.02	ns	0.02	0.03	ns	0.02	0.03	ns
Organic matter (%)	1.86	1.83	ns	1.5	1.39	ns	2.10	1.67	ns	1.89	1.58	ns	1.53	1.69	ns	1.26	2.01	ns
Phosphorus (mg/kg)	5.01	3.76	ns	3.96	1.98	ns	5.81	2.52	ns	3.91	2.76	ns	3.7	4.18	ns	2.01	2.56	ns
Potassium (mg/kg)	56.8	72.86	ns	44.69	39.32	ns	63.44	34.03	ns	46.41	31.92	ns	58.89	54.96	ns	40.76	39.37	ns
Calcium (mg/kg)	60.68	65.43	ns	18.75	22.46	ns	128.24a	72.11b	*	51.55	62.01	ns	120.38a	58.35b	*	57.47a	30b	*
Magnesium (mg/kg)	21.13	20.54	ns	7.52	7.19	ns	24.76	17.33	ns	13.09	13.01	ns	29.55	15.52	ns	15.21	5.99	ns

Note: ns=not significant, * = significant at $p \leq 0.05$, Different litters indicate a significant difference of the same litter in each rubber plantation system. RF = rubber with fruit, RMF=rubber monoculture compared with RF, RP = rubber with pakliang, RMP = rubber monoculture compared with RP, RT = rubber with timber and RMT = rubber monoculture compared with RT.

4. Conclusion

Three intra-couple differences were examined in this study and did not yield statistically significant differences in litterfall, litter decomposition, soil macrofauna, and nutrient contents among rubber monoculture and rubber-based agroforestry plantations. However, rubber-based agroforestry plantations, particularly RT sites, had litterfall levels higher than other rubber plantations and had the tendency for higher plant part decomposition. Macrofauna was also found at higher density levels in rubber-based agroforestry plantations. Nutrient content was also higher in the topsoil at these sites, and calcium dominated the intra-couples. This information provides evidence that therefore supports rubber farmer decisions to grow rubber-based agroforestry due to the carbon returns from litterfall in rubber-based agroforestry plantations. This will help regulate physical properties and processes of the soil, and further support C and nutrient cycles that benefit against low external input fertilizer management. Therefore, government agencies should support these practices, and recommend that other farmers in the region also consider the broader soil dynamics. This study has shown that these more fine-grained ecosystem dynamics can also help to make a difference in changing agrarian practices.

Acknowledgments

Thanks to the Hevea Research Platform in Partnership and LUSES, An International Joint Laboratory on the Impacts of Rapid Land Use Change on Soil Ecosystem Services who supported this research financially. Special thanks to the rubber farmers that allowed us on their rubber plantations and took the time to discuss with us for this study.

References

Austin A. T., Luc V., Adelia G. A. & Luis I. P. (2014). There's no place like home? An exploration of the mechanisms behind plant litter–decomposer affinity in terrestrial ecosystems. New Phytologist. 204: 307–314.

Bradford, M.A., George M. T., Till E., Jones T. H. & John E. N. (2002). Microbiota, fauna, and mesh size interactions in litter decomposition. OIKOS. 99 (2): 317-323.

Brown, G.G., Amarildo P., Norton P.B., Adriana M.A. & Maria E.F.C. (2001). Diversity and functional role of soil macrofauna communities in Brazilian no tillage agroecosystems: A preliminary analysis. Paper based on an oral presentation at the "International Symposium on Managing Biodiversity in Agricultural Ecosystems" Montreal, Canada, 8-10 November, 2001.

Bruyn, L. A. L. (1997). The status of soil macrofauna as indicators of soil health to monitor the sustainability of Australian agricultural soils. Ecological Economics. 23 (2): 167-178.

Chen, C., W. L. Xiaojin & J. J. Wu. (2017). Effects of rubber-based agroforestry systems on soil aggregation and associated soil organic carbon: Implications for land use. Geoderma. 299:13-24.

Cizungua, L., Jeroen S., Dries H., Jean W., Daniel M., Oswald V. C. & Pascal B. (2014). Litterfall and leaf litter decomposition in a Central African tropical mountain forest and eucalyptus plantation. Forest Ecology and Management. 326: 109–116.

Kaspari, M., Milton N. G., Kyle E. H., Mirna S., Joseph S. W. & Joseph B. Y. (2008). Multiple nutrients limit litterfall and decomposition in a tropical forest. Ecology Letters. 11 (1): 35-43.

Ke, X. Karin W. & Juliane F. (2005). Effects of soil mesofauna and farming management on decomposition of clover litter: a microcosm experiment. Soil Biology and Biochemistry. 37 (4): 731-738.

ORRAF (Office of Rubber Replanting Aid Fund). (2006). A study of problems and occupational development needs of rubber plantation farmers. Research and Planning Office of the Rubber Aid Fund. Ministry of Agriculture and Cooperatives. (In Thai).

Pausch, J., Zhu B., Kuzyakov Y. & Cheng W. (2013). Plant inter-species effects on rhizosphere priming of soil organic matter decomposition. Soil Biology & Biochemistry. 57: 91-93.

Sarah, A. (2016). There's no place like home an exploration of the mechanisms behind plant litter–decomposer affinity interrestrial ecosystems. https://besplantsoileco.wordpress.com accessed on 28 September 2016.

Stroesser, L., Penot E., Michel I., Tongkaemkaew U. & Chambon B. (2018). Income diversification for rubber farmers through agroforestry practices: How to withstand rubber price volatility in Phatthalung Province, Thailand. Revue internationale des études du développement. 235(3): 115-143.

Villamor, G.B., Quang B. L., Utkur D., Meinevan N. & Paul L.G.V. (2014). Biodiversity in rubber agroforests, carbon emissions, and rural livelihoods: An agent-based model of land-use dynamics in lowland Sumatra. Environmental Modelling & Software. 61: 151-165.

Wang, Q., S. Wang & Y. Huang. (2008). Comparisons of litterfall, litter decomposition and nutrient return in a monoculture Cunninghamia lanceolata and a mixed stand in southern China. Forest Ecology and Management. 255: 1210–1218.

Yang, Y. S., J. F. Guo, G. S. Chen, J. S. Xie, L. P. Cail & P. Lin. (2004). Litterfall, nutrient return, and leaf-litter decomposition in four plantations compared with a natural forest in subtropical China. Annals of Forest Science. 61(5): 465-476.

Pabsta, H., Kuhnelb A., Kuzyakova Y. (2013). Effect of land-use and elevation on microbial biomass and water extractable carbon in soils of Mt. Kilimanjaro ecosystems. Applied Soil Ecology. 67: 10-9.

6

Forest, water and people: The roles and limits of mediation in transforming watershed conflict in Northern Thailand

Ahmad Dhiaulhaq[1,2,*], **Kanchana Wiset**[1], **Rawee Thaworn**[1], **Seth Kane**[1], **David Gritten**[1]

[1] RECOFTC – The Center for People and Forests, Kasetsart University, Bangkok, 10903, Thailand; e-mail: kanchana.wiset@gmail.com, rawee@recoftc.org, seth.kane100@gmail.com, david.gritten@recoftc.org
[2] Crawford School of Public Policy, Australian National University. Stanner Building, 37 Lennox Crossing, ANU, Canberra, ACT 2601; e-mail: ahmad.dhiaulhaq@anu.edu.au
* Corresponding author

Abstract: This study focuses on watershed management in Northern Thailand, where conflict over forest, land and water-use is a prevailing problem. A characteristic of watershed conflicts is that they are often multifaceted and involve multiple stakeholders with different interests and values, consequently requiring conflict management approaches that are sustainable in their outcomes, including addressing the underlying causes of the conflicts. Drawing from a case study in Mae Tia Mae Tae watershed in Northern Thailand, this study explores how mediation by external third party can contribute to the transformation of conflicts in the watershed and how the broader institutional contexts in which the conflict is embedded shapes the mediation outcomes. The study suggests that co-creation of mutual understanding and recognition of each party's socio-cultural differences, including land-use practices, are critical in building trust and in how conflict transformation processes moved forward. Moreover, the ability of the mediator in facilitating the establishment of a deliberative institution (i.e. a watershed network committee) and agreed rules on forest utilization were also critical in maintaining long-term collaboration in the watershed and potentially preventing other conflicts arising in the future. Some issues, however, may threaten the continuity of the cooperation and sustainability of peace in the watershed, including the lack of structural reform that formally recognizes local people's rights, insecure land tenure, and the absence of legal recognition for the watershed network committee as a legitimate mechanism for watershed decision making. The paper discusses these findings by comparing it with those from our previous studies in other locations (Cambodia, Indonesia and Western Thailand) to strengthen the insights from Northern Thailand. Finally, the research puts forward some recommendations for reforms and to strengthen the use of effective mediation, to achieve transformative outcomes, in conflicts of this nature.

Keywords: Watershed management; forest conflict; mediation; conflict transformation; Thailand

---

1. Introduction

Northern Thailand is known for its rich natural environment, including dense forests and high levels of biodiversity. The mountainous region is also known for its cultural diversity, as it is home to both ethnic northern Thais, who mainly live in the lowland areas and minority hill-tribes (e.g. Karen and Hmong) who live mainly at higher elevations (Uhlig, 1980; Wittayapak and Dearden, 1999). There is a great deal of pressure on these ecologically and culturally rich landscapes, as well as a large amount of contestation between various stakeholder groups. Numerous publications have shown that conflict often occurs in watersheds in Northern Thailand between the hill-tribes minority groups living at higher elevations and forest authorities (Laungaramsri, 2000; Tungittiplakorn, 1995) as well as between these minority groups and communities (e.g. farmers) residing lower down the watershed over land-use practices and water management (Wittayapak and Dearden, 1999; Laungaramsri, 2000). The tension has also been driven by other social and political issues such as conservation movement, social prejudices among different ethnic groups, perceived illegal immigration and efforts to eradicate opium poppy cultivation in the Northern highland areas (Hares, 2009). For example, upstream ethnic minorities are largely perceived by forest authorities and the lowland communities to be responsible for uphill forest degradation due to their settlement and land-use practices (e.g. swidden agriculture), negatively affect downstream communities (Laungaramsri, 2000; Hares, 2009).

Conflict over watershed management is a common issue in many countries around the world (e.g. Steinberg and Clark, 1999; Imperial and Kauneckis, 2003; Church, 2009; Tungittiplakorn, 1995), reflecting the plethora of stakeholders with differing views, values and interests (e.g. socio-cultural, political, environmental and economic) over the management of forest, land and water resources. These characteristics can also explain why watershed conflicts are usually complex; they often involve multi-parties or multiple stakeholders, often with a large number of constituents with multiple objectives, diverse socio-economic status, culture, ethnicity and values. These types of conflicts require conflict management approaches that are sustainable in their outcomes, including addressing the underlying causes of the conflicts.

While studies have shown that natural resource conflicts in certain circumstances can bring positive impacts such as strengthening collective action, opening a space for communication and negotiation between the conflict parties, and increasing awareness of the need to clarify and address tenure problems (de Koning et al., 2008; Yasmi et al., 2009, 2013; Dhiaulhaq et al., 2014b, 2015), in most cases forest conflicts predominantly bring negative consequences (economic, environmental and social) such as triggering destructive behaviors, anxiety, fear, forest depletion, and reduced income (Yasmi et al., 2009, 2013). Consequently, effective conflict transformation mechanisms are crucial to ensure that conflicts have constructive outcomes.

This study is concerned with watershed in Northern Thailand, where forest and land conflict is a prevailing watershed management problem. While several studies have been conducted on watershed conflicts in Northern Thailand (e.g. Buergin and Kessler, 2000; Laungaramsri, 2000; Delang, 2004; Hares, 2009), their focus, however, is mainly on examining the causes, intensity, impacts, and actors in watershed conflicts. Research on how these conflicts in Northern Thailand and watershed conflicts in general are addressed is underdeveloped (Dhiaulhaq et al., 2015). This study intends to address these gaps. Drawing from a case study in Mae Tia Mae Tae (MTMT) watershed in Northern Thailand, this study explores whether third-party mediation can contribute to the transformation of conflicts in the watershed. This entails examining what processes lead to positive outcomes and how broader institutional structures or contexts shape the mediation outcomes. Additionally, as this study is part of a regional study (in Cambodia, Indonesia and Thailand) analysing the role and effectiveness of conflict mediation in transforming forest and land conflicts, insights from our other cases (Dhiaulhaq et al., 2014a, b, 2015) are also presented and discussed to assess similarities and differences.

2. Analysing conflict mediation through a transformative mediation framework

Mediation can be understood as an assisted negotiation process facilitated by a third party (or parties) which aims to assist the conflicting parties to find mutually acceptable solutions, including by addressing the underlying causes of the conflict (e.g. Wall et al., 2001; Engel and Korf, 2005). Many scholars believe that the use of mediation, as a form of alternative dispute resolution (ADR), can enhance procedural justice by devolving decision-making to the conflicting parties, encouraging equality among parties, as well as minimizing coercion, for example through legal measures (e.g. Rubenstein, 1999). This paper follows those scholars who view the mediation field as diverse and pluralistic—not all mediators follow the same practices and objectives (e.g. Augsburger, 1992; LeResche, 1992; Bush and Folger, 2005). As an example, some authors have pointed out that mediation in a Western and non-Western culture can take different pathways or strategies in addressing conflicts (Augsburger, 1992; Callister and Wall, 2004; Lee and Teh, 2009).

Within the mediation field, transformational approaches are increasingly prominent for both practitioners and scholars. Conflict transformation aims not only to end or settle conflicts but also to create a major shift in the quality of interaction and promote conditions that create long-term constructive relations (Mitchell, 2002; Reimann, 2004; Bush and Folger, 2005). The transformative approach is often put in contrast with 'problem-solving' or 'settlement' approach which emphasizes generating a 'resolution' in order to settle the conflict. In contrast, the transformative mediation model does not put resolution or consensus as their immediate priority.

Mitchell (2002) provides an excellent review of different views within the 'conflict transformation school'. Although different scholars stress different aspects of what conflict transformation transforms, she argues that most seem to agree that the core aspect that needs to be transformed includes changes at the micro (i.e. personal and interpersonal or relationship level) and macro level (i.e. structural level). Similarly, Vayrynen (1991) argues that conflict transformation may involve: 1) *Actor transformation*, which involves major internal changes within the conflict parties and their relations with others; 2) *Issue transformation*, which involves changes in the political issue of the conflict, such as finding common ground among the parties; 3) *Rule transformation*, which involves changes in the norms shaping the parties' interactions; and 4) *Structural transformation*, which involves changes in the structural issues that cause the conflict. We use these frameworks to structure our analysis in this article.

Additionally, in investigating the 'micro-level' conflict mediation process and outcomes, we also use Bush and Folger's (1994; 2005) 'transformative mediation' framework to guide our analysis. The role of the mediator in this model is expanded to assist conflicting parties to shift towards more positive interactions through, and characterized by two key aspects: empowerment and recognition. Empowerment here defined by Bush and Folger as improving the capability of the conflict parties to handle and make better decisions on their conflict problems. This includes the ability to clarify their goals, options and preferences as well as to communicate and negotiate these with other parties. Regarding recognition, it includes the willingness of a party to acknowledge, respect and empathize with the perspectives, views, situation and experiences of others. Bush and Folger (2005) argue that empowerment and recognition are the keys to transformative mediation because when these occur in conflict, the quality of the interaction will transform from destructive to constructive.

Moreover, in this article, we also examine the "macro" perspective of the socio-political context that shapes the conflict, mediation processes and outcomes, thereby providing a more comprehensive examination of the mediation. The socio-political context may include power relations, institutional structure governing forests, discourses, economic interests, and political and cultural environments (e.g. Peluso and Watts, 2001; Raitio, 2012; Zachrisson and Lindahl, 2013). This is mainly because mediation and conflict do not occur in a vacuum. Lederach (1997) suggested that the goals of conflict transformation include enhanced wellbeing, interdependence, and justice, and where possible strive to change and address systemic factors and policies that underlie conflicts. This study will assess whether the processes and outcomes of mediation in the case study site satisfy these various definitions and goals of conflict transformation.

3. Methodology

The empirical focus of this article is on Mae Tia Mae Tae (MTMT) watershed in the Northern Thailand. The findings presented in this study are drawn from data gathered by authors in 2012 as part of a regional research project examining forest-conflict mediation in Southeast Asia (Dhiaulhaq et al., 2015). The study was designed as a qualitative research and employed several methods of data collection, including: 1) semi-structured interviews; 2) focus group discussions (FGDs); and 3) field observations in the MTMT watershed. Semi-structured interviews were conducted in August 2012 involving 54 key informants (37 males and 17 females) representing the conflict and mediation stakeholders, including members of highland and lowland communities, mediator team (i.e. the Sustainable Development Foundation, SDF), government staffs and national park officers. Additionally, three FGDs [with representatives of the mediator team, the lowland and highland communities, and members of the watershed committee] were also conducted. The field observations were also undertaken to understand local land use and livelihoods pattern, as well as conservation activities in the watershed. The analysis also included the collection and review of reports, newspapers, and journal articles. Some of the findings from the 2012 fieldwork were also included in the earlier regional-level article (Dhiaulhaq et al., 2015). Further interview with the SDF team was conducted in June 2017 to get some updates on the findings from the earlier fieldwork, including examining the impacts of the mediation.

MTMT watershed is a sub-watershed of larger Mae Krang watershed. It is located in Chom Thong District, about 2-3 hours drive from the provincial capital of Chiang Mai. It is a large (approximately 14,800 hectares) mountainous landscape with an altitude range of 300 – 1,740 meters. Approximately 84 percent of the land in this watershed (148 km^2) is forest: hill evergreen, dry dipterocarp, and mixed deciduous. Most of the watershed is government-designated protected areas, which includes Ob Luang National Park, Doi Inthanon National Park and National Forest Reserved area. It is a source for the Mae Tia and Mae Tae Streams which feed the Ping River. The area is inhabited by various communities in 23 villages scattered in three sub-districts (Doi Kaew, Sob Tea, and Mae Soi) in Chom Thong District. It is home to three major ethnic groups: Hmong (one large community), *Paka-Kyaw*--also known as Karen (five communities), and Northern Thai communities who live in the downstream of the watershed (17 communities). MTMT watershed is also home to at least 147 wildlife species which can be divided into four groups: amphibians (9 species), reptiles (23 species), birds (97 species) and mammals (12 species). Some of them are rare wildlife species protected by the Government based on 1992 Wildlife Preservation and Protection Act B.E. 2535.

4. Results

4.1. The conflict in Mae Tia Mae Tae watershed

The case study exemplifies the complex nature of conflict over watershed management in Northern Thailand; it involves multiple parties often with differing interests and values. Several social, environmental and economic changes in the area directly and indirectly affected relationships between the upland communities, lowland communities and forest authorities, leading to conflict between these actors (Figure 1).

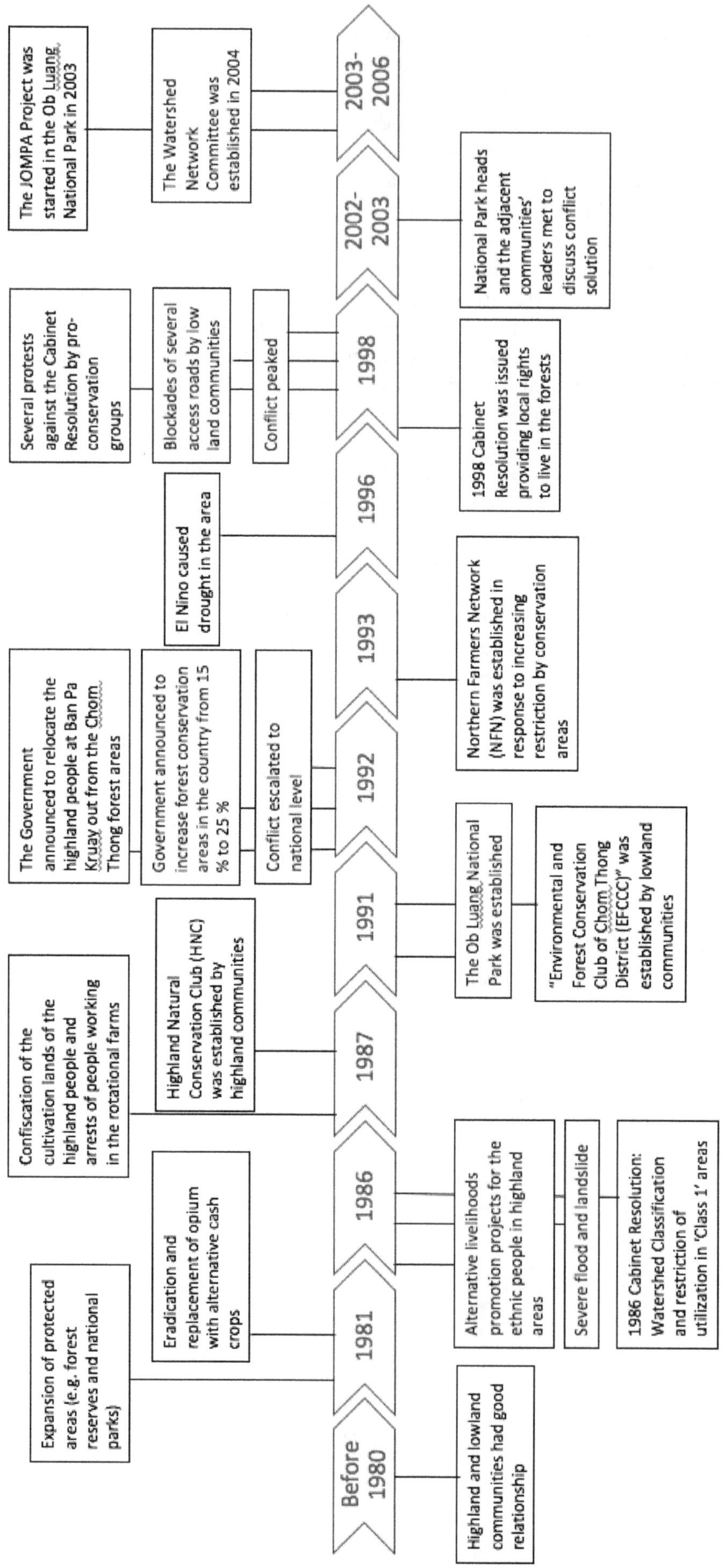

Figure 1. Timeline of key events related to conflict in Mae Tia Mae Tae Watershed

During our fieldwork, both upland and lowland communities, especially the elders, were able to recall that both groups of communities had good relationships in the past (before the 1980s), indicated by joint-agricultural activities and substantial trade in forest and agricultural products. Changes in livelihood, water and land use patterns (Figure 1) gradually change this relationship and contribute to the emergence of the MTMT watershed conflict (Bonell and Bruijnzeel, 2005), especially after the State-led opium substitution programs starting in the 1980s encouraged cash, mono-crop agriculture (e.g. cabbage and other vegetables for upland communities). This coincided with many lowland communities converting rice fields to Longan (*Dimocarpus longan*) plantations to earn more income. The changes in these land-use practices, such as growing cabbage and longan, requires more water creating competition for resources, particularly in the dry season. This dynamic was particularly pronounced in the 1996 – 2001 period, which saw regular droughts and increased inter-communal tensions.

Additionally, state territorialization (Vandergeest and Peluso, 1995) and exclusionary conservation policies were also responsible for the escalation of conflict in MTMT watershed. During the 1980s and 1990s, partly due to the heightened concern for environmental conservation as a result of rapid forest degradation, state actors zoned large areas of forest and land in the Northern Thailand as protected areas. In 1981 and 1991, for example, the Thai Government extended the boundaries the Doi Inthanon National Park and established Ob Luang National Park (OLNP) respectively. These protected areas covered large parts of the MTMT watershed, including parts of cultivation lands of the highland communities. Additionally, in 1986, the government also released a Cabinet Resolution establishing the Watershed Classification System, which designated most mountainous areas in Northern Thailand as Watershed Area Class 1 (WSC1), prohibiting natural resource extraction and human settlements[1]. These regulations directly affected highland communities who have been living in the mountainous forest areas for generations. Conflict incidences arose when the state territorialization was strictly enforced, with the exclusion of local inhabitants. Upland communities reported that there were incidences where many highland farmers were arrested due to their cultivation activities and destruction of the community member's crops by National Park officials.

The exclusionary approach was also proposed by some 'dark green' conservation NGOs operating in the area. Interviewed communities, especially in the upland, said that Conservation movement promoted by some NGOs exacerbated the conflict due to its tendency to recommend resettlement of the upland ethnic minorities to other locations as a solution to preserve the headwater and overall watershed area, and to maintain the water supply to the lower plain (see also Laungaramsri, 2000). The reason for relocation was largely influenced by the perception that the ecological problems in the downstream area were mainly caused by the upland ethnic minority communities settling and cultivating the headwater areas in the upstream (Laungaramsri, 2000). The movement rapidly received significant support from downstream villagers who increasingly needed water for their expanding mono-crop agriculture, and have complained about water shortages in their orchards and paddy fields (Forsyth and Walker, 2008). In the 1990s some lowland communities also formed a conservation group called Environmental and Forest Conservation Club of Chom Thong District (EFCCC) that also advocated for relocating highland communities from the forests. These NGOs were able to gain significant social support amongst lowlanders. In that period, the government passed the 1998 Cabinet Resolution acknowledging certain hill tribe forest rights, including rights to remain staying in the forest which the 'dark green' conservation NGOs and lowlanders perceived it as a deforestation and forest encroachment risk (Yamaguchi, 2006).

In response, highland communities also established their own group called the Highland Natural Conservation Club in Chom Thong District (HNC). Their movement was supported by local

---

[1] The watershed classification in Thailand is identified into 5 zones with the first (WSC1) and the second (WSC2) preserved as protection and commercial forests respectively and the rest (WSC3-WSC5) can be used for agricultural and other purposes with appropriate conservation and mitigation measures (See Tangtham 1996).

NGOs, such as the Inter Mountain Peoples Education and Culture in Thailand Association (IMPECT), which worked primarily with ethnic minority groups.

In the late 1990s, the conflict between highland and lowland communities escalated, peaking in 1998 when lowland communities protested by obstructing several main roads to restrict highland community transportation. The roads blockade, which lasted about three months, caused serious inconvenience, especially for the highland communities. Economic impacts were felt by both parties, as travel and trade were impeded. Some schools were closed on occasion, and protesting took time away from the fields and other livelihood activities. In response to the blockade, the highland group, represented by HNC, undertook several mass protests at the local district and provincial government offices. The conflict eventually garnered significant attention, including from national news media. This conflict case was also used by the 'dark green' conservation NGOs to support their cause against the Community Forest Bill in the national level debate (Delang, 2004).

The conflict had many negative impacts, including disharmony and division within and between communities, including depression, fear, and violence, especially when the conflict was at its peak; with livelihood implications for both highland and lowland community members. During the conflict, for example, some lowland communities erected fences painted with red, white and blue colors—representing Thailand's national flag, causing the highland communities to perceive that they were considered unlawful non-Thai immigrants. One of upland villagers, for example, explained what he felt when he saw the fences: "I felt that we are not considered Thai people anymore. I was so upset because the poles' fences were made and painted with the color of Thai flag. At that moment, I felt that they thought that we are alien residents, like Burmese immigrants." As a result of the tensions, highlanders who live in the lowland areas could not visit relatives in the highlands. Highland communities also always prepared weapons, as they feared they would be ousted from their lands by force.

4.2. The process of conflict transformation

4.2.1 How the process was initiated

There were a number of informal attempts to address the conflict in the watershed. In 2002, for example, the head of OLNP together with the leaders of communities living in and next to the National Park area, for example, organized a discussion to find possible solutions to the conflict. In 2003, the Office of Natural Resource and Environmental Policy and Planning (ONEP), a government agency, organized a forum, with the support of key agencies in the area, i.e. Doi Kaew Tambol Administration Office and Sustainable Development Foundation (SDF). The forum invited the leaders of both upland and lowland communities to discuss the conflict issues and gave an opportunity to the community leaders to express their concerns and exchange information.

The attempt to address the conflict in a more systematic way through mediation was initiated when a local NGO, the Sustainable Development Foundation (SDF), started a project promoting joint management in protected areas (JOMPA project) in the OLNP in 2003. Initially, SDF did not focus on managing the conflict because their main objective was to promote participatory and collaborative processes in natural resource management. However, SDF discovered that the conflict intensity in the area hindered cooperation between the upland and lowland communities in managing their forests. SDF also realized there was a lack of actors who could act as an intermediary or a neutral party who can mediate the conflict.

In this regard, as part of the JOMPA project, SDF team conducted a preliminary assessment of the conflict, including undertaking a stakeholder analysis, identifying the main issues, and assessing the potential for mediation. SDF strategically used its prior relationships with the local communities from their extensive experience in the area and throughout Chiang Mai province. To approach the conflict parties, SDF used both formal and informal methods, such as talking individually to community leaders. Then the SDF identified key persons who would work with them in the processes of addressing the conflict and build collaboration among MTMT stakeholders. The upland and lowland communities eventually agreed to participate in the processes facilitated by the SDF in order to achieve these objectives and address their conflicts together.

In this regard, different to conventional mediation where a mediator is normally appointed by the conflict parties, in MTMT case, the SDF staffs were the ones who proactively approached the conflict parties, as part of their efforts to develop collaborative forest management in the area. The leader of SDF explained that the aim of the mediation was not only to resolve the conflict but also to build long-term collaboration in the watershed management; so that both parties could manage and mutually benefit from the natural resources in the MTMT watershed. He also explained that: "We focused on facilitating collaboration processes and the setting of long-term common goals together among the villagers, [because] when collaboration increases, conflict will also decrease." In total, the conflict mediation process facilitated by the SDF took more than three years (Figure 1), including some stages presented in the following sections.

4.2.2. Actor and issue transformation

The study suggests that co-creation of mutual understanding between the conflicting parties and recognition of each party's socio-cultural differences, including land-use practices, are a critical foundation for building trust and serve as a prerequisite for conflict transformation processes to move forward. To achieve this, at the initial stage, the mediator team, supported by the local government (the sub-district administrative organization, TAO) organized several meetings attended by conflict stakeholders. In these initial meetings, the mediator facilitated the representatives of upland and lowland communities to share their concerns and problems. The aim was to clarify the issues and create a mutual understanding and trust between parties. One of the villagers from the lowland recalled that in this first meeting "many exchanges of arguments and blaming to each other occurred in that meeting which SDF tried to mediate."

Additionally, exchange visits were also conducted between representatives of lowland and upland communities. The mediators also used videos as a tool to collect information about each party's culture and land-use practices. The videos showing the livelihood systems of both upland and lowland communities, including interviews with influential people in the area, were also shown to both communities to improve mutual understanding about other parties' culture and land-use practices.

As results of these processes, a change in understanding and perception about other parties' culture and livelihood patterns constituted important milestones for moving forward in addressing the conflict. These also led to a change of attitudes towards one another—from one of recrimination to mutual understanding and respect. One of lowland villagers, for example, describes this change in their interaction: "The relationship now has improved. The lowland people now can go up to the upland areas, and the upland people will greet and invite us for meals. During the conflict, they would not even talk to the lowland communities, and they would not even give water to us."

One of the most important processes facilitated by the mediators was the creation of the MTMT watershed network committee, as a platform for discussing conflict and other watershed-management issues. During one of the joint-meetings between upland and lowland communities, the mediator team proposed an idea of forming an MTMT watershed network committee as a platform for lowland and upland people to exchange information and discuss their problems in managing the watershed. The idea of establishing MTMT watershed network committee was also to ensure that any potential issues that might lead to conflict in the future could be addressed and therefore could contribute to the sustainability of a peaceful environment and maintain long-term cooperation. Although the initial idea to establish the network committee came from the mediator team, the decision was agreed and made by both groups of communities, and endorsed by the NP officials. The network consisted of representatives from 23 villages (two representatives from each village) in the upland and lowland areas and was supported by the head of OLNP. One of committee members from an upland village reflected: "If we still blame each other, we cannot work well, and if all people cooperate and are discipline in using forests and water, the problems will not happen. At least, if a problem comes up, we can still solve it through our cooperation."

However, the mediators consider that although the watershed committee offers a deliberative and problem-solving mechanism and participation in watershed management, it is currently still an informal local platform without a legal basis, raising a question of the legitimacy of the decisions made. This issue will be discussed in the discussion section.

4.2.3. Rule transformation

Once the watershed network committee was established, in order to address the contested boundaries, a participatory land demarcation process was conducted to map the land utilized by communities, including the settlement, agriculture and protected areas. The demarcation of the boundaries involved representatives of both upland and lowland communities, the National Park and the mediator team. To ensure the effectiveness of the participatory processes, before this demarcation process, the mediator team trained all representatives and the watershed network committee members to use GPS as well as survey and mapping techniques, so all parties could work and communicate in the same technical 'language'. Based on the results of the participatory demarcation process, the mediators converted the result of land demarcation into computerized maps (i.e. GIS). Considering the wide area, the number of issues and stakeholders involved, the demarcation process took over three months. The boundary maps were edited three times until all parties agreed.

Once the boundaries were clarified, the parties started a discussion on establishing rules and regulations on land and forest resource utilization in the watershed, facilitated by the mediator team. These rules were established through participatory processes—the mediators invited the communities' representatives to discuss the rules and regulations so they would fit within the existing local rules and context of each community, including their local or traditional practice of forest management. In order to gain recognition of the results of the land demarcation and the agreed rules, the mediator team proposed that these agreed rules be presented to relevant organizations such as the NPs and local governments for endorsement.

The agreement on these rules helped both upland and lowland communities understand their rights and responsibilities in managing and utilizing the natural resources in the area. As the rules were the result of participatory process, it contributed to the compliance to the rules, including preventing encroachment. For the government, the mediation supported the work of the NP officers in protecting the forest. A park ranger reflected that the success of mediation and compliance with the rules and regulations by the communities made the work of the national park officers easier. He gave an example that now, when the local communities meet the NP officers they are willing to talk and sometimes offer them meals. In cases where someone breaks the rules, the officers let their leaders address the issue first according to the communities' rules, with the national park officers only acting if the communities cannot manage the cases themselves.

Following the conflict's de-escalation, the mediators started building collaborative and institutional capacity of all communities in managing the watershed, including conserving forest and developing alternative livelihoods sources. The communities and mediator team ensured that conservation activities aligned with the local culture and beliefs of both groups of communities. For example, the lowland communities conducted forest ordinations and water ceremonies, and the highland communities worshipped forest and water spirits while inviting representatives to participate in each other's activities. Moreover, both communities also worked together to make forest fire breaks supported by the local government.

4.2.4. Limited structural transformation

Interviewed villagers said that the boundary demarcation process, including clarification and demarcation of land functions (e.g. agriculture, protected areas) resulted in a better sense of security and recognition of their rights towards land use, at least at the local level. This was partly the result of participatory boundary demarcation being acknowledged by the NPs and local government staffs. The agreed boundaries also contributed to the reduced access restriction to forest resources in agreed locations. After the land demarcation and clarification of the boundaries, villagers said that they were able to continue utilizing the land for agriculture and other income generation activities without disruption. The upland communities, for example, now can grow their crops throughout the year without worrying about being arrested by NP officers. Additionally, the agreed rules, which include agreement on forest and water management, also ensures that both communities have a better distribution of water for household consumption and agricultural activities contributing to income security.

However, despite these improved situations, the mediators and communities realized that the agreed land demarcation and maps could not be used as a legal basis for communities' land rights, causing insecurity for forest dwellers. This is because the State Law, particularly the NP Act (1962) strictly prohibited settlement and agricultural activity in NP areas. In this regard, change at a larger scale (e.g. at the national level), in terms of securing legal recognition of local land rights is still limited. This also shows the limits of mediation in addressing conflict which involve structural and public policy problems, particularly in the forestry context. Up until the time of writing of this article, the SDF is still facilitating community efforts in gaining a legal right from the Department of National Parks, Wildlife and Plant Conservation (DNP).

A stronger, formal recognition of both the local land rights and the watershed network platform are considered important by the mediators and communities to provide a more secure situation, especially regarding land tenure. The mediator mentioned that in a situation where, for example, there is a leadership change in the protected areas, the policy might also change. Moreover, a new leader may not have the same level of awareness of the conflict issues in the area, which may make him/her make a policy that goes to a different direction. The formalization of recognition is to ensure that whoever lead the NP, s/he will promote similar participatory policy and recognition of local land rights.

5. Discussion and concluding remarks

5.1. The need to enhance recognition

In this study, we have sought to understand the role of external third-party mediation to facilitate the process of addressing a watershed conflict, its implications for conflict transformation and the challenges it faced. The findings presented in earlier sections suggested that an external third-party mediation has the potential to play an important role in opening up dialogue among the conflict parties and changing conflict interactions among watershed stakeholders, especially when the parties could not address the conflict by themselves.

The study, however, found that despite some improvements in the quality of interaction among the conflict parties during and after the mediation process (actor and local rule transformations), the mediation suffers several limitations, especially at the higher and structural level. An important limitation in the mediation in MTMT is that some of the goals to change the structural issues relating to recognition of rights, that became the underlying cause of the conflict, have not been fully achieved. For example, although after the participatory land demarcation the local communities were able to continue to live and do their agricultural activities, they still do not have a strong legal basis for their land rights. The fact that some of their agricultural plots are within the NP area, it means they are still regulated under the 1961 NP Act (Government of Thailand, 1961) which prohibits settlement, forest product extraction and agricultural activity in NP areas. Patel et al. (2013) argue that access to forest resources is essential in meeting subsistence needs of local stakeholders, and therefore any policies or practices that strictly limit local access to forest resources will lead to conflict. In this regard, the study suggests that change at a larger scale (e.g. at the national level), in terms of securing legal recognition of rights (e.g. rights to access, use, manage, live in the forest) is still needed and should be facilitated by Thai law. If the rights of local people over forest resources is not secured, the conflicts may reoccur in the future.

This study therefore shows the limit of mediation in addressing conflict which involves structural and public policy problems, particularly in the forestry context. A similar issue was also found in other research sites including in Kanchanaburi, Western Thailand (Dhiaulhaq et al. 2014b, 2015), where the agreement between local communities and National Park could not be used as a strong legal basis for their land tenure although it eventually allowed the communities to stay and continue their agricultural activities (Dhiaulhaq et al., 2014b, 2015). However, as de Koning et al. (2008) point out, processes of getting recognition of rights in the context of forest management are political rather than technical. They are subject to a growing competition and may also face opposition from vested interests in land and forest resources (e.g. from authorities in charge of forest and conservation).

Based on the insights from the MTMT mediation, at the conceptual level, the paper would suggest expanding Bush and Folger's (2005) concept and definition of recognition, to also include institutional recognition, for example, recognition of rights by the state. Bush & Folger's definition of recognition, which includes the willingness of a party to acknowledge, respect and empathize with the perspectives, views, situation and experiences of others, may not be enough when the conflict involves structural or public policy issues. This paper would argue that in the case such as MTMT watershed, recognition should be beyond the level of actors (i.e. interpersonal) but also include the institutional and structural level. Additionally, as Martin et al. (2016) suggest, part of improving recognition in conservation is to also ensure equitable engagement and influence over decision-making, for example by integrating local actors and their ways of knowing, valuing and managing resources in conservation management.

5.2. The need to strengthen the capacity

Bush and Folger (1994; 1995) highlight the importance of 'empowerment', defined as improving the capability of the conflict parties to handle and make better decisions on their conflict problems, as an important condition for conflict transformation. This includes the ability to clarify their objectives, options and preferences as well as to communicate and negotiate these with other parties. One of the important achievements of the mediation process in the MTMT case was the ability of parties to agree on and to put in place an institution or procedure that can be used as a platform to address, maintain and prevent conflict in the future (i.e. watershed network committee). This type of institution, according to Mitchell (2002) is one of the important mechanisms needed to ensure that conflict transformation can be sustainable. Based on studies on watershed partnership in various countries, Leach and Pelkey (2001) argue that forms of partnership like the watershed network committee can create opportunities for all stakeholders to communicate before problems or conflicts occur. Moreover, by enhancing communication and building consensus, such partnerships may help prevent unnecessary disruptions and tensions in water and forest management as well as community activities and productivity (e.g. agriculture, business) while generating more technically sound management plans and policies that draw upon the variety of expertise and knowledge within the community and beyond (Leach and Pelkey,

2001). This discussion platform, although it is not a formal institution, can also potentially promote informal governance processes in the area and bolster local forest management with positive and sustainable outcomes (Mansourian, 2017). For example, in this case, some impacts included enhanced local awareness of forest conservation, the revision of rules and regulations on forest use which appropriate for the multi-stakeholders, and joint conservation activities between highland and lowland communities.

The study suggests that there is still a need for a formal, legal recognition (e.g. from the government) of the MTMT watershed network committee as an institution and platform for decision making over watershed management. This legal recognition would hopefully be able to strengthen the legitimacy and sustainability of this mechanism in addressing any issues and conflict in the future. A legal status recognition would also help ensure their long-term existence, legitimacy and credibility, for example in finding financial support for their programs. Moreover, in order to ensure the continuity of the network committee this paper would recommend that the relevant government agencies who work in the area to establish support systems or mechanisms that maximize the function of the watershed network in the watershed management. With sufficient support, the communities and mediators hope to create it as a platform for conflict management.

The mediation in MTMT case has some special characteristics in terms of the nature of conflict parties (e.g. multi-party, indigenous people) and the nature of the conflict issues (e.g. protected area, conservation movement) involved. One of the differences between the MTMT case and other case studies that were part of this research project (Dhiaulhaq et al., 2015) is that conflict mediation in MTMT watershed was comparatively more complex and consequently require a longer time (more than three years). It is understandable considering the conflict involved various issues (i.e. direct and underlying causes of conflict), actors and interests that needed to be addressed. The conflict in MTMT watershed involved multiple stakeholder groups (the upland and lowland communities in 23 villages spread over the upland and lowland areas, as well as NP authorities). This is relatively different to, for example, other case studies of forestry and oil palm plantation cases in Indonesia (Dhiaulhaq et al., 2014a, 2015) where the primary conflict parties (and mediation process) involved two primary parties—between a plantation company and local community. Analysing mediation with this conflict characteristic offers valuable insights, especially in comparison to mediation in two-party conflicts.

As mentioned in other literature (e.g. Lesnick & Ehrmann, 1987), mediators in multi-party disputes face challenges of conducting mediation processes mainly due to the large number of parties and interests that may be dispersed across a broad geographic area. The long process also means that the mediation requires more resources (human, financial). It is fortunate that the mediation of the MTMT conflict is part of a participatory forest management project funded by international donor organizations. Without such support, it is difficult to imagine the long mediation efforts could be effectively undertaken.

Additionally, this study suggests that mediation in MTMT conflict is different to a conventional form of mediation. In this case, for example, rather than a stand-alone conflict mediation process facilitated by a professional mediation organization or mediators, mediation of the conflict in MTMT is part of a participatory forest management project facilitated by an NGO. Our earlier studies found that this (NGO facilitating mediation) is not uncommon in forest conflicts in Southeast Asia. In Indonesian case (Dhiaulhaq et al., 2014a), an NGO who facilitated the mediation process between a plantation company and local community was NGO activists who at first assisted and advocated in the interest of the community, and then with agreement of the company, eventually become mediators of the conflict.

The SDF raised some recommendations for improving the mediation process. First, the mediator should gain formal, legal support from the authorized government agencies for effectively addressing forest conflicts; so, the impact would be more concrete and sustainable, especially when the conflict involves State forest areas. Second, besides promoting a participatory process in forest management, improving the quality of life of local people should be conducted simultaneously (in addition to conflict transformation efforts). This would help to provide alternative livelihood opportunities for local communities instead of depending solely on forest, reducing tension between local livelihood and conservation objectives. Third, the mediator should try to link or incorporate the implementation of the results of the mediation process into local government plans so that all activities would be more sustainable. Finally, if the watershed network is promoted as a platform for conflict management, local communities especially the watershed network committee, should be supported in improving their natural resource and conflict management capacities.

At the practical level, our results suggest that in order to effectively address the complex and multidimensional nature of watershed conflict, mediators need to take a variety of roles, therefore requiring skills and capacities which enable them to take these roles. This includes facilitating effective communication and negotiation, building trust and when necessary, proposing alternative solutions. The mediators also need to understand the local and national contexts of watershed management. According to the SDF mediators, these skills were acquired from their previous experiences in facilitating and promoting participatory processes in natural resource management in other areas, not from a formal training on conflict management or mediation, which were still rarely available at that time. However, while the mediators' skills are one of the key factors in successful mediation, the number of capable mediators to address conflict over natural resources is still limited. This is in line with the earlier study findings that the number of mediators in Asia is still relatively low (Yasmi et al., 2010; Dhiaulhaq et al., 2015). In this regard, this paper would suggest the importance of developing the capacity not only of the mediators but also the conflict transformation capabilities of key watershed stakeholders in general (i.e. community, government, National Park authorities) on how to prevent, address and deal with conflicts through targeted training programmes on conflict transformation. This is particularly important because conflict mediation is sensitive and if done improperly can exacerbate conflict situations (e.g. Kerr, 1954).

Finally, the study shows that the use of 'conflict transformation' framework can help practitioners understand the opportunities and gaps in the mediation process. This study would also recommend that future research should also cover cases in which mediation attempts failed and why. Learning from failure may help us understand more about the fundamental attribution of errors (whether mediators' strategies or external factors outside their control led to the failure) and more about the limitation of mediation in transforming forest and land conflict. Additionally, a careful assessment of the impacts of mediation on conflict transformation is also needed, in order to provide stronger evidence on the role of mediation for transforming watershed conflicts.

Acknowledgements

The authors are grateful to the Swedish International Development Cooperation Agency (SIDA), the Norwegian Agency for Development Cooperation (NORAD) and the Swiss Agency for Development and Cooperation (SDC) for funding this research. Views in this article belong to the authors. We are indebted to the participants of the experts' workshop, those who took part in the interviews and FGDs as well as the anonymous reviewers for their constructive feedback on the earlier version of the article.

References

Augsburger, D. W. (1992). *Conflict mediation across cultures: Pathways and patterns*. Westminster: John Knox Press.

Bonell, M., & Bruijnzeel, L. A. (Eds.). (2005). *Forests, water and people in the humid tropics: past, present and future hydrological research for integrated land and water management*. Cambridge: Cambridge University Press.

Buergin, R., & Kessler, C. (2000). Intrusions and exclusions: Democratization in Thailand in the context of environmental discourses and resource conflicts. *GeoJournal*, 52(1), 71-80.

Bush, R. A. B., & Folger, J. P. (1994). *The promise of mediation: Responding to conflict through empowerment and recognition*. San Francisco: Jossey-Bass.

Bush, R. A. B., & Folger, J. P. (2005). *The promise of mediation: The transformative approach to conflict*. San Francisco: John Wiley & Sons.

Callister, R. R., & Wall Jr, J. A. (2004). Thai and US community mediation. *Journal of Conflict Resolution*, *48*(4), 573-598.

Church, J. (2009). Avoiding Further Conflict: A Case Study of the New York City Watershed Land Acquisition Program in Delaware County, NY. *Pace Envtl. L. Rev.*, 27, 393.

De Koning, R., Capistrano, D., Yasmi, Y., & Cerutti, P. (2008). *Forest-related conflict: Impact, links, and measures to mitigate*. Washington D.C., USA: RRI.

Delang, C. (2004). *Living at the edge of Thai society: The Karen in the highlands of northern Thailand*. London; New York: Routledge.

Dhiaulhaq, A., Gritten, D., De Bruyn, T., Yasmi, Y., Zazali, A., & Silalahi, M. (2014a). Transforming conflict in plantations through mediation: Lessons and experiences from Sumatera, Indonesia. *Forest Policy and Economics*, 41, 22-30.

Dhiaulhaq, A., De Bruyn, T., Wiset, K., Thaworn, R., Gritten, D., Yasmi, Y., Soontornwong, S., Kritsanarangsan, S. (2014b). Transformative mediation, a tool for maximising the positives out of forest confict: A case study from Kanchanaburi, Thailand. *Forests under pressure: Local responses to global issues*, IUFRO World Series, vol. 32, Vienna: International Union of Forest Research Organizations (IUFRO).

Dhiaulhaq, A., De Bruyn, T., & Gritten, D. (2015). The use and effectiveness of mediation in forest and land conflict transformation in Southeast Asia: Case studies from Cambodia, Indonesia and Thailand. *Environmental Science & Policy*, 45, 132-145.

Engel, A., & Korf, B. (2005). *Negotiation and mediation techniques for natural resource management*. Rome, Italy: FAO.

Forsyth, T., & Walker, A. (2008). *Forest guardians, forest destroyers: the politics of environmental knowledge in northern Thailand*. University of Washington Press.

Government of Thailand (1961). National Park Act, B.E. 2504 (1961). Government of Thailand. 5 p.

Hares, M. (2009). Forest conflict in Thailand: Northern minorities in focus. *Environmental management*, 43(3), 381-395.

Imperial, M.T., & Kauneckis, D. (2003). Moving from conflict to collaboration: watershed governance in Lake Tahoe. *Natural Resources Journal*, 1009-1055.

Kerr, C. (1954). Industrial conflict and its mediation. *American Journal of Sociology*, 230-245.

Laungaramsri, P. (2000). The ambiguity of" watershed": the politics of people and conservation in northern Thailand. *Sojourn: Journal of Social Issues in Southeast Asia*, 15(1), 52-75.

Leach, W.D., & Pelkey, N.W. (2001). Making watershed partnerships work: a review of the empirical literature. *Journal of water resources planning and management*, 127(6), 378-385.

Lederach, J. P. (1997). *Building peace: Sustainable reconciliation in divided societies*. Washington

DC: United States Institute of Peace Press.

Lee, J., & Teh, H. H. (Eds.). (2009). *An Asian perspective on mediation*. Singapore: Academy Publishing.

LeResche, D. (1992). Comparison of the american mediation process with A Korean-American harmony restoration process. *Conflict Resolution Quarterly*, *9*(4), 323-339.

Lesnick, M. T., & Ehrmann, J. R. (1987). Selected strategies for managing multiparty disputes. *Conflict Resolution Quarterly*, *1987*(16), 21-29.

Mansourian, S. (2017). Governance and forest landscape restoration: A framework to support decision-making. *Journal for Nature Conservation*, *37*, 21-30.

Martin, A., Coolsaet, B., Corbera, E., Dawson, N. M., Fraser, J. A., Lehmann, I., & Rodriguez, I. (2016). Justice and conservation: The need to incorporate recognition. *Biological Conservation*, *197*, 254-261.

Mitchell, C., 2002. Beyond resolution: what does conflict transformation actually transform?. *Peace and Conflict Studies*, *9*(1), pp.1-23.

Patel, T., Dhiaulhaq, A., Gritten, D., Yasmi, Y., De Bruyn, T., Paudel, N.S., ... & Suzuki, R. (2013). Predicting future conflict under REDD+ implementation. *Forests*, 4(2), 343-363.

Peluso, N. L., & Watts, M. (Eds.). (2001). *Violent environments*. Cornell University Press.

Raitio, K. (2012). New institutional approach to collaborative forest planning on public land: Methods for analysis and lessons for policy. *Land Use Policy*, *29*(2), 309-316.

Reimann, C. (2004). Assessing the state-of-the-art in conflict transformation: Reflecting from a theoretical perspective. In A. Austin, M. Fischer & N. Redpers, eds. *Transforming ethno-political conflict: The Berghof handbook.* Berlin: VS Verlag fur Sozialwissenschaften.

Rubenstein, R. E. (1999). Introduction: Conflict Resolution and Social Justice. *Peace and Conflict Studies*, *6*(1), 1-7.

Steinberg, P. E., & Clark, G. E. (1999). Troubled water? Acquiescence, conflict, and the politics of place in watershed management. *Political Geography*, *18*(4), 477-508.

Tangtham, N. (1996). Watershed Classification: The Macro Land-use Planning for the Sustainable Development of Water Resources. Paper presented at the International Seminar Workshop on "Advances in Water Resources Management and Wastewater Treatment Technologies", 22-25 July 1996, Suranee University of Technology, Thailand.

Tungittiplakorn, W. (1995). Highland-Lowland conflict over natural resources: A case of Mae Soi, Chiang Mai, Thailand. *Society & natural resources*, 8(4), 279-288.

Uhlig, H. (1980). Problems of land use and recent settlement in Thailand's highland-lowland transition zone. In *Conservation and development in Northern Thailand,* ed. J. D. Ives, S. Sabhasri, and P. Vorurai, 33-42. Tokyo: United Nations University Press.

Vandergeest, P., & Peluso, N. L. (1995). Territorialization and state power in Thailand. *Theory and society*, *24*(3), 385-426.

Vayrynen, R. (1991) To Settle or to Transform? Perspectives on the Resolution of National and International Conflict, in R. Vayrynen, ed., *New Directions in Conflict Theory*. Thousand Oaks: Sage Publications.

Wall, J.A., Stark, J.B., & Standifer, R.L. (2001). Mediation a current review and theory development. *Journal of Conflict Resolution*, 45(3), 370-391.

Wittayapak, C., & Dearden, P. (1999). Decision-making arrangements in community-based watershed management in northern Thailand. *Society & Natural Resources*, *12*(7), 673-691.

Yamaguchi, K. (2006). *Scarcity and conflict of resources: Chom Thong water conflict*. Bali, Indonesia: International Association for the Study of the Common Property.

Yasmi, Y., Guernier, J. and Colfer, C.J.P., (2009). Positive and negative aspects of forestry conflict: lessons from a decentralized forest management in Indonesia. *International Forestry Review*, *11*(1), pp.98-110.

Yasmi, Y., Kelley, L., & Enters, T. (2010). *Conflict over forests and land in Asia: Impacts, causes, and management*. Bangkok, Thailand: RECOFTC.

Yasmi, Y., Kelley, L.C., & Enters, T. (2013). Community-outsider conflicts over forests: Perspectives from Southeast Asia. *Forest Policy and Economics*, 33(21-27).

Zachrisson, A., & Lindahl, K. B. (2013). Conflict resolution through collaboration: Preconditions and limitations in forest and nature conservation controversies. *Forest policy and economics, 33*, 39-46.

7

Potential Economic and Development Prospects of Non Timber Forest Products in Community Agroforestry Land around Sibolangit Tourism Park

Oding Affandi[1,*], Anita Zaitunah[2], Ridwanti Batubara[3]

[1] Forestry Faculty, University of Sumatera Utara; Email: affandio_usu@yahoo.com
[2] Forestry Faculty, University of Sumatera Utara; Email: azaitunah@gmail.com
[3] Forestry Faculty, University of Sumatera Utara; Email: ridwantibb@yahoo.com
* Correspondence author: affandio_usu@yahoo.com ; oding@usu.ac.id

Abstract: The communities who live around Sibolangit Tourism Park have developed nontimber forest products (NTFP) in their own agroforestry lands. This research evaluates the potential economic and development prospects from NTFP development in the Park by examining: (1) type of NTFP and economic value from community agrofrestry land, (2) contribution of NTFPs on household income, (3) development prospects of NTFP-based agroforestry around Sibolangit Tourism Park. The research was conducted in two selected villages around Sibolangit Tourism Park: Sembahe Village and Batu Mbelin Village. The research took place over a period between June and August 2016. Research data was obtained from in-depth interviews and observations. A descriptive method was used to analyze and describe facts related to the research aims. The type of NTFPs cultivated by communities at the research sites include mangosteen, durian, garcinia, candlenut, lanzones, lansium, bitter bean, and areca nut (as their forestry component) and ginger, turmeric, chili, papaya, etlingera, and banana (as the agriculture component). Most NTFPs are cultivated as a comercial product. The economic value of NTFPs in Batu Mbelin Village has reached IDR. 547,275,000/year or contribute 80.07% of total family income. Meanwhile, the economic value of NTFPs in Sembahe Village has reached IDR 682,100,000/year, contributing to 78.75% of total household income. Therefore, the prospects for supporting and expanding NTFP in agroforestry plots in and around Sibolangit Tourism Park has high potential for supporting household income.

Keywords: Economical value; NTFP; NTFP development; Agroforestry; the Tourism Park of Sibolangit

---

1. Introduction

From a national economic context, forests have multiple functions that provide both direct and indirect benefits. Economic studies however, rarely examine these multiple functions holistically and in an integrated fashion. In an ideal condition, sustained forest cultivation could produce both timber forest products (TFPs) and non timber forest products (NTFP[1]). In most cases however, forest resource utilization tends to be conducted partially, oriented to single commodity extraction that is timber production. The singular timber production approach however, is restricted by the knowledge, capability, technology, and local development interests and presents barriers for the potential benefit of forest cultivation in a given area. The singular timber extraction model however, is no longer fit for the context of supporting national development priorities given the expansive logging that has occurred in Indonesia during the past two decades. Government priorities have shifted in the interest of seeking potential multi-dimensional solutions to focus on

---

[1] Based on Explanation of section 4 subsection (1) Law No.41 Tahun 1999 on Forestry, NTFP can be biological entities (plants and animal products), non biological entities (conservation and service functions, not including mining products) and direct produtcs from processing or called underlying product or see Regulation of Ministry of Forestry No. 35 year 2007 on Non Timber Forest Product

people's welfare, and therefore desire to build the capabilities to support local community livelihood priorities.

NTFPs are part of trees or plants that have special characteristics that can become products to support people's needs, sold as export commodities, or as raw material for industries. NTFP is a byproduct from a tree such as sap, leaves, bark, fruit or plants which have special character such as rattan, bamboo and numerous others. NTFP collection, in general, is a traditional activity conducted by people living in and around the forest. In some locations, the collection of NTFPs are the main activity of people's daily lives. Some examples include rattan collection, timber sap (such as agathis sap), and shorea, also called *damar* (Djajapertjunda, 2001).

Forest resource management systems that support NTFP cultivation are generally conducted by local people as part of their agroforestry plots. There are numerous long-standing practices that continue to function until today in various regions in Indonesia, mainly in Sumatera, such as Repong in Pesisir Krui, Lampung; Parak in Maninjau, West Sumatra; mixed rubber gardens in Jambi and South Sumatera (de Foresta, et al., 2000); and, the Kemenyan gardens in Tapanuli Utara, North Sumatera (Suhardjito, et al, 2000). In the past, forest resource management systems that included these often complex NTFP arrangements were overlooked by government programs and other development institutions. More recently, the government has acknowledged NTFP as an important complement and alternative from classical forest resource management practices.

Several decades ago agroforestry began to develop around Sibolangit Tourism Park. The system is known as *Reba Juma* (from the local Karo language; *Reba* means house, *Juma* means field). In other words, this means that the fields are easily accessed from cultivator homes (Affandi and Siregar, 2010). When examining *Reba Juma* more closely, it can also be described as an agrisilviculture pattern. According to Sardjono, et al. (2003), an agrosofrestry system combines the forest component (or timber) with the agricultural component (or non-timber plants). Timber refers to long cycle crops (*tree crops*) and non-timber plants are any seasonal crops (*annual crops*). Based on the research site, *Reba Juma* consists of forestry plants such as mangosteen, durian, garcinia, candlenut, lanzones, lansium, bitter bean, areca nut, combined with agriculture plants, including ginger, turmeric, chili, papaya, etlingera, and banana.

The economic potential of NTFPs are from both state forest areas and from private land ownership ares. The potential coverage area is assumed to be quite vast, even though the information on values is imprecise due to the lack of data availability. NTFPs are also assumed to provide a key component for realizing the potential growth of the 'green economy' concept (Andayani, 2015).

Local people acknowledge the benefit of NTFP management in agroforestry areas around Sibolangit Tourism Park, in terms of economic, ecological, and socio culture forms. The important role of NTFPs in agroforestry areas around the park continues to be maintained until now. The agroforestry practices around Sibolangit Tourism Park is also acknowledged by numerous stakeholder to serve as an important function as a buffer zone. Not only do the agroforestry areas around the Park provide an ecological buffer, they also serve as a safeguard against encroaching population expansion to the Sibolangit region. Population expansion into these areas would have a negative impact on both the forests and the existing communities. However, such land management practices rarely receive the attention and acknowledgement from parties with formal management responsibilities. Furthermore, there is still limited information on how agroforestry management systems are sustained, the relationship to local cultural practices, and especially lacking information on the way local forest management practices contribute to people's welfare. On the other hand, although NTFPs in agroforestry practices provide additional value for local incomes, custom, and subsistence, it is still far from optimal in achieving welfare standards for people around the forest. Therefore, in this research we present comprehensive information on the economic potential of NTFPs in agroforestry management in order to support the development of agroforestry systems that are in accordance with local interests.

AS such, the problem this research aims to address are multiple, and include: (1) examining the type and economic value of NTFPs and the uses among agroforestry practitioners at the case study sites, (2) assessing NTFP and their relationship to household income, (3) analyzing the prospect of NTFPs development in agroforestry system around Sibolangit Tourism Park.

2. Materials and Methods

2.1. Time and Location

This reseach was conducted in two selected villages around Sibolangit Tourism Park, which are Sembahe Village and Batu Mbelin Village in June - August 2016. Other people find their livelihoods as workers, trader/entrepreneurs, government officers, army, police officers, and others. The ethnicity of both villages is dominated by Karo, including some people who identify as Batak from the neighboring region, and also others that are Javanese.

2.2. Data collection methodology and data analysis

The research was explorative, and data collection was conducted using a combination of documentation approaches from various secondary data sources. Direct data collection was through primary data collection in the field through interviews, focus group discussions, and field observation. Direct data collection in the field consisted mainly of collecting data directly at the study sites and crosschecking with secondary data in the field and elsewhere.

Primary data collection included inventorying different species, surveying the number of forest products being utilized, gathering baseline socio-economic data; cataloging the period and time of forest product collection; compiling cost and forest product utilization information; and, returning for several iterations of data collection. Secondary data, such as the general condition of the research area or general data from the village office and regency was also included for analytical considerations.

The field data on forest product value of each species per year was calculated to estimate the price of the product using a market approach. There are information on quantity of product per species, total collection per year, percentage of economical value and contribution of the product.

The sample in this study is the community who live around the Sibolangit Tourism Park area in the Batu Mbelin village and Sembahe Village of Sibolangit Sub District, Deli Serdang District as many as 30 samples of household heads, 5-8 persons for FGDs and key informants from each village. The sample in this study is selected by purposive sampling, which means the subject is not based on strata, random, or region distribution, but rather conducted with a specific purpose (Arikunto, 2006).

3. Results and Discussion

3.1. Type and category of NTFP

Agroforestry systems at the research site produce two categories of results including the forestry component and agriculture component. The forestry component consists of TFPs and NTFPs. Therefore, the management of each component is important to defining the overall agroforestry system. Management of plots determine the value of agroforestry products to be sold in the market. Appropriate technologies, requisite processing systems, access to markets, and product diversification will support the optimalization of forest products.

In the forestry context, the main product of modern agroforestry is TFP and NTFP from wood plants. The Sibolangit Tourism Park case is different because local communities have developed agroforestry systems traditionally utilizing various plant types, mainly NTFPs that have a close relationship with local socio-cultural meanings passed down and practiced from generation to generation by individual land owners and under other land tenure regimes (Sardjono, et al., 2003; Widiyanto, 2013).[2]

NTFP species that are utilized by people around the Park mainly cultivate their own land under a common agroforestry pattern. Forest products vary from fruit trees to vegetables plants. Collection of NTFPs by people is an ongoing traditional activity. NTFP utilization by people can be grouped into two categories: (a) productive, being sold at market (commercial), and (b) consumptive, that is self consumed and not being sold (subsistent) (Ngakan, et al., 2006). Data on NTFP species utilized in the study area is shown in Table 1.

Table 1. Species and category of NTFP utilized by communities in agroforestry practices around Sibolangit Tourism Park

No.	Species of NTFP	Category	Notes
1	Areca nut (*Areca catechu*)	Productive, consumptive	Cultivation, unseasonal
2	Garcinia (*Garcinia atroviridis*)	Productive, consumptive	Cultivation, unseasonal
3	Mangosteen (*Garcinia mangostana*)	Productive, consumptive	Cultivation, seasonal
4	Durian (*Durio zibethinus*)	Productive, consumptive	Natural[3], seasonal
5	Lansium (*Lansium domesticum var*)	Productive, consumptive	Natural, seasonal
6	Lanzones (*Lansium domesticum var domesticum*)	Productive, consumptive	Natural, seasonal
7	Bitter bean (*Parkia speciosa*)	Productive, consumptive	Natural, seasonal
8	Candlenut (*Aleurites moluccana*)	Productive, consumptive	Natural, seasonal
9	Ginger (*Zingeber officinale*)	Productive, consumptive	Cultivation, unseasonal
10	Turmeric (*Curcuma domestica*)	Productive, consumptive	Cultivation, unseasonal
11	Banana (*Musa paradisiaca*)	Productive, consumptive	Cultivation, unseasonal
12	Chili *(Capsicium annum L.)*	Productive, consumptive	Cultivation, unseasonal
13	Papaya *(Carica papaya* L.)	Productive, consumptive	Cultivation, unseasonal
14	Etlingera *(Etlingera elatior)*	Productive, consumptive	Cultivation, unseasonal
15	Corn *(Zea mays)*	Productive, consumptive	Cultivation, unseasonal

Based on research results, Table 1 highlights NTFP utilization by people around the Park. The categories also relate directly to the main source of the household economy. The household economy is characterized by all the NTFP categories listed above that are mainly to be sold as commercial products (productive) and also for its local uses (categorized as consumptive).

3.2. Economic value of NTFP

The economic value of NTFPs are obtained from a multiplication between "total collection" per species per year and "price" of forest product per species. Based on research results in the two villages around Sibolangit Park the economic value from utilization of NTFPs from agroforestry is 547,275,000 rupiahs per year for Batu Mbelin Village and 682,100,000 rupiahs per year for Sembahe Village. The specific breakdowns from the calculations are provided in Table 2 and Table 3.

---

[2] Traditional Agroforestry or classic agroforestry is 'every agriculture system, where trees from planting or maintenance of stand/plant is integrated as part of the socio-economic system and forms the ecological whole of the system (*agroecosystem*)' (Sardjono, et al., 2003)

[3] existing in or caused by nature

Table 2. NTFP economic value per year from agroforestry practices in Batu Mbelin Village (Year 2016)

No.	NTFP species	Number of collection (Unit/year)	CF	Total collection	Unit	Price (IDR)	Economic value (IDR)
1	Areca nut	790	2	1,580	Kg	15,000	23,700,000
2	Garcinia	5,830	3	17,490	Kg	5,000	87,450,000
3	Mangosteen	10,650	1	10,650	Kg	20,000	213,000,000
4	Durio	10,350	1	10,350	Piece	7,000	72,450,000
5	Lansium	3,350	1	3,350	Kg	10,000	33,500,000
6	Lanzones	10,200	1	10,200	Kg	5,000	51,000,000
7	Bitter bean	850	1	850	Bundle	20,000	17,000,000
8	Candlenut	2,550	4	10,200	Kg	5,000	51,000,000
9	Ginger	4,900	1	4,900	Kg	8,000	39,200,000
10	Turmeric	880	1	880	Kg	12,000	10,560,000
11	Banana	190	1	190	bunch	35,000	6,650,000
12	Chili	25	15	375	Kg	30,000	11,250,000
13	Papaya	79	36	2,844	Piece	5,000	14,220,000
14	Etlingera	158	24	3,792	Kg	10,000	37,920,000
15	Corn	4,400	1	4,400	Kg	3,000	13,200,000
		TOTAL					**547,275,000**

Notes: CF = Collection Frequency

Table 3. NTFP economic value per year from agroforestry practices in Sembahe Village (Year 2016)

No.	Type of NTFP	Number of collection (Unit/year)	CF	Total Collection	Unit	Price (IDR)	Economical value (IDR)
1	Areca nut	1,214	2	2,428	Kg	15,000	36,420,000
2	Garcinia	4,110	3	12,330	Kg	5,000	61,650,000
3	Mangosteen	6,550	1	6,550	Kg	20,000	131,000,000
4	Durio	18,200	1	18,200	Piece	7,000	127,400,000
5	Lansium	4,050	1	4,050	Kg	10,000	40,500,000
6	Lanzones	4,230	1	4,230	Kg	5,000	21,150,000
7	Bitter bean	1,050	1	1,050	Bundle	20,000	21,000,000
8	Candlenut	1,500	4	6,000	Kg	5,000	30,000,000
9	Ginger	4,900	1	4,900	Kg	8,000	39,200,000
10	Turmeric	1,050	1	1,050	Kg	12,000	12,600,000
11	Banana	57	1	57	Bunch	35,000	1,995,000
12	Chili	24	15	360	Kg	30,000	10,800,000
13	Papaya	58	36	2,088	Piece	5,000	10,440,000
14	Etlingera	13	24	312	Kg	10,000	3,120,000
		TOTAL					**682,100,000**

Note: CF = Collection Frequency

Table 2 and Table 3 show that NTFP economic value per year among agroforestry practices in Batu Mbelin Village and Sembahe Village are derived largely from the forestry component. As for the NTFP with the highest economic value, mangosteen holds the highest value, followed by durian and garcinia. In general, economic value of NTFP is influenced by the number collected, collection frequency, and price of each type of NTFP. The number of collection depends on productivity from each species, which is related to land management intensity in agrofrestry practices. Collection frequency depends most on harvest time and fruiting system of each species. Furthermore, the price depends on market chain (*tengkulak,* or middlemen/collectors, or direct sourcing), difference of selling time and quality of product sale (Awang, et.al., 2002).

Research results show that most NTFPs are still sold in its raw material form, so the nominal financial value for producers is still low. It is also assumed that to increase product sale in the form of an intermediate product there are still disconnects between the commercialized forms for the downstream product. Therefore, the strategy and opportunities to influence price determination needs to accomodate the existing market system and ways to anticipate the probability of product export potential.

The economic value of various types of NTFPs in agroforestry practices in general also yield various values in accorandce to the area of land and comodity type. Therefore, there is a variation of harvesting for each agroforestry product that also results in variation time of collecting income from the given agroforestry product. Findings about cost variation are detailed more specifically in Widiarti and Prajadinata (2008) that shows mixed garden planting patterns provide a variety of incomes that are routine, daily, weekly, monthly, seasonal and yearly. Therefore, the mixed garden provides for approaches to sustainable incomes at certain intervals for every farmer.

Mangosteen is one of the main commodity in agroforestry practices in Batu Mbelin and Sembahe villages. It has a high economic value because of its relatively high selling price in the market. According to Qosim (2013) mangosteen has high economic value and good prospects to be developed as a commodity export crop because of the lack of competitors. As a main commodity, mangosteen has been managed intensively by people in various agroforestry settings. This is in accordance with the research by Setiawan (2014) that shows high income in agroforestry practices is a function of the variation of plant types and the main commodity crops having high commercial value in the market. According to interviews at the research sites, respondents noted that mangosteen grew naturally without any special treatment and although previously had low harvesting yields, but the increasing prospect and value for mangosteen, farmers in Batu Mbelin and Sembahe began to cultivate for more intensive treatments that provided improved contributions to their income.

3.3. NTFP contribution to household income

As noted, people of Sembahe and Batu Mbelin Villages have various professions. Therefore, they do not solely rely on their income from NTFPs in their agroforestry plots. Cultivators also count on income from other occupations such as more conventional farming in other areas, construction work, entrepreneur, broker, government official, wetland agriculture systems, and others. Household income per year from respondents beside NTFP from agroforestry plots in Desa Batu Mbelin and Desa Sembahe are presented in Table 4.

By comparing between total economic value of people from NTFP and total economic value of people from apart of NTFP, it is evident that economic value from NTFPs are higher than economic value from those apart from NTFP (NTFP: 79,41% > apart of NTFP: 20,59). A more general overview is also presented in Table 5. The high economic value of NTFP utilization shows that NTFP provides a very real contribution to people's livelihood around Sibolangit Tourism Park. In other words, communities have a high dependency on the existence of forest resources and agroforestry practices.

Table 4. Household economic value per year apart from NTFP utilization in agroforestry practices among respondents in Batu Mbelin and Sembahe (Year 2016)

	Source of income	Batu Mbelin Village		Sembahe Village	
		Total (IDR)	%	Total (IDR)	%
1	Entrepreneur	47,400,000	34.80	44,400,000	24.11
2	Construction Worker	28,800,000	21.14	26,000,000	14.12
3	Own business	49,800,000	36.56	76,000,000	41.30
4	Farm Labor	10,200,000	7.50	37,700,000	20.47
	Total	136,200,000	100	184,100,000	100

Table 5. Comparison of economic value between NTFP and apart from NTFP at the research sites (Year 2016)

Economic value	Batu Mbelin Village		Sembahe Village		
	Sum (IDR)	%	Sum (IDR)	%	Average %
NTFP					
• Forestry component	469,120,000	68.64	549,100,000	63.39	66.02
• Agriculture component	78,155,000	11.43	133,000,000	15.36	13.39
Sub total of NTFP	**547,275,000**	**80.07**	**682,100,000**	**78.75**	**79.41**
Apart from NTFP	**136,200,000**	**19.93**	**184,100,000**	**21.25**	**20.59**
Total	683,475,000	100	866,200,000	100	

The high contribution value from NTFP utilization in agroforestry for household income in general is due to agroforestry practices that can produce continuously, and which have various harvesting cycles (Indriyanto, 2008). This is in accordance with the results from Ngakan, et.al. (2006) study, which indicates that forest products (mainly NTFP) from natural forest resources present one of the most valuable economic activities for people from around the area. Having the additional economical value, in this way can be viewed as being much more valuable than timber. Furthermore, the collection of NTFP do not contribute to extractive forest destruction, which has grown in public recognition in Indonesia in recent years. Therefore, key environmental services will also continue to be maintained, such as the ecological and hydrological cycles. In other words, including agroforestry as a natural resource and livelihood strategy, presents multiple functions, and forest utilization in these was can contribute to establishing improved welfare for local communities.

The high contibution of agroforestry-based NTFPs (meaning, in plots located outside the TWA forest area) to household income helps to push down the drivers of forest loss, especially in areas where people have a direct impetus in sustaining the local resources. This is clearly the case in the boundary areas around Sibolangit Tourism Park.

It is important to contextualize that the Sibolangit Tourism Park area is a conservation area in North Sumatra Province that has functioned as natural sanctuary and wildlife reserve to protect biodiversity, ecosystems, and natural beauty for the aims of educational development, recreation and tourism. Nurapriyanto, *et.al.* (2011) research also show that the development of NTFPs can reduce losses of forest function.

3.4. *Development prospect of agroforestry based NTFP*

NTFP are plant species that can grow well inside or outside the forest. With this capability, the role of NTFPs has been accepted by people as one source of income. As we have shown in this research, the area of Sibolangit Tourism Park and surrounding areas function as conservation areas that also provide an important role for local livelihoods. Therefore, in order to provide additional benefits in the forestry sector and around the area of Sibolangit Tourism Park co-development arrangements of certain plant sepecies as NTFPs that can be supported in particular agroforestry systems. This is an alternative approach to move away from extractive TFP approaches while also supporting local livelihoods. In order to increase productivity and production quality, approaches can be developed to further complement incomes around Sibolangit Tourism Park.

Development of NTFPs in agroforestry systems around Sibolangit Tourism Park has a very high prospect. This can be seen in the practice of NTFP based agroforestry, a practice that has been in place for a long time by people around Sibolangit Tourism Park, continuing their traditional practices from generation to generation. Furthermore, it is also evident that there is significantly additional economic value and contribution to household income from NTFPs (such as mangosteen, durian, garcinia, lansium, candlenut, etc) and from other agricultural commodity crops (such as ginger, turmeric, chili, papaya, etlingera, corn, and others). The high economic value of NTFPs within agroforestry systems can help to diversify local incomes.

Beyond NTFPs role as a market commodity opportunity, they also function as a sustained source of local subsistence needs. For example, rice, corn, tubers, vegetable and fruits also function as NTFPs in agroforestry areas. Such crop examples can be consumed directly, or any surplus can also be sold to market. Such subsistence functions also allow for greater resilience during shocks related to food security considerations.

In the context of conservation and the environment, NTFPs among the agroforestry patterns managed by local people have various compositions and combinations. In general, the species varieties support greater ecological resilience against pests and climatic shocks. According to de Forestra, et.al. (2000), in some cases, agroforestry practices of mixed gardens that do not recieve too much owner treatment, have reached more than 50 plant types. This high species diversity has an important role to conservation.

According to Von Maydell (1986: in Hairiah, et.al., 2003) land management practices using agrofrestry patterns have higher prospects in developing improved commodity opportunities for NTFPs. This has been confirmed through this research for the following reasons:

1) Productivity: In this research, it is evident that the total economic value of products from mixed agroforestry systems provides higher returns than mono-systems in the area. This is due to the output from various returns from NTFPs from the land, and also the continued returns that can be harvested throughout the year. Mixed plant systems provide benefits because failure of one component/plant species can be compensated by other component/plants.
2) Diversity: The existence of a combination of two components or more in agroforestry systems provide high diversity in types of products or services. Therefore, from an economic perspective it can avoid permanent fatal failures that can wipe out livelihoods from single cultivation approaches.
3) Self regulation: High diversification in agroforestry systems can meet local community needs and also provide a cushion against dependency from particular products. System independence function to reduce requirements from intensive outside inputs (for example fertilizer, pesticide).
4) Stability: Agroforestry practices provide diversity and optimal productivity and can provide a balance across the land use management system, which can assure stability (and continuity) of farmers' income.

4. Conclusions and Recommendations

4.1. Conclusions

The communities around Sibolangit Tourism Park have long utilized and developed their agroforestry systems for NTFP on their own land. Most NTFPs are for commercial production and others help to fulfil subsistence requirements. Economic value of NTFP across two locations indicate that there is a higher economic value from NTFP, which also garner high values that contribute to 79.41% of total family income.

The prospects for NTFP development based on agroforestry in Sibolangit Tourism Park area is high. As we have shown herein, NTFP expansion can support social-economic needs and ecological functions. Socially, NTFP in agroforest systems are part of a longterm continuity of local traditional practices passed down through generations. Economically, we have highlighted the high returns from NTFP economic value, which provide for the main source of income to local community households. Finally, from an ecological perspective, numerous research also shows that agroforestry systems provide for enhanced biodiversity.

4.2. Recommendations

1) In order to develop NTFPs based on agroforestry, support programs can help to enhance local prospects by providing clear and exact price information, potential avenues for achieving market and specie diversification of NTFPs, support for understanding the overall market structure, and support effective marketing strategies that help to shorten market chains and strategy commodity promotion.
2) In order to develop NTFP in agroforestry plots to meet market demand in the future a better understanding of farmer production and consumer needs can help involve communities in the decision making process to prioritize mutual actions. This can then help to enhance the capacity and target the appropriate programing and technological transfer.

Acknowledgments

We acknowledgements the support from the Directorat General of Research Strengthening and Development of the Ministry of Research and Technology and Higher Education where this research was part of the 2016 Fundamental research grants. This research was supported by the Ministry of Research and Technology in Higher Education, under a program of the Directorat General of Strengthening Research and Development. The title of the research was: "Study on Agroforestry Management Practice in order to Maintain Biodiversity by People's Around the forest" and the research took place in 2016.

References

Affandi, O. and Siregar, E.B.M. (2010). Reba Juma: Kelestarian Praktek Agroforestry Lokal pada Masyarakat Karo, Propinsi Sumatera Utara. *AGRICA. Jurnal Agribisnis Sumatera Utara.* 3 (1): 1-11

Andayani, W. (2015). Strategi Penetapan Harga And Pengembangan Pemasaran Hasil Hutan Bukan Kayu (HHBK) Dalam Rangka Meningkatan Pendapatan Sektor Kehutanan. *Prosiding Seminar Nasional "Peranan dan Strategi Kebijakan Pemanfaatan Hasil Hutan Bukan Kayu (HHBK) Dalam Meningkatkan Daya Guna Kawasan Hutan, 6-7 November 2014, Fakultas Kehutanan Universitas Gadjah Mada. Yogyakarta*

Arikunto, S. (2006). *Prosedur Penelitian.* Penerbit Rineka Cipta. Edition VI. Jakarta.

Awang, S.A., Andayani, W. Himmah, B. Widayanti, W.T. and Affianto, A. (2002). *Hutan Rakyat Sosial Ekonomi dan Pemasaran.* BPFE- Yogyakarta. Yogyakarta.

de Foresta, H., Kusworo, A., Michon, G., & Djatmiko, W. A. (2000). Ketika kebun berupa hutan: Agroforest khas Indonesia sebuah sumbangan masyarakat. *Institut de Recherche pour le Development (IRD) and International Centre for Research in Agroforestry (ICRAF). Bogor.*

Djajapertjunda, S. (2001). *Studi Kolaborasi Pengelolaan Repong Dama Kruing Lampung.* Syafa'at Advertesing and Forum Komunikasi Kehutanan Masyarakat.

Hairiah, K., Sardjono, M.A. and Sabarnurdin, S. (2003). *Bahan Ajar 2. Pengantar Agroforestry.* World Agroforestry Center (ICRAF). Bogor

Indriyanto. (2008). *Silvikultur Pada Sistem Wanatani. Pengantar Cultivation Hutan.* PT Bumi Aksara. Jakarta

Ngakan, P. O., Komarudin, H., Achmad, A., & Wahyudi, T. (2006). *Ketergantungan, persepsi dan partisipasi masyarakat terhadap sumberdaya hayati hutan: studi kasus di dusun Pampli kabupaten Luwu utara, Sulawesi selatan.* Center for International Forestry Research (CIFOR). Bogor

Nurapriyanto, I., Tuharea, A., & Arifin, N. (2003). Sistem Pengusahaan Beberapa Hasil Hutan Bukan Kayu dan Alur Tataniaganya di Jayapura, Papua. *Penelitian Sosial Ekonomi dan Kebijakan Kehutanan. Jayapura.*

Qosim, W. A. (2013). Pengembangan Buah Manggis sebagai Komoditas Ekspor Indonesia. *Jurnal Kultivasi. Fakultas Pertanian Universitas Padjadjaran, 12*, 40-45.

Sardjono, M.A., T. Djogo, H.S. Arifin, and N. Wijayanto. (2003). *Bahan Ajaran 2. Klasifikasi and Pola Kombinasi Komponen Agroforestry.* World Agroforestry Center (ICRAF). Bogor

Setiawan, A. (2014). Analisis Sosial Ekonomi Pola Agroforestri Di Sub Das Lahumbuti Kabupaten Konawe Propinsi Sulawesi Tenggara. Thesis. Universitas Hasanuddin. Makasar.

Suhardjito, D., Khan, A. Djatmiko, W.A. Sirait, M.T. and Evelyna, S. (2000). *Karakteristik Pengelolaan Hutan Berbasiskan Masyarakat*. Aditya Media. Yogyakarta

Widiarti, A., & Prajadinata, S. (2016). Karakteristik hutan rakyat pola kebun campuran. *Jurnal Penelitian Hutan dan Konservasi Alam, 5*(2), 145-156.

Widiyanto, A. 2013. Beberapa Catatan Mengenai Hasil Hutan dalam Sistem Agroforestry. *Retrieved from* https://www.researchgate.net/publication/299980458

8

Managing Forest Conflicts: Perspectives of Indonesia's Forest Management Unit Directors

Larry A. Fisher [1,*], **Yeon-Su Kim** [2], **Sitti Latifah** [3], **Madani Makarom** [4]

[1] School of Natural Resources and the Environment, University of Arizona, Tucson, AZ 85721, USA; lafisher@email.arizona.edu
[2] School of Forestry, Northern Arizona University, Flagstaff, AZ 86011, USA. e-mail: ysk@nau.edu
[3] Foretsry Study Program, University of Mataram, Indonesia; e-mail: sittilatifah@gmail.com
[4] West Rinjani Forest Management Unit, Lombok, Indonesia; e-mail: aganampuh@gmail.com
* Correspondence author: lafisher@email.arizona.edu

Abstract: Recent expansion of the forestry and plantation sectors in Indonesia has intensified agrarian and natural resource conflicts, and created increased awareness of the social, economic and environmental impacts of these disputes. Addressing these disputes is a critical issue in advancing Indonesia's commitment to sustainable forest management. The Forest Management Units (*Kesatuan Pengelolaan Hutan*, or KPH*)*, have become the pivotal structural element for managing all state forests at the local level, with responsibility for conventional forest management and policy implementation (establishing management boundaries, conducting forest inventory, and developing forest management plans), as well as the legal mandate to communicate and work with indigenous people and local communities. This paper presents the results of a national survey of all currently functioning KPH units, the first of its kind ever conducted with KPH leadership, to obtain a system-wide perspective of the KPHs' role, mandate, and capacity for serving as effective intermediaries in managing forest conflicts in Indonesia. The survey results show that the KPHs are still in a very initial stage of development, and are struggling with a complex and rapidly evolving policy and institutional framework. The most common conflicts noted by respondents included forest encroachment, tenure disputes, boundary conflicts, and illegal logging and land clearing. KPH leadership views conflict resolution as among their primary duties and functions, and underscored the importance of more proactive and collaborative approaches for addressing conflict, many seeing themselves as capable facilitators and mediators. Overall, these results juxtapose a generally constructive view by KPH leadership over their role and responsibility in addressing forest management conflicts, with an extremely challenging social, institutional, and political setting. The KPHs can certainly play an important role as local intermediaries, and in some cases, as facilitative mediators in resolving local conflicts, but only with a more concerted effort from central and provincial government authorities to provide greater consistency in policies and regulations, improved policy communication, and a sustained commitment to strengthening the capacity of individual KPHs.

Keywords: Sustainable forest management; conflict resolution; Indonesia

---

1. Introduction

Conflicts over the management of natural resources are common phenomena everywhere in the world (Ayling and Kelly, 1997; Hellstrom, 2001; Yasmi et al., 2006; Gritten and Mola-Yudego, 2011). This has been well documented in the forestry sector, where national and regional government priorities clash over jurisdictional control, and compete with local communities over their livelihood needs (De Koning et al., 2008; Harwell, 2011; Yasmi et al., 2012). More recently, localized forest management conflicts have increasingly become nested within the broader national and international debates over timber extraction, harvesting of valuable non-timber forest products, rapid land conversion to palm oil and pulp and paper plantations, biodiversity conservation, and global carbon sequestration (Yasmi et al., 2012; Dhiaulhaq et al., 2015).

Various authors have discussed the complexity of natural resource management conflicts, and have developed various typologies, using actors, causes, intensity, tractability, sources of power, and management strategies (FAO, 2000; Gritten and Mola-Yudego, 2011; Krott et al., 2014). When managed effectively, these conflicts can serve as constructive forces, driving social change for continuous evolution of society (Lee, 1997; Deloges and Gauthier, 1997; Castro and Nielsen, 2001; Hellstrom, 2001). However, in many cases, conflicts escalate into damaging situations with substantial social and economic costs, including the disenfranchisement of local communities, social protests, and the engagement of security forces and the use of lethal violence (Bannon and Collier, 2003). For the private sector, these conflicts can represent significant financial costs and risks; in an analysis of projects in the extractive sector, researchers reported losses of roughly US$20 million per week in individual project settings, largely due to lost productivity and sales (Davis and Franks, 2014).

The number and intensity of forest and land use conflicts are particularly acute in developing countries blessed with rich natural resources and cultural diversity, but plagued with weak governance systems and economic pressures for resource exploitation (de Koning et al., 2008). Indonesia is a prime example of such countries (Gritten and Mola-Yudego, 2011).

The recent expansion of the forestry and plantation sectors in Indonesia has intensified agrarian and natural resource conflicts, especially those related to land tenure, forest boundary and classification, and illegal logging and land clearing, and created increased awareness of the social, economic and environmental impacts of these disputes (Forest Watch Indonesia, 2011; Dhiaulhaq et al., 2014). Persch-Orth and Mwangi (2016) report that Indonesia has the highest number of industrial tree plantation conflicts worldwide. In 2013, the Consortium for Agrarian Reform (*Konsorsium Pembaharuan Agraria,* or KPA), reported 396 cases of agrarian conflict (affecting more than 1.2 million ha), up from 198 cases (300,000 ha) during the previous year. A World Bank study (2014) reported that nearly 25 million ha of all designated state forest lands (or *kawasan hutan)* are in conflict (more than 20 percent of the total forest area, encompassing nearly 20,000 villages), primarily due to competing land claims and governance issues. The Institute for Policy Analysis of Conflict (IPAC), has underscored the fact that "land and resource conflicts are becoming a major source of lethal violence in Indonesia and there are no good mechanisms for resolving them" (IPAC, 2014).

Indonesia has the third largest tropical forest in the world, with one of the world's fastest rates of deforestation – more than 1,000 km^2/year (Hansen et al., 2013). At the same time, Indonesia has emerged as one of the major beneficiaries of global negotiations to mitigate climate change through improved forest management. Substantial financial gains through international climate mitigation initiatives provide strong motivation for Indonesia to reform its forestry sector (Brockhaus et al., 2012). Indonesia's Forest Management Units (*Kesatuan Pengelolaan Hutan*, or KPH*)*, have become the pivotal structural element for managing all state forests at the local level (MoF, 2014; Hernowo and Ekawati, 2014). By May, 2012, Indonesia had projected the designation of 600 KPHs nationally, covering an area of 130,680,000 ha, with multiple KPHs established in each District, and oversight and management primarily at the provincial level.[1] Each KPH is expected to carry out conventional tasks for forest management, such as establishing its management boundaries, conducting forest inventory, and developing forest management plans (ROI, 2007). The KPHs also implement various forest management activities at the site level, including the application of national, provincial and district forest policies (MOF, 2014). In addition, the KPHs have become the first government entity managing natural resources with a legal mandate to communicate and work with indigenous people and local communities (Kartodihardjo et al., 2011; Bae et al., 2014).

---

[1] See: http://www.forclime.org/documents/Brochure/Bahasa/FAQ%20KPH%20_Bahasa.pdf

[2] Partnership agreements (*kemitraan*) are a relatively recent program initiative from the Minister of Forestry (Ministerial Regulation No P.39/2013) that augments existing community-based forestry initiatives such as village forests (*hutan desa*), social forestry (*hutan kemasyarakatan*, or HKM), and Partnership agreements

Although the central government created the KPH system, jurisdiction over its operation is now vested with the provincial government where the KPH is located (Sahide et al., 2016a,b). As an emerging institutional actor bridging multiple jurisdictions and stakeholders within the forestry sector, they are explicitly viewed as a potential mediator in systematically addressing forest management disputes. Individual KPHs are expected to make sense of policies and regulations from different levels of governments, manage ambiguity, and negotiate, bargain and exercise discretion in implementing these policies within local contexts (Kim et al., 2015).

Given these expectations for the KPHs' role in managing forest conflicts in Indonesia, it is important to understand KPH perspectives of both their mandate and capacity for managing these conflicts. This paper presents the results of a national survey, the first of its kind ever conducted with KPH leadership, to obtain a system-wide perspective of all KPH units by eliciting KPH directors' perspectives on their role, mandate, and capacity for serving as effective intermediaries in managing forest conflicts in Indonesia.

2. Methods

The research team developed an initial draft questionnaire with primary input from selected KPH directors (*kepala*) and national experts in the forestry sector. Discussions with several KPH leaders reinforced their interest in gaining a national perspective on the nature of forest management conflicts, and on KPH attitudes, capability, responses, and priorities. The survey instrument was further refined after preliminary testing with a smaller group of resource persons (see Acknowledgments). Revisions included editing of the questions' phrasing and vocabulary for more ready understanding, and adjustment of some of the question formats. The survey was formatted and distributed in Bahasa Indonesia using the on-line service Survey Monkey (www.surveymonkey.com). Initial discussions with KPH directors suggested that the use of an on-line survey was practicable, given general email accessibility, and their strong interest in pioneering the use of this rapid response tool.

The survey was distributed via email link in June 2016 to all KPH directors, through the email mailing list of the National Association of KPHs (the professional organization representing all KPHs in Indonesia). The email included an introductory message (also in Bahasa Indonesia) from the national coordinator explaining the purpose, importance, and need for participating in the survey. Three subsequent reminders were sent to increase the survey response rate.

Invitations to participate in the survey were sent to 125 KPH directors, as identified by the Association's most current membership and email distribution list (note that at the time of the survey, 130 KPHs had been officially designated as operational). The introductory email asked specifically that the KPH directors personally fill out the surveys, in order to obtain direct responses from the highest level of KPH leadership. However, survey respondents were allowed to remain anonymous, to encourage candid participation, without concern that responses would be traced back to individuals.

The survey consisted of 19 questions, covering the following topics:

- General information about the KPH (location, area, status, staffing)
- Respondents' personal information (age, years of service, educational background)
- Perspectives about KPH roles and functions
- Perceptions of conflict (type and extent of conflicts, actions to date, outcomes, needs)
- Roles of other actors (perceived responsibilities, potential contributions)

An English translation of the survey questionnaire is presented in Appendix 1. All survey responses were received during June - July, 2015.

3. Results

We received responses from 87 KPHs (69.6%), although not all respondents filled out every question of the survey. Analysis of results was therefore adjusted for individual questions, accounting for variable response rates.

3.1. Overview of KPHS represented in the survey

Responses were received from all regions, with the majority of responses from three main islands: Sumatera (35.6%), Sulawesi (20.7%) and Kalimantan (16.1%) (see Table 1). These figures are generally representative of the current national distribution of KPHs and their sizes. KPH responses were somewhat evenly distributed in terms of size, with more than half of the KPHs managing less than 100,000 ha.

Table 1. Responses by Location and Size

Location\size	Less than 50K ha	50K-100 K ha	100K-15 0K ha	150K-250K ha	Over 250K ha	Blank	Total (%)
Java	1						1(1.1)
Maluku	1			1			2(2.3)
Papua		1		2			3(3.4)
Bali-Nusa Tenggara	9	3					12(13.8)
Kalimantan	2	2	4	1	5	1	14(16.1)
Sulawesi	4	3	6	4	1	2	18(20.7)
Sumatera	9	13	4	1	4		31(35.6)
Blank						3	3(3.4)
TOTAL (%)	26 (29.9)	22(25.3)	14(16.1)	9 (10.3)	10(11.5)	6(6.9)	87

Since the KPHs have all been established since 2007 (Government Regulation 6/2007), respondents' experience working officially within the KPH system is relatively recent, with 24.7% reporting less than one year of working experience, 42.9% from 1 – 2 years, 28.6% from 3 – 5 years, and just 3.9% with more than five years experience. The number of respondents by age: 6.4% between 20 – 30 years, 35.9% 31 – 40 years, 34.6% 41 – 50 years, and 23.1% between 51 – 60 years (no respondents identified as being older than 60 years). Thus it appears that although experience within the KPH system is rather limited, KPH directors are generally more mature (57.7% above 41 years), and they do bring considerable professional experience to their more recent work with the KPHs.

KPH staffing varies significantly in both size and status (see Table 2). Most of the KPHs reported less than 20 civil service staff (35.8% less than 5, 34.6% between 6 and 10, and 21.0% between 11 and 20), with 4.9% (or 4 KPHs) between 21 and 50, 2.5% (2 KPHs) between 51 – 100, with only one KPH (1.23%) reporting more than 100 civil service staff. Given the challenges of managing these extensive units with so few permanent staff, many of the KPHs augment their staffing needs with short-term hires of newly graduated forestry students (*Kontrak Bakti Sarjana Rimbawan*), and with contract labor hired for specific programs. This may tend to restrict management capacity, professionalism, and continuity within the KPH system, as many of the KPHs tend to rely on short-term and temporary staff. The number of staff, particularly the number of civil servants, does not correlate with the size of KPHs; however, the larger KPHs do tend to rely more on project-based temporary staff.

Table 2: Average number of staff per unit, by employment status and KPH size

Size\KPH Staff Status	Civil Servants	Contract (forestry students)	Contract -project based	TOTAL
All Size	12.34	9.84	9.24	31.4
Less than 50K ha	15.8	10.5	11.9	38.2
50K-100K ha	12.6	8.4	5.1	26.0
100K-150K ha	7.6	12.0	4.4	24.0
150K- 250K ha	9.4	9.8	5.4	24.7
Over 250K ha	13.9	8.7	22.3	44.8

In terms of educational background, most respondents reported receiving Bachelors' (59.5%) or Masters' (36.7%) degrees, with only two respondents (2.5%) reporting a high school level education, and one (1.3%) having completed a doctorate degree. The great majority completed their studies in traditional fields such as forestry (69.7%) and agriculture (10.5%), with fewer numbers reporting degrees in law (6.6%), social sciences (5.3%), and economics (3.95%). Once again, it is clear that KPH directors do bring significant education and experience to their work with the KPH, despite their limited experience working within these new management units.

Nevertheless, little more than half the respondents (55.13%) reported having participated in any coursework or training related to conflict management, and of these, most (56.8%) cited a single introductory training event. These included basic skills training in conflict mapping, conflict management, tenurial conflict, partnership building, and facilitation training. Many cited the basic "Conflict Mapping and Conflict Management" or "Managing Tenurial Conflict" training courses they receive during their introductory in-service training for officials nominated to become KPH directors.

3.2. Perceptions of KPH duties and Functions

Asked about the primary duties and functions of the KPH, respondents cited as highest priorities: 1) forest management (92.73%), 2) long-range planning (87.27%), 3) forest inventory (83.64%), 4) capacity building for KPH staff (72.73%), 4) conflict resolution (69.09%); and 5) forest protection and conservation (69.09%) (see Figure 1). Other duties noted as important included enforcement, rehabilitation and reclamation, networking, community capacity building, administration and financial management, forest investment and business development, clarification of forest policies, forest utilization, and monitoring and evaluation. Many underscored the critical, cross cutting importance of staff capacity building and building good working relationships with local communities, including efforts at improving local economies. As one respondent commented: "We have to develop an approach to secure adequate incomes for people on the forest margins; if we can achieve this, *Insya Allah*, then we will have a much better chance of resolving forest management conflicts."

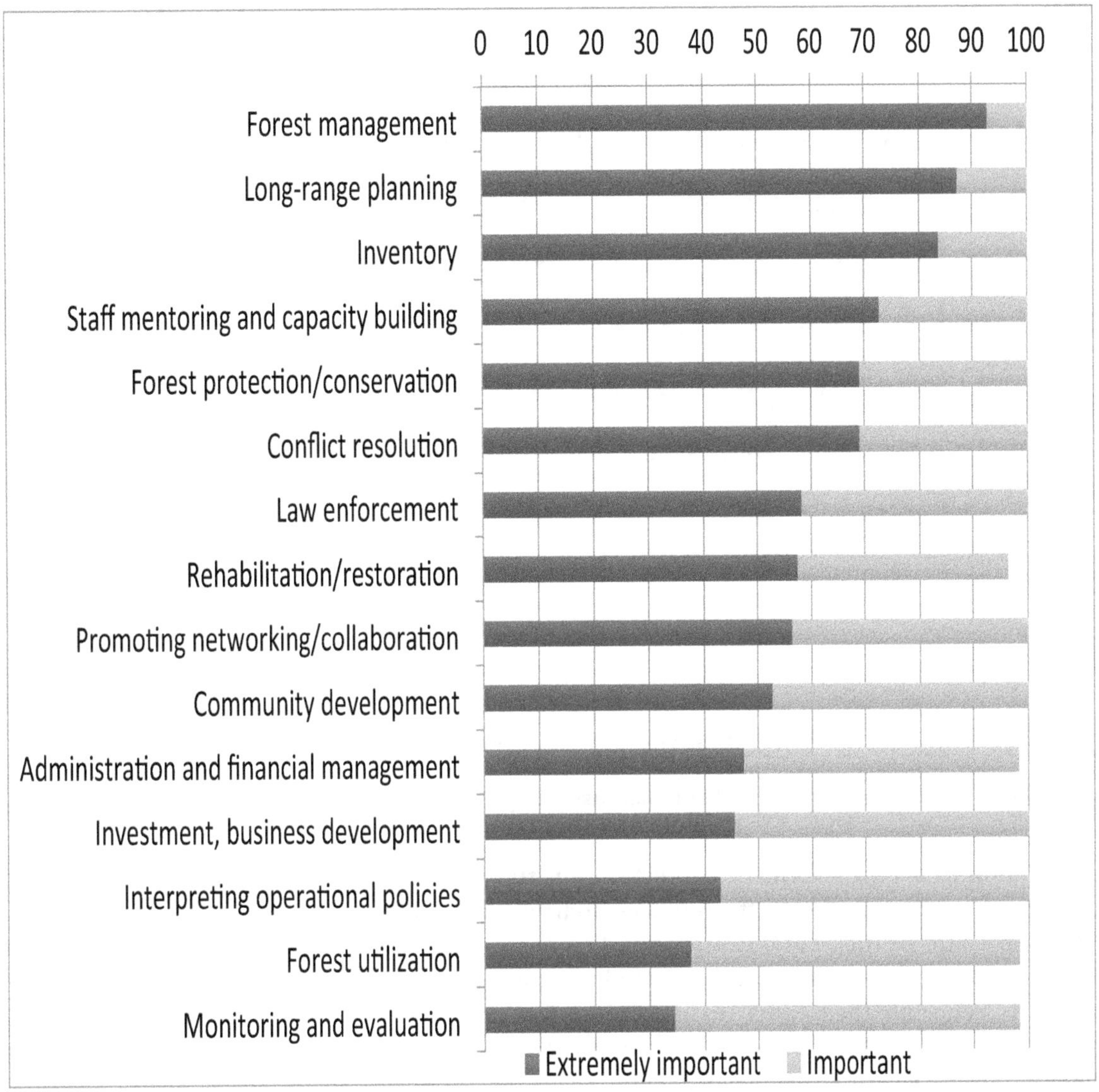

3.1. KPHs, conflict, and conflict management approaches

The most common conflicts (extremely serious or serious) noted by respondents included forest encroachment (89.09%), tenure disputes (86.79%), boundary conflicts (84.91%), illegal logging (75.48%), and illegal land clearing (69.82%). Also noted: customary claims, disputes over permitting, community-based conflicts (within and between communities, between local residents and migrants) and conflicts with concession holders (see Figure 2). This data underscores the extra-legal nature of the majority of forest management conflicts, and the challenges KPHs face as the front-line implementing unit for national forest policies. Several respondents noted this point in their comments, e.g., "Conflict resolution should be based on adhering to existing laws and policies; in the end, it all depends on political will and the commitment of all stakeholders to follow regulations."

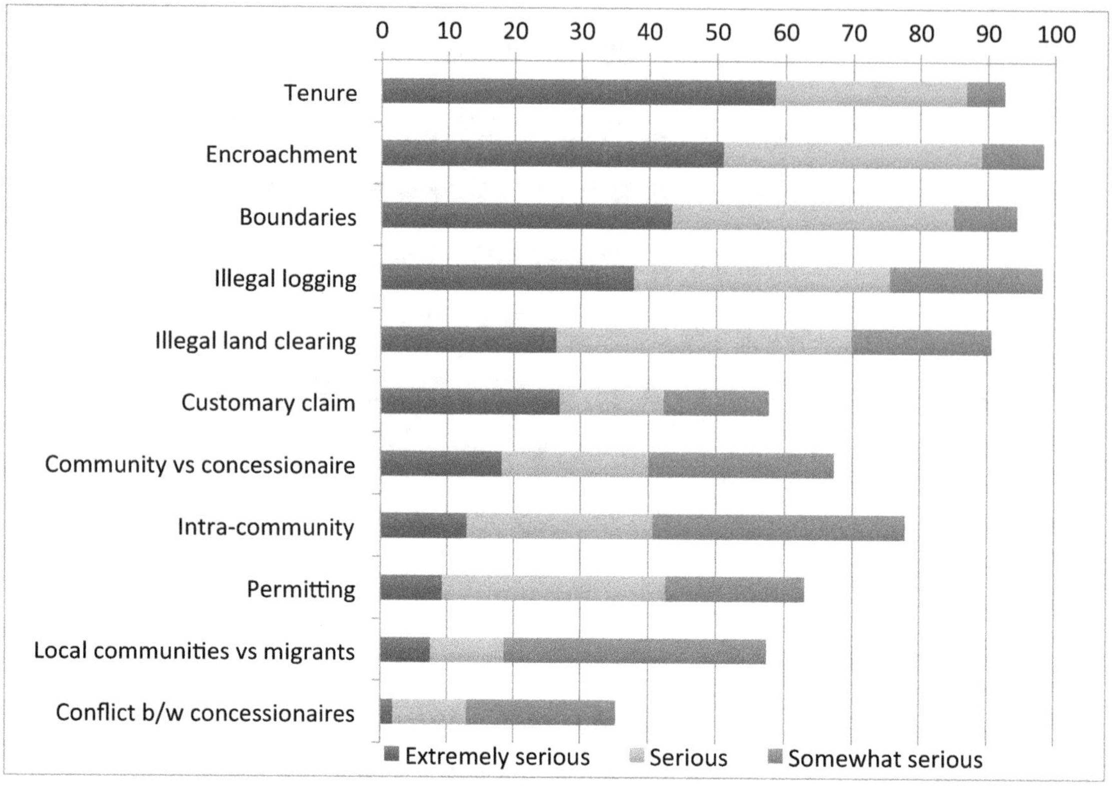

Figure 2. Type of conflict and level of seriousness

Commenting on the extent of conflict within the KPHs, almost half of respondents (48%) estimated that active conflicts affected one quarter of the KPH area, 28% estimated conflicts impacted 26 – 50% of their KPH, 16% reported conflicts in 51 – 75% of the area, and 8% reported that 75 – 100% of their sites were affected by conflicts (Table 3). Conversely, only 25% of respondents reported their KPHs to be conflict free in more than half the total area. KPHs reporting that most of their area (75% or more) is in current conflict, also noted that managing conflicts is a primary responsibility of the KPH.

Table 3: KPH and area in conflict

Extent of Conflict	1 - 25%	26 - 50%	50 - 75%	76 - 100%
Currently in conflict	48%	28%	16%	8%
Potential Conflict	24%	47%	16%	14%
Conflict free	48%	27%	17%	8%

KPH directors voiced strong support for more constructive views about addressing forest management conflicts (Figure 3): 98.18% agreed with the statements that they needed to learn more about conflict and were willing to work with others to resolve disputes, 92.72% said that conflict resolution is an important part of their responsibilities, 89.09% said they were willing to seek assistance in resolving disputes, 87.27% said they seek compromise in addressing conflicts, 85.46% described themselves as capable facilitators or mediators, and 74.55% said they were effective negotiators. Among the representative comments: "Conflict resolution begins with an incremental approach to the communities, understanding their issues and concerns and the challenges they face, and mapping these out so that they are better understood and more readily addressed." Many other comments underscored the importance of building stronger working relationships among stakeholders, for example: "In addressing tenurial conflict in my area, we have formed an integrated team with local government, communities, and law enforcement, to negotiate through consensus building and seek alternative means of resolving the conflicts." These are important and encouraging statements, as it is widely recognized that acknowledging conflict, using effective communication, working collaboratively, and taking joint responsibility for solutions are keys to effective resolution (Moore, 1986).

At the same time, 65.46% agreed with the statement that conflicts must be resolved through extreme measures (e.g., through more direct involvement of law enforcement, stronger application of laws and policies), while 30.91% said they try to avoid conflicts as much as possible, 21.82% said that conflict is the primary responsibility of law enforcement agencies; only 5.46% said they preferred not to think about conflicts. These latter attitudes would be important to explore further in follow up research, as they are inconsistent with the KPHs' legislative mandate, and may not serve as an effective strategy for resolving these disputes.

Figure 3. Attitudes about managing forest conflict

Asked about which actors have the greatest influence over forest management in their areas, respondents listed the Ministry of Environment and Forestry, KPHs, forest margin communities, and Regency-level governments, followed by provincial government, concession holders, and NGOs (see Figure 4). Many emphasized the MoEF's primary role in setting forest boundaries, establishing use classifications, and in setting overarching policies, and underscored the KPHs' role in implementing these policies and providing local level planning, management, and monitoring. Respondents also highlighted the important role of customary (*adat*) leaders in ensuring effective forest management. Many respondents underscored the need for synergy and coordination among actors, i.e., the importance of collaboration across jurisdictions, and among all actors, including government, communities, concession-holders, and NGOs. As one respondent commented: "The KPH is stuck in the middle, between government policies and practical realities; resolving conflicts requires the commitment of all government agencies, from national to local, and certainly the NGOs and forest margin communities share a common interest and mutual influence in achieving effective forest management."

Respondents identified the KPH and MoEF as the institutions having the greatest responsibility for resolving forest conflicts (both at 57.41%), followed by District-level governments (44.44%) and the Ministry of Agrarian Affairs and Spatial Planning (MAASP), which has authority for regulating and coordinating overall land use policies (39.22%) (Figure 4). This is consistent with views expressed in other responses, that many of the disputes primarily relate to forest management and boundary issues, and that many of these disputes have national policy dimensions, but reflect local social and economic dynamics. However, respondents' comments acknowledged the multiple roles and responsibilities, and the complex interaction among stakeholders in addressing forest management conflict, for example, in this succinct summary of the role of various actors: "The three levels of government, National, Provincial and District governments, have the greatest influence on forest management, because they have responsibility for planning and administration of all forest lands. The Ministry, particularly, is the primary initiator of all these aspects of forest management and administration. The KPH, in contrast, is merely an implementer/operator of these management policies. Meanwhile, the NGOs serve a key role in extension and communication, as a bridge to the communities in helping convey improved understanding of our programs. Concession holders can exert both positive and negative influences – positive if they manage their site effectively, but negative if they simply obtain their permit but don't have the necessary management capability, which results in illegal activities like illegal logging, encroachment, occupation, etc. And obviously the community has the most immediate and direct relationship to the forest, so we have to continue to seek ways to improve their understanding and capacity, change their way of thinking about forest management toward greater social benefits and greater sustainability."

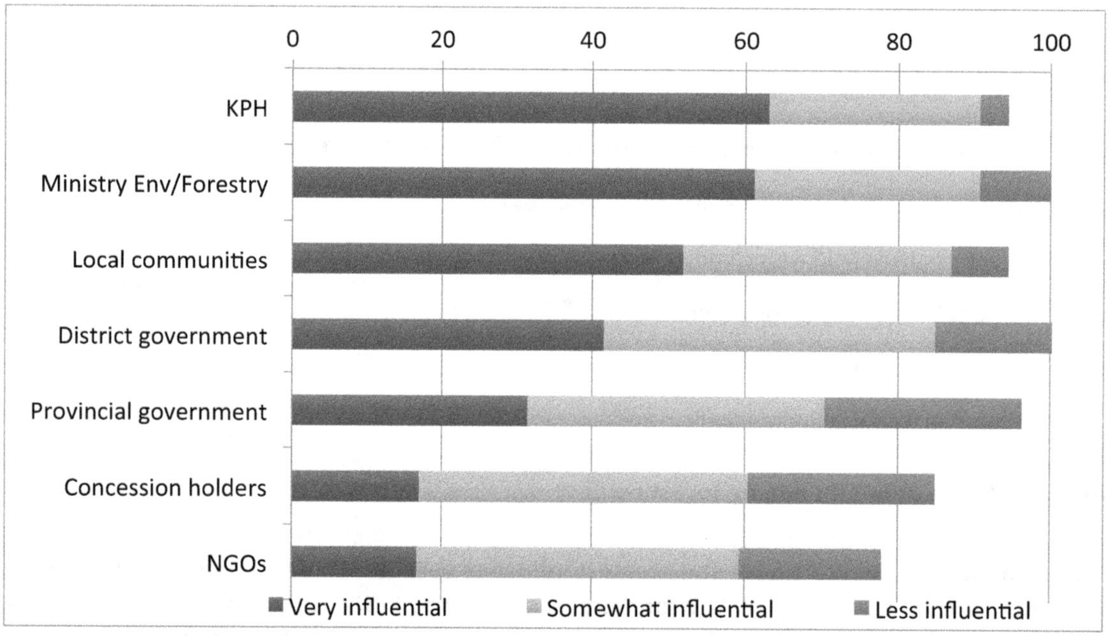

Figure 4. Actors and their influence over forest management

Commenting on the most effective approaches they have used to date in resolving conflicts, respondents generally rated pro-active measures as most effective: promoting collaboration among parties (92.31%) developing partnership agreements[2] (88.24%), supporting staff training and capacity building (86.27%), including conflict resolution plans within their long-term planning documents (76.47%), collecting data (71.16%), engaging a third-party neutral/mediator (67.34%), and conflict mapping and analysis (67.31%) (Figure 5). Least effective actions noted: using Free

---

[2] Partnership agreements (*kemitraan*) are a relatively recent program initiative from the Minister of Forestry (Ministerial Regulation No P.39/2013) that augments existing community-based forestry initiatives such as village forests (*hutan desa*), social forestry (*hutan kemasyarakatan*, or HKM), and Partnership agreements allow for local discretion and more active participation of community groups in managing and sharing benefits of forest management (see Adnan et al. 2015).

Prior and Informed Consent3 (FPIC) protocols (54.35%), enhanced law enforcement (52.00%), and development of grievance/complaint procedures (48.98%). These results suggest a positive emphasis on more positive, pre-emptive approaches to mitigating conflicts through cooperation and staff skill building, but it is interesting to note the low scores for FPIC and the application of grievance procedures, both of which would be considered proactive, systems approaches to reducing conflict and its escalation. This may be a reflection of the more limited understanding and use of these two approaches, which are as yet uncommon within the KPH system.

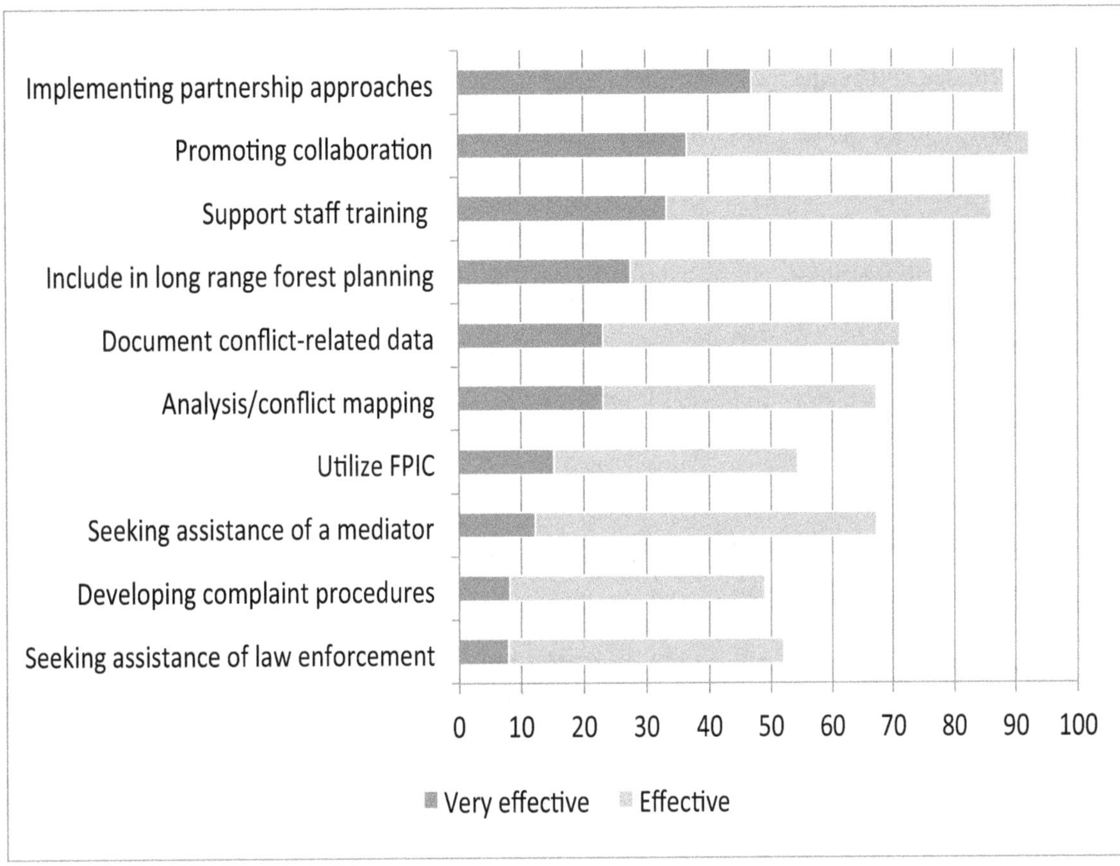

Figure 5. Most Effective Approaches for Managing Conflict

Asked about the support deemed most helpful to the KPH in resolving conflicts, respondents again emphasized more pro-active capacity building measures: partnership building (77.78%), staff training (74.07%), active engagement of community leaders (74.07%), local adaptation of national policies (70.37%), greater engagement of Regency government (66.04%), increased budget for the KPH (62.96%), regular forest patrols (57.41%), conflict mapping and documentation (55.56%), and increased KPH staff numbers (48.15%) (Figure 6). Worth noting here are the relatively low scores for greater engagement of law enforcement agencies (28.30%) or the intervention of professional mediators (26.42%), suggesting that many KPH directors feel they have the capacity, or could develop the capacity, for managing disputes.

---

[3] Free, Prior, and Informed Consent (FPIC) is the principle, outlined in the UN Declaration on the Rights of Indigenous Peoples, that a community has the right to give or withhold its consent to proposed projects that may affect the lands they customarily own, occupy or otherwise use Colchester, 2010). In Indonesia, FPIC is referred to by the acronym PADIATAPA (*Persetujuan atas Dasar Informasi di Awal Tanpa Paksaan*).

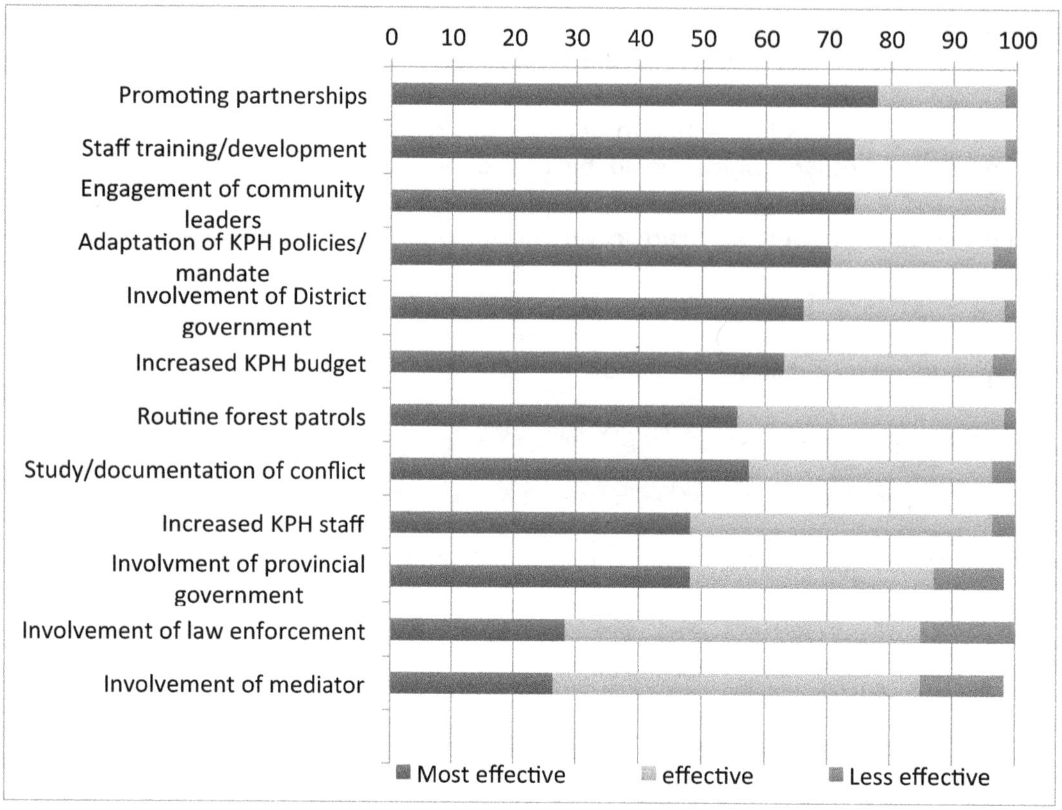

Figure 6. Support or interventions regarded as most helpful in resolving conflicts

4. Discussion

The results of this initial, broad-scale survey offer important insights into the current status of the KPHs, their leadership's perceptions of forest conflicts, and their sense of the KPHs' mandate, opportunities, and needs in addressing forest management disputes. The response rate (69.6%) and the many thoughtful and insightful comments received within the survey, suggest a high level of interest in the topic, and, as one respondent suggested, the sense that "KPHs will never be free of conflict with local communities, so we will have to continue seeking ways to manage conflict effectively."

The profile of KPHs and conflict that emerges from the survey underscores the following key points:

- The KPHs are still in a very early stage of development, and are struggling with a complex and rapidly evolving policy and institutional framework, one in which, as many respondents noted, the KPHs are "stuck in the middle", and in the position of "implementer/operator of...management policies". Overall, they are constrained by limited professional staff, and with leadership and staff who lack on-the-job experience. Note that less than 4% of KPH leaders have been in their positions for more than 5 years; fewer than half of the KPHs have more than 20 permanent staff, and many rely heavily on contract labor. In terms of basic skills, only a little more than half (55%) have participated in any training related to conflict management, and this only at the very introductory level.
- KPH leadership's sense of priorities very much focuses on the fundamental aspects of forest management: inventory, planning, forest protection, conservation, and restoration. However, staff development, conflict resolution, routine law enforcement, networking and collaboration, and socializing national forest policies, followed closely in terms of perceived priorities and responsibilities.
- Forest management conflicts are primarily attributed to encroachment, illegal logging and land clearing, and tenure and boundary disputes, reflecting a combination of both limited management capacity, as well as an emboldened sense of peasant empowerment (Maryudi et

al., 2016). But respondents also noted that customary claims, permitting/licensing, and intra- and inter-community conflicts constituted frequent causes of forest management disputes, and this again suggests the need for greater clarity of national policies and more consistent implementation at the local level.

- KPH leaders generally offered support for more constructive comments about dealing with conflict: an acknowledgement of the need to learn more about approaches to conflict management, a willingness to work with others to resolve disputes, the recognition that dealing with conflict is an important part of their responsibilities, and their openness to seeking compromise. In short, they tend to embrace their role and responsibility in managing conflict, and the majority of respondents described themselves as capable facilitators, mediators, or negotiators.
- Respondents underscored the importance of more collaborative approaches for addressing conflict: partnership and network building, staff development, engaging community leaders and other key partners, adaptation of policies to accommodate local context and needs. Considered as lesser importance were more coercive means, such as increasing law enforcement and forest patrols, and seeking the outside assistance of professional mediators.

Overall, these results juxtapose a generally constructive view by KPH leadership over their role and responsibility in addressing forest management conflicts, with an extremely challenging social, institutional, and political setting – multiple actors and issues, changing policies and authority, and uncertainty in their role and degree of autonomy (Sahide et al., 2015). The KPHs' willingness to embrace their responsibility and acknowledge the need for enhanced staff capacity, greater flexibility in adapting policy, and financial and analytical support, offers a necessary, but certainly insufficient, foundation for positive change. However, while these survey results are helpful in articulating the perspectives of the KPH leaders, they must be tempered with an understanding of the limits of self-assessment, and they are necessarily constrained by the harsh realities of a forest governance system in need of significant reform (Contreras-Hermosilla and Fay, 2005; World Bank, 2014). Certainly additional research could offer deeper understanding of the challenges faced by KPHs in addressing the changing policy framework (e.g., the ongoing shift from national to regional control, balancing economic and environmental priorities), the social and economic context (e.g., by region, by forest type), the relationship with local communities and other key actors, and the capacity of individual KPHs and their staffs.

In their review of 118 conflict cases, Yasmi et al. (2006) underscored the fact that natural resource management conflicts necessarily involve multiple stakeholders and complex, interdependent issues and interests. Many of these site-based disputes are situated within, and informed by more macro-level governance structures (Wulan et al., 2004; Sahide and Giessen, 2015). It is significant, therefore, that KPH leaders report that tenure and boundary disputes, and a range of illegal activities (encroachment, land clearing, and illegal logging), constitute the majority of forest management disputes in Indonesia. While the KPHs represent the potential for decentralizing and improving forest governance, the system still requires stronger integration within provincial and district-level governments (Kartodihardjo et al., 2011; Bae et al., 2014), greater authority for KPHs to settle disputes over community access rights, and clearer guidance for KPHs in monitoring and supervising concession operations (Ota, 2015). There remain real sources of tension for the KPHs (with their implicit emphasis on community-based forestry approaches) in positioning themselves within these national, provincial and district level bureaucracies (Sahide et al., 2016a,b).

The collaborative, stakeholder-based approaches emphasized in many of the survey responses do offer promise, and underscore a commitment to a more socially and environmentally appropriate means of increasing local participation in forest management decision making. Embracing collaborative processes can be a truly liberating and empowering process for communities and other stakeholders, but the experience with co-management regimes in many places suggests that such arrangements can result in the cooptation or further marginalization of local interests (Castro and Neilsen, 2001). Indeed, while the merits of community forest approaches (such as partnership agreements) lie, potentially, in the simplification of permit issuance procedures and reduced transaction costs, national legislative reform is still needed to expand both the area of potentially usable forest for local communities, as well as the KPHs' authority to include non-concession areas (Jang and Bae, 2014).

While skill building, partnership development, and collaboration with stakeholders are all important elements of improved governance, the plethora of boundary disputes and illegal activities still requires significant legal reform and enhanced law enforcement methods. The KPHs represent an important element of this reform, and further research on individual KPHs' capacity can provide insights for KPHs to share their early successes and experiences with project implementation, including lessons learned in engaging local communities and indigenous groups. Similarly, understanding individual KPHs' ability to analyze conflicts and distinguish causality, including socio-economic and environmental aspects, as well as location and context-specific dynamics, would be an important contribution for improving management of these conflicts (Gritten and Mola-Yudego, 2011).

Dhiaulhaq et al. (2015) have emphasized that successful mediators take on a variety of roles. They facilitate communication among parties and help explore, and when necessary propose, alternative solutions. They often require the technical capacity and negotiation skills of the parties involved to address imbalances, as well as providing resources and technical assistance. Theoretically, at least, the KPHs have the mandate, and in some cases the experience and skill, to play this important role. Survey results have emphasized KPH directors' general confidence in their ability to serve this function. However, while mediation can certainly be an effective tool for resolving forest management disputes (FAO, 2000; Dhiaulhaq et al., 2015), it remains questionable whether the KPH – as an institution, since it is an active stakeholder in many of these conflicts – can serve as an effective, neutral mediator (Persch-Orth and Mwangi, 2016).

More importantly, the nature and extent of conflicts described by KPHs are clearly more chronic in nature, and require broader systemic change, rather than site-based mediation – to resolve existing conflicts, and prevent future conflicts from occurring. In the absence of this broad-scale systemic change, the central question for the KPHs, if they are to be considered as effective intermediaries, is the degree of discretion and authority they are given for promoting effective, collaborative solutions that are suitable to local conditions.

5. Conclusions

The KPHs continue to be viewed as a fundamental element of forest governance reform in Indonesia. A projected 600 KPHs is expected to be fully functional by 2020, and they are increasingly being viewed as the frontline presence for the Ministry of Environment and Forestry in managing the forest estate in Indonesia (Kartodihardjo et al., 2011; Hernowo and Ekawati, 2014). However, as Ota (2015) has noted, a mere increase in numbers is an unlikely solution to the fundamental challenges represented by the complex forest management conflicts described above. For the KPHs to function effectively at the operational level, clear legal and policy mechanisms, reconciliation of jurisdictional roles and responsibilities, and enhanced enforcement, are needed to support improved implementation in forest management.

The KPHs can certainly play an important role as local intermediaries, and in some cases, as mediators to local conflicts, but only with a more concerted effort from central, provincial, and district-level government authorities to provide greater consistency in policies and regulations, improved policy communication, and a sustained commitment to strengthening the capacity of individual KPHs (Kim et al., 2015).

As one respondent commented in a final summary response: "to be successful, [KPHs] will have to address three fundamental issues: building staff capacity (in terms of both quality and quantity), securing adequate budgets that provide for independence, and appropriate regulations that truly acknowledge and seek to accommodate a multi-stakeholder perspective."

Acknowledgments

This research was supported by a Global Innovation Initiative grant from the United States Department of State and the Office of Foreign Affairs, United Kingdom. Additional support was provided by the Ford Foundation and the Climate and Land Use Alliance. We benefited greatly from early input received from several members of the National Association of Forest Management Units in Indonesia. We especially thank KPH directors for taking the time to complete the survey questionnaire and provide comments on preliminary findings

References

Adnan, H., H. Berliani, G. Hardiyanto, Suwito, and D. K. Sakti. (2015). *Pemberdayaan Masyarakat melalui Kemitraan Kehutanan*. The Partnership for Governance Reform, Jakarta.

Ayling, R.D., and K. Kelly. (1997). Dealing with conflict: natural resources and dispute resolution. *Commonwealth Forest Review,* 76 (3): 182–185.

Bae, J.S., Y.S. Kim, L.A. Fisher, M. Moeliono, and J. DeShazo. (2014). Promises and Perils of Decentralized Forest Governance: The Case of Indonesia's Forest Management Units in Reducing Emission from Deforestation and Forest Degradation (REDD+). *Society and Natural Resources*, 27:1346–1354

Bannon, I., and P. Collier. (2003). *Natural Resources and Violent Conflict.* World Bank, Washington DC.

Brockhaus, M., K. Obidzinski, A. Dermawan, Y. Laumonier, and C. Luttrell. (2012). An overview of forest and land allocation policies in Indonesia: Is the current framework sufficient to meet the needs of REDD+?. *Forest Policy and Economics,* 18:30-37.

Castro, A.P., and E. Nielsen. (2001). Indigenous people and co-management: implications for conflict management. *Environmental Science and Policy,* 4: 229–239.

Colchester, M. (2010). Free, Prior and Informed Consent: Making FPIC work for forests and peoples. *The Forest Dialogue Research Paper* No. 11, Yale University, New Haven.

Collier, P., and A. Hoeffler. (2002). Greed and Grievance in Civil War. *Oxford economic papers,* 56(4): 563-595.

Contreras-Hermosilla, A., and C. Fay. (2005). *Strengthening Forest Management in Indonesia Through Land Tenure Reform: Issues and Framework for Action*. World Agroforestry Center, Bogor, Indonesia.

Davis, R., and D. M. Franks. (2014). Costs of Company-Community Conflict in the Extractive Sector. *Corporate Social Responsibility Initiative Report* No. 66. Harvard Kennedy School, Cambridge, MA.

De Koning, R., D. Capistrano, Y. Yasmi, and P. Cerutti. (2008). *Forest-Related Conflict: Impacts, Links and Measures to Mitigate.* Rights and Resources Initiative, Washington, D.C., USA.

Deloges, C., and M. Gauthier. (1997). *Community forestry and forest resource conflicts: an overview*. Paper prepared for the XI World Forestry Congress, Antalya, Turkey.

Dhiaulhaq, A., D. Gritten, T. De Bruyn, Y. Yasmi, A. Zazali, and M. Silalahi. (2014). Transforming conflict in plantations through mediation: Lessons and experiences from Sumatera, Indonesia. *Forest Policy and Economics,* 41 (0):22-30.

Dhiaulhaq, A., T. De Bruyn, and D. Gritten. (2015). The use and effectiveness of mediation in forest and land conflict transformation in Southeast Asia: Case studies from Cambodia, Indonesia and Thailand. *Environmental Science and Policy*, 45 (0):132-145.

Food and Agriculture Organization (FAO). (2000). *Conflict and Natural Resource Management*. FAO, Rome.

Forest Watch Indonesia. (2011). *Keadaan Hutan Indonesia*. Forest Watch Indonesia, Bogor Indonesia.

Gritten, D., and B. Mola-Yudego. (2011). Exploration of the relevance of geographical, environmental and socio-economic indicators regarding forest conflict types. *International Forestry Review*, 13 (1): 46 – 55.

Hansen, M. C., P. V. Potapov, R. Moore, M. Hancher, S. A. Turubanova, A. Tyukavina, D. Thau, S. V. Stehman, S. J. Goetz, T. R. Loveland, A. Kommareddy, A. Egorov, L. Chini, C. O. Justice, and J. R. G. Townshend. (2013). *High-Resolution Global Maps of 21st-Century Forest Cover Change. Science* 342 (6160), 850-853

Harwell, E. (2011). *Forests in Fragile and Conflict-Affected States.* World Bank, Washington, D.C.

Hernowo, B., and S. Ekawati (eds.). (2014). *Operasionalisasi Kesatuan Pengelolaan Hutan (KPH): Langkah Awal Menuju Kemandirian.* Kanisius, Jakarta.

Hellstrom, E. (2001). *Conflict cultures: Qualitative comparative analysis of environmental conflicts in forestry.* Silva Fennica Monographs 2.

Institute for the Policy Analysis of Conflict (IPAC). (2014). *Indigenous Rights vs. Agrarian Reform in Indonesia: A Case Study from Jambi.* IPAC, Jakarta. Retrieved from http://file.understandingconflict.org/file/2014/04/IPAC_Report_9_Case_Study_from_Jambi_web.pdf in February 10 2016

Jang, S.K., and J.S. Bae. (2014). New Forest Governance in Indonesia: A Forest Partnership between Forest Management Units and Local Communities. *Southeast Asia Study* 24 (4): 95-144

Kartodihardjo, H., Nugroho, B., Putro, H.R. (2011). *Forest Management Unit Development (FMU): Concept, Legislation and Implementation.* Ministry of Forestry, FORCLIME Forest and Climate Change Programme, Jakarta, Indonesia.

Kim, Y.S., J.S. Bae, L. A. Fisher, S. Latifah, M. Afifi, S.M. Lee, and I.A. Kim. (2015). Indonesia's Forest Management Units: Effective intermediaries in REDD+ implementation?. *Forest Policy and Economics*, 62:69-77.

Krott, K., A. Bader, C. Schusser, R. Devkota, A. Maryudi, L. Giessen, H. Aurenhammer. (2014). Actor-centered power: The driving force in decentralised community based forest governance. *Forest Policy and Economics*, 49: 34–42.

Lee, K. (1997). Compass and Gyroscope: Integrating Science and Politics for the Environment, Island Press, Covelo, California.

Maryudi, A., Citraningtyas, E. R., Purwanto, R. H., Sadono, R., Suryanto, P., Riyanto, S., & Siswoko, B. D. (2016). The emerging power of peasant farmers in the tenurial conflicts over the uses of state forestland in Central Java, Indonesia. *Forest Policy and Economics, 67*, 70-75.

Ministry of Forestry (MoF). (2014). Executive Summary About Forest Management Units (February 2014). Ministry of Forestry, Jakarta, Indonesia.

Moore, C. W. (1986). *The Mediation Process: Practical Strategies for Resolving Conflict.* Jossey-Bass, San Francisco.

Ota, M. (2015). Can Forest Management Units improve the management of state forests in Indonesia's outer islands? Institutional analysis and case studies from Lampung province of Sumatra. *International Forestry Review*, 17(1), 76-85.

Persch-Orth, M., and E. Mwangi. (2016). Company–community conflict in Indonesia's industrial plantation sector, *CIFOR info brief*, No. 143, May 2016

Republic of Indonesia (ROI). (2007). Government Regulation No. 6/2007 Forest Arrangement and Establishment of Forest Management Plan as well as Forest Exploitation (in Bahasa Indonesia). Republic of Indonesia, Jakarta, Indonesia.

Sahide, M.A.K., and L. Giessen. (2015). The fragmented land use administration in Indonesia: Analysing bureaucratic responsibilities influencing tropical rainforest transformation systems. *Land Use Policy*, 43: 96–110.

Sahide, M. A. K., Maryudi, A., Supratman, S., & Giessen, L. (2016a). Is Indonesia utilising its international partners? The driving forces behind Forest Management Units. *Forest Policy and Economics, 69*, 11-20.

Sahide, M.A.K., S. Supratman, A. Maryudi, Y.S. Kim, and L. Giessen. (2016b). Decentralisation policy as recentralisation strategy: forest management units and community forestry in Indonesia, *International Forestry Review*, 18 (1): 78 - 95

World Bank. (2014). *Towards Indonesian Land Reforms: Challenges and Opportunities. A Review of the Land Sector (Forest and Non-forest) in Indonesia*, World Bank, Jakarta.

Wulan, Y.C., Y. Yasmi, C. Purba, and E. Wollenberg. (2004). *Analisa Konflik Sektor Kehutanan di Indonesia 1997 – 2003*. Center for International Forestry Research, Bogor.

Yasmi, Y. (2003). Understanding conflict in the co-management of forests: The case of Bulungan Research Forest. *International Forestry Review*, 5(1): 38-44.

Yasmi, Y., H. Schanz, and A. Salim. (2006). Manifestation of conflict escalation in natural resource management. *Environmental Science and Policy*, 9:538 – 546.

Yasmi, Y., L. Kelley, D. Murdiyarso, and T. Patel. (2012). The struggle over Asia's forests: an overview of forest conflict and potential implications for REDD+, *International Forestry Review*, 14(1): 2012

9

From voluntary private to mandatory state governance in Indonesian forest certification: Reclaiming authority by bureaucracies

Agung Wibowo[1] and Lukas Giessen[2,3,*]

[1] Department of Forestry, Faculty of Agriculture, University of Palangka Raya, Indonesia
[2] Chair of Forest and Nature Conservation Policy, University of Göttingen, Germany
[3] European Forest Institute (EFI), Bonn Office
* Corresponding author: agungwi@for.upr.ac.id; Lukas.Giessen@efi.int

Abstract: Forest certification has been introduced by non-state actors as a voluntary and market-based instrument addressing forest problems, which state policies failed to resolve. Lately, however, state-driven forest-related certification schemes can be observed, e.g. in Indonesia, through the EU FLEGT-VPA negotiation process. It is argued, specific state agencies in a struggle for power and authority develop mandatory certification schemes which are directly competing with private ones. Before this background, the aims of this study are: (i) describing the current trend from voluntary private to mandatory state certification schemes in Indonesia, (ii) mapping the main actors involved in certification politics, and (iii) explaining this trend with the interests of the main actors. The results confirm a trend from voluntary private to mandatory state-driven certification of forest management. The Ministry of Forestry, the Ministry of Trade, the Ministry of Industry, wood producer and processing associations, European Union, local funding organizations, environmental organizations, certification bodies and international buyers are detected as the main coalitions and actors in the certification politics. The stronger coalition develops a mandatorily-timber legality verification system as strategies to counter their voluntary private competitor schemes.

Keywords: competition; forest certification; international governance; private governance; state governance; bureaucratic politics; forest policy

---

1. Introduction

As a response to the failed negotiations among governments on a global forest convention in 1992, forest certification has subsequently been introduced by non-governmental actor coalitions for promoting sustainable forest management (Boström, 2003; McDermott, 2014). There are two ideas underlying the emergence of the non-state-actor driven certification schemes, such as the Forest Steward Council (FSC). First, it was an effort by environmental non-governmental organizations to voluntarily certify responsible-managed forest in Europe and North America (Bartley, 2007). Second, such non-state, market-driven, and voluntary forest certification schemes also aimed to certify tropical and mega-diversity forests including problems such as vast deforestation, degradation and illegal logging (Gulbrandsen, 2004; Blackman et al., 2014).

Shortly after the FSC was launched in 1993, other voluntary, non-state certification schemes that were better known locally and that had more support from industry groups were founded during the 1990s to better accommodate local forestry practices. These included e.g. the Sustainable Forestry Initiative (SFI) and the American Tree Farm System (ATFS) in North America, the Programme for the Endorsement of Forest Certification (PEFC, initially the Pan-European Forest Certification) in Europe, the Sustainable Forest Management System in Canada, and Lembaga Ekolabel Indonesia (LEI) in Indonesia (Tosun, 2012; Maryudi 2006, 2009). All these forest certification initiatives have been largely driven by non-state, private actors and voluntary subscription and are consequently regarded as private institutions (Pattberg, 2007). They, yet, mirror an underlying competition between public and private actors over the authority to regulate forest management practices (Sikor, 2013).

However, the success of private certification schemes in tackling specific forestry problems like illegal logging, forest encroachment and deforestation, and in improving forest governance in developing countries thus far has been limited (Durst et al., 2006; Tacconi, 2007; Espach, 2006). As a consequence, a state-driven, inter-governmental cooperation between timber producer and consumer countries was initiated by the European Union (EU) in 2003, known as Forest Law Enforcement, Governance and Trade (FLEGT), aiming to certify the legality of harvested timber (Wiersum and Elands, 2013). Such state-based forest certification initiatives now counter-compete with private ones and are assumed having the potential to dominate and even replace private certification schemes (Nurrochmat et al., 2014; Malets, 2015).

Recent empirical findings support this latter perspective and suggest even a crucial role for state agencies in forest certification and related politics (Gale and Haward, 2011; Gulbrandsen, 2014; Sahide et al., 2015). According to these literatures state agencies can obstruct (Gale and Haward, 2011) or support certification initiatives (Hysing, 2009; Bell and Hindmoor, 2012). More importantly, however, specific state bureaucracies, due to their formal mandate and ability to develop binding regulatory policy instruments, may even develop own certification schemes, as currently observed under the FLEGT process and regarding timber legality verification in Indonesia (Nurrochmat et al., 2014). In so doing, it is argued, state agencies develop mandatory certification schemes which are directly competing with private ones, mainly with foreign initiatives, and by this reclaim authority over forest certification through legitimized public actors (Giessen et al., 2016). Before this background, the aims of this study are (i) describing the current trend from voluntary private to mandatory state certification schemes in Indonesia, (ii) mapping the main actors involved in certification politics, and (iii) explaining this trend with the interests of the main actors.

2. Conceptual framework

2.1. Voluntary forest private certification

The FSC and PEFC forest certification systems have grown significantly and have become the largest in world, with a share of about 98% of all certified forest and chain of custody (CoC) certificates (FSC and PEFC, 2013). This amounts to 417 million hectares, or 10.3% of global forests area and a share of about 28% of the total round wood production (UNECE/FAO, 2013). This approach has successfully built a new mechanism in timber product trade by requiring all traded goods to meet a balance of ecological, social and economic requirements. By creating its own rules and simultaneously increasing public awareness of certified products, forest certification initiative governs the interaction between actors in forest product trade, and replaces the role of governments which failed to sign a global convention on forestry at the Earth Summit in 1992 (Bernstein and Cashore, 1999; van Kooten et al., 2005). This private forest certification has transformed into real governance (Pattberg, 2007). Depending on their viewpoints in specific studies, forest policy analysts refer to this phenomenon using various terms, like non-state global governance (Bernstein, 2011), non-state market driven (NSMD) governance (Cashore, 2002; Cashore et al., 2007; Burns et al., 2016), non-state authority (Cashore et al., 2005), transnational (business) governance (Bartley, 2010), private forest governance (Pattberg, 2005), global forest governance (Gan et al., 2013), and global private meta-governance (Derkx and Glasbergen, 2014).

2.2. Mandatory state certification

At the 1998 G8 foreign ministers meeting, there was an agreement to cooperate with timber-producing countries in the development of their own measures to counter illegal logging and trading in illegally harvested timber (G8 Action Programme on Forests, 2002). Following this agreement, in 2003 the European Union (EU) declared the Forest Law Enforcement, Governance and Trade (FLEGT) action plan, which enabled producers from partner countries that had already signed FLEGT-voluntary partnership agreements (VPAs) to sell freely their legal products in EU territory (Brack, 2005, 2012). The agreement gave authority to partner countries to define "illegal timber" based on their own regulations, so that it acknowledged partner countries' sovereignty over their forests. To date, six countries, including Indonesia, have signed FLEGT-VPA, while nine others are in negotiations (EU FLEGT Facility, 2014). The FLEGT-VPA is seen as a more equitable type of cooperation, since it binds both producers and consumers to sell and to buy only legally harvested timber. In addition, the direct participation of consumer countries in the penalization of the sale of illegal wood-based products in their territory, as promised by the US through the Lacey Act (2008) and by European Parliament through EU Timber Regulation (2010), is expected to diminish the space for illegal timber trade (Bartley, 2014). This is what cannot be addressed by voluntary forest certification initiatives such as FSC, PEFC, ISO 14000 or other more domestic initiatives (Solutions, 2003).

The history of Indonesia-EU VPA started in 2007 while Indonesia saw that EU FLEGT could be the trigger for improving its national forest governance as well as to get international legitimacy on Indonesia wood products. A series negotiation had been done until 2011, and both parties agreed to implement VPA commitment in 2014. Two years later EU confirmed that Indonesia had fulfilled all requirements of VPA, and therefore entitle for unrestricted export into EU. Along with the negotiation, government of Indonesia has been mandating all forest logging companies and timber manufacturers to be certified under SVLK, a national timber legality assurance agreed upon in FLEGT-VPA.

2.3. Bureaucratic politics and the power of actors

We use bureaucratic politics theory in order to cover more broadly those actors that may be influencing policies related to forest certification at the national level. Bureaucracies are defined as "agencies that have been created by governments or other public actors with some degree or permanence and coherence and beyond formal direct control of single nation governments and that act in the international arena to pursue a policy" (Bauer e t al., 2012:28). Bureaucracies are different from organizations, which are institutional arrangements built upon normative frameworks, members, and bureaucracies as administrative core institutions. They are also different from institutions, which are merely sets of principles and norms (Bauer et al., 2012). Studies on bureaucracies' behavior (e.g. Krott, 1990; Peters, 2002, Giessen & Krott 2009, Giessen et al. 2014) reveal that although bureaucracies have a formal mandate to serve public interests, they will informally prioritize their own interests and compete with other bureaucracies for more resources, political domain and influence. To develop and direct specific policies, bureaucracies are equipped with power, which is defined as the ability to shape a programme according to own interests, even against resistance from other actors (Krott et al., 2014). According to Krott et al. (2014), power consists of three elements, namely (i) coercion, associated mostly with the ability to force an outcome, (ii) (dis-)incentives, the ability to offer (dis-)advantages to subordinate actors, and (iii) dominant information, selected information that is difficult or impossible to verify by another party. These power elements may be acquired through sanction mechanisms, transfer funds, and the expertise of bureaucracies.

Given the rise of mandatory state forest certification systems and the obstructions faced by voluntary private systems, state bureaucracies now have the following options: (i) increasing their power vis-à-vis other bureaucracies by developing own certification schemes, and (ii) reclaiming authority from private certification schemes to regulate forest management. Before this theoretical background this article will scrutinize these propositions using the following methods.

3. Methods

We employed document analysis, interviews, and observations to acquire solid data to meet our objectives. First, trends in the numbers of certified forest areas and certified companies under private (FSC and LEI) and state (Ministry of Forestry) certification schemes were drawn from the website of those bodies and were placed in charts to gain an overview of each initiative's progress and to make comparisons possible. Any information from professional journals or professional mailing lists, and position papers from international organizations and associations were also considered to be valuable sources.

Second, semi-structured interviews (Halperin and Heath, 2012) to identify the main actors within this issue and their positions were conducted in Jakarta and Bogor, Indonesia, during August 2013-February 2014, and were followed by online interviews until April 2014. The interviewees were staff from the Ministry of Forestry (MoF), the Ministry of Trade (MoT), and the Indonesian Woodworking Association (ISWA), the Forest Concessionaires Association (Asosiasi Pengusaha Hutan Indonesia or APHI), the Multistakeholder Forestry Programme (MFP), FSC Indonesia, certification auditors, and environmental non-governmental organizations (ENGOs). We were also involved as observers in the respective meetings during that period to listen to each actor's ideas and to find out what expectations the actors had from others.

We treat sustainability certification and legality verification in same wave since both are quite similar in term of cost incurred and the process to go through. Some researchers (eg. Mcdermott et al., 2015; Brack, 2014) also reckon them are the same.

Actors' positions and their prospective coalitions were raised from theory of interest in environmental politics study. In analyzing actors' positions we assumed that every actor behaved rationally, meaning that actors tried to maximize their own benefit rather than to promote a general goal, like maintaining environmental sustainability and equality. To contest norm- and interest-driven is acceptable (see: Van Schaik and Schunz, 2012; Gulbrandsen, 2003) since each actor (which can be an individual, organization or state) has its own interest, adding to their effort to meet common desire. We follow expert judgment (Wibowo and Giessen, 2015) and our experiences in the field to rise up main issues and discussing them based rational behavior of actors.

4. Results

4.1. Voluntary private and mandatory state certifications in figures

Until 2014 there have been three forest certification schemes in Indonesia, namely the LEI and FSC, both of which are voluntary-private initiatives, and the PHPL (*Pengelolaan Hutan Produksi/* mandatory state scheme). Forest certification assignment under the FSC was first conducted in 1998 and successfully certified three out of five assessed FMUs in Java, which was then followed by seven others within the next two years. However, all of these certificates were suspended and withdrawn in 2003 due to unsatisfactory management improvement and non-compliance with FSC standards (Muhtaman and Prasetyo, 2006). A new era of FSC standards-based certification was introduced in 2008, when six FMUs, with a total managed area of 707,709 ha, obtained sustainable certificates (FSC, 2008). This grew to 1,089,942 ha in 2009, then dropped to 638,455 ha in 2011, and peaked again at 2,002,710 ha in November 2014, representing the certified area of 29 FMUs (FSC, 2009, 2011, 2014).

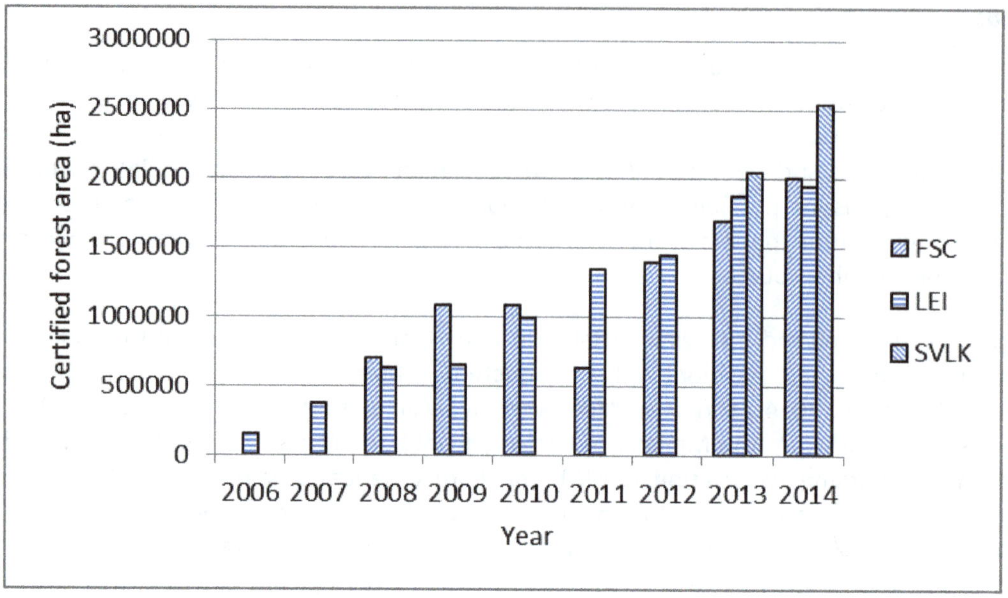

Source: FSC (various years), LEI (various years) and MoF (various years).
Figure 1. Size of certified forest area under FSC, LEI and SVLK schemes in the period 2006-2014

Similarly, the first certification process using LEI standard was also conducted in 1998, and putting PT Diamond Raya Timber who managed 90,956 ha of natural production forest as the first cooperation obtained the certificate a year later (Muhtaman and Prasetyo, 2006). Since then, there has been no serial data available to the public until the LEI released FMU and manufacture LEI certified document (LEI, 2013), which mentioned that there were 1,873,428 ha of certified-forests, consisting of 411,690 ha of natural production forests (22%) from two FMUs, 1,429,055 ha of plantation forests (76%) managed by 15 FMUs, and 32,683 ha of community forests (2%) from 22 community groups. Furthermore, the LEI (2014) informed that the certified natural forest area had dropped about 80% within the period 2009-2014, and that the plantation forest area had increased 400% during that time.

Unlike the two previous voluntary schemes, PHPL is a mandatory state scheme enacted by the government in 2012, and effectively applied to all forest concessionaires and FMUs in 2013. The PHPL (SVLK)-certified forest area until May 2014 was 2,542,091 ha, representing an area managed by 40 FMUs (calculated from MoF, 2014). The certified forest area under the PHPL scheme was higher than those under FSC and LEI, even though the PHPL started in 2013 (Figure 1). It is likely that FSC- or LEI-certified FMUs are also certified under SVLK, since the processes to obtain such certificates are quite similar, and same certification body may carry out the late assessment. Consequently, the total certified forest area under the three schemes cannot represent the total sustainably managed forest area in Indonesia.

In respect of the chain of custody (CoC) certification, there were six companies which held LEI certificates in 2009, and one of them did not reengage with the LEI in the second round (LEI, 2014). Conversely, the number of FSC CoC certificates doubled, from 87 in 2009 to 175 in 2012, and reached 193 by November 2014 (FSC, various years). A significant number of SLKs (sertifikat legalitas kayu or timber legality certificate; a form of certificate under the SVLK) were also issued by certification bodies, as the government started to force all exporters to provide V-legal documents for their wood products. By 2013, this number reached 726 companies (MoF, 2013) or almost four times the FSC holders, and grew continuously to 894 firms in 2014 (Table 1)

Table 1. Number of companies holding FSC, LEI and SVLK certificates

Year	Number of certified companies under each scheme		
	FSC	LEI	SVLK
2009	87	6	-
2010	136	6	-
2011	161	6	-
2012	175	6	-
2013	188	6	726
2014	193	5	894

Source: FSC (various years), LEI (2013, 2014), MoF (2014).

4.2. Costs of sustainable forest certification and legality verification

The cost of forest certification and timber legality verification vary, depending on the size of forest area, investment capital, and production capacity of the company. This cost applied at first assessment to gain the certificate and at yearly surveillance within certificate's validity period. Cost component charged by certification body to the company including fee for auditors, data and information collection, field assessment, reporting fee, documentation, and security assistance for auditors if needed. For SVLK scheme the Minister of (Environment and) Forestry released the Ministry Regulation P.13/2013, which amended by P.96/2014, on the standard cost for sustainable forest management assessment and timber legality verification. It is mentioned that the total cost for legality assessment for small and medium scale-wood processing industry could reach 6.6 million IDR, and up to 28.8 and 170 million IDR for bigger industry and large forest management units, respectively. In the meantime, the cost for forest sustainability assessment is around 222-280 million IDR and up to 132 million IDR for its yearly surveillance. All these costs are excluding travelling cost for auditors from Jakarta to the field. From private certification side, there is no official data on the cost of assessment and surveillance processes available for public. Pratiwi et al. (2015) reported that the costs paid by the companies for gaining FSC, LEI and PHPL-SVLK certificates are at least 31, 60, and 75 million IDR, respectively, depending on the scale of company.

4.3. Mapping main actors and their interests

Actors involved in forest certification in Indonesia are found along the production-consumption chain, starting with the actors initiating certification schemes and ending with buyers in consumer countries. The following are three groups of influential actors and their roles in shaping recent construction based on their tendencies and interests.

4.3.1. The coalition on state certification

- The Ministry of Forestry

The Ministry of Forestry (MoF) was the main state agency responsible for establishing the national timber legality assurance system called SVLK. The SVLK installation process began in 2003, when the MoF organized some civil social organizations, i.e., Telapak, Environmental Investigation Agency, Indonesian Ecolabel Institute, and the Nature Conservancy to hold a series cross sectoral dialog in developing SVLK (MFP, 2014). SVLK, which enacted through the Minister of Forestry Regulation No P.38/2009, and up to December 2014 was already five times amended (MoF, 2011; 2012; 2013; 2014a; MoEF, 2014a), has become the core of Indonesia-EU FLEGT-VPA negotiation. The MoF tried to apply the SVLK comprehensively to all forest industry and timber manufacturers immediately, in order to maintain the credibility of the SVLK and the MoF. However, due to external pressure, the MoF (which was transformed into the Ministry of Environment and Forestry or MoEF in October 2014) extended the deadline for obtaining the certificates and simplified some provisions in the SVLK so that small and medium-sized enterprises could join it (see MoF, 2014b; MoEF, 2014b).

- The Ministry of Trade

The Ministry of Trade (MoT) has the mission of increasing the added value and volume of exported goods by setting up regulations to support progressive export activities. Regarding wood legality, the MoT issued a regulation on the export of forestry industrial products, whereby one group of export timber products had to have V-Legal documents in 2013, and another group had to have these in 2014 (MoT, 2012). However, since the small-scale enterprise group was not ready by the deadline, the MoT delayed the inclusion of V-Legal documents in export requirements until 2015 (MoT, 2013a), and postponed it again until 2016 (MoT, 2014a). The latest delay was taken to meet the MoT's ambition in increasing the export value of wood products by 300% within the next 5 years (Kompas, 2014). Aside from setting up regulations for ease of export, the MoT was also actively involved in SVLK negotiation with the EU and in promoting SVLK so that Indonesian furniture demand grew up by 15-20% in 2014 (MoT, 2014b).

- The Ministry of Industry

The Ministry of Industry (MoI) supports the SVLK as the way to expand overseas markets for wooden industrial products. However, the small-scale industry's unreadiness for certification and the high cost associated with it forced the MoI to ask the MoF and MoT for a delay in SVLK implementation of one year and for simplification of the SVLK verifier, mainly for wooden furniture and handicraft products (AMKRI, 2014). Previously, in March 2013, the MoI had also accommodated SMEs' scruple on SVLK and vowed to discuss a relief for them with the MoF (Agroindonesia, 2013).

- Indonesian Forest Concessionaires Association (APHI) and the Indonesian Pulp and Paper Association (APKI)

The Indonesian Forest Concessionaires Association (APHI) and the Indonesian Pulp and Paper Association (APKI), as representatives of big players in the Indonesian wood industry, actually support the SVLK, in order to improve their product image in the global market (Agrofarm, 2013). Nevertheless, since end buyers ask for FSC-certified products, local industries have to opt exclusively for FSC-certified timber in their production line. Consequently, SVLK-certified intermediary goods cannot be sold locally, let alone in the global market (The Jakarta Post, 2014b) and the producers have to bear a double cost to tackle the problem (Hutan Indonesia, 2013). Since the SVLK is designed to comply with an agreement with the EU (Pohnan and Stone, 2013), the policy of making the SVLK obligatory for all logging companies and wood-processing industries is considered detrimental to local businesses and industries not oriented towards the European market. Both associations have already asked the government to open more local markets for SVLK-certified products and to promote the SVLK more broadly.

- Indonesian Sawmill and Woodworking Association (ISWA), Association of Indonesian Furniture and Crafts Producers (Asmindo) and small tree growers

Small wood industries and small tree growers support the SVLK, associating it with better market access in Europe and other countries (The Jakarta Post, 2014a; Agroobserver, 2015). However, they warn that the requirements for small industries should be simpler than those for big ones, since small tree growers work in private land parcels, and they do their own planting and harvesting (Agroindonesia, 2014). They also may change the use of the land or sell it in a traditional way, without permits from anybody else. Hence, they don't have to "formalise" their business by means of legal permits, taxation, environmental assessment and other administrative documents, which are costly (Obidzinski et al., 2014).

- European Union (EU)

EU efforts to halt illegal timber circulation inside and outside the region are evident in the two main regulations under the FLEGT Action Plan, namely the FLEGT-VPA and the EU Timber Regulation (EUTR). The EUTR can be seen as an economic tool to strengthen the position of European timber against its competitors, since any wood products to be exported to Europe have to pass a due diligence investigation by operators (van Heeswijk and Turnhout, 2013). In contrast, the FLEGT-VPA, which includes a timber legality assurance system (TLAS) from the producer country, ignores the due diligence process for FLEGT-licensed products. In the case of the SVLK, both Indonesia and EU parties signed and ratified the VPA in 2014 and acknowledged SVLK as Indonesian TLAS. However, the EU support for the SVLK cannot be seen simply as a passport of sorts for Indonesia's wood products entering EU territory, since it could be suspended once a SVLK-licensed product is discredited (Hawin et al., 2010).

- The Multistakeholder Forestry Programme (MFP) and Kehati Foundation

The MFP is a UK-funded program that assists the MoF in developing and promoting the SVLK, and in helping FMUs and small enterprises to obtain SVLK-based certificates through its coaching program (Raharjo, 2013). Together with the MoF, the MFP was engaged in SVLK negotiations with the EU. The MFP also strongly encourages the government of Indonesia to implement SVLK and rejects its delay, which the industry sector has requested often. The Kehati foundation, on the other hand, is a national funding agency that supports activities related to biodiversity conservation. Both the MFP and Kehati reckon that SVLK is in line and could be the way to succeed REDD+ program (Dharmawan et al., 2012).

4.3.2. The coalition on private certification

Most leading ENGOs support private certification initiatives to achieve better forest governance. These include: the Borneo Initiative, the Tropical Forest Foundation and the WWF-Global Forest and Trade Network, which work on the LEI and FSC schemes, and Greenpeace and the Nature Conservancy, which work on the FSC (Greenpeace, 2014). These ENGOs promote schemes they support and decline others for reasons of lack of accountability, poor sustainability standards and double costs consequence, among others. A network of ten ENGOs, called the Anti Forest-Mafia Coalition has criticized the SVLK, judging it to be flawed and a new form of "green washing" illegal timber (Anti Forest-Mafia Coalition, 2014). However, their members also help small enterprises to obtain SVLK certificates (Antara, 2015). Some of them publish SVLK's weaknesses repeatedly, even though they do not show their partiality to private forest certification schemes.

4.3.3. Other actors

- Certification Bodies

Currently there are 18, 2 and 4 certification bodies working on the SVLK, the FSC and the LEI, respectively (MoF, 2013; various sources) where some of them work with more than one scheme. Certification bodies have the right to issue certificates of sustainable forest management practice or chain of custody and the right to withdraw such a certificate during an annual inspection if the holders no longer comply with the standards of a given certification scheme. Due to a disproportional ratio between the certification body and number of areas/companies to be certified, the certification cost is increasing and is becoming unaffordable for small industries (Obidzinski et al., 2014). The principle of transparency in the audit process makes it possible for one certification body to know what other bodies do, what auditor works for which body, and the certification fees applied to particular audit process. Furthermore, once a certification body has conducted an audit, it has to announce the plan or the results on determined website so that it is open to input from independent observers, NGOs, and other stakeholders. The transparency principle allows the work of the auditor to be criticized (see also Maletz and Tysiatchniouk, 2009); hence, the certificates already issued may be withdrawn for many reasons. Frequently revocations of certificates will affect not only the certification body, but also the certification scheme they use.

- International buyers

After FLEGT-VPA signation, timber certified for sustainability and timber certified for legality compete for consumer demand for eco-friendly products. International buyers, as the end users of such products, will choose between state-based certified products and private-based certified woods. Accordingly, it is important to provide potential buyers with appropriate information on own products' superiority (Cashore, 2002). Potts et al. (2010) point out that major retailers in North America have decided to choose forest products derived from sustainable sources, whereas EU countries require importers, manufacturers, and retailers to have chain of custody (CoC) certification that clarifies the origin of their timber. Cai and Aguilar (2013) compiled 19 studies on consumer's willingness-to-pay for certified wood products, and found that 1- 39% of consumers were willing to pay premium prices for these products, no matter what kind of label it had. This means that the number of consumers willing to pay a higher price for certified products was less than that of those unwilling to pay. Consumers do not care whether products are certified or not, let alone whether the certification initiatives are private or governmental.

We summarize actors' major power under each power element type (Table 2) to make a way in predicting their fragmentation and future possibly alliance. The power elements are used rationally, depends on actor interest; and not always used or appeared in their action although they are able to do so.

Table 2. Major power element of powerful actor

Type of power element	Powerful actor	Strategic power
Coercion	Ministry of Environment and Forestry	Permit for forest management unit and industries
	Ministry of Trade	Permit to export wooden goods
	EU	FLEGT licencing for wood products to enter EU
Dis(incentive)	Ministry of Environment and Forestry	Providing financial support for legality verification
	International buyers	Creating demands for certified wooden goods
Dominant information	Ministry of Environment and Forestry	Revising and evaluating SVLK standard
	Ministry of Industry	Holding information on industry competitiveness and healthy
	Private certification owners	Good image
	MFP and Kehati Foundation	Negotiation experts and facilitators
	Environmental NGOs	International influencers and opinion makers for environmental issues
	Certification body	Issuing sustainability and legality certificates

5. Discussion

5.1. Fragmented certification schemes and their alliances

Based on policy-making authority, Cashore (2002) divides environmental governance into three categories, namely, "non-state market driven governance", characterized by the lack of government involvement in the policy-making process; "shared private/public governance", in which policy-making is shared with the government; and "traditional governance", where policy-making authority belongs fully with the government. Gulbrandsen (2004) uses the term "private governance" to refer to NSMD governance, and "public governance" (meaning state governance) for traditional governance. Gulbrandsen (2004) adds hybrid private-public governance to denote industry-dominated involvement in mixed private and public governance. Following categorization by Cashore (2002) and Gulbrandsen (2004) we found that there are now four clusters of forest certification governance present in Indonesia, namely environmental-private governance, hybrid private-public governance, producer-private governance, and state governance, with the FSC, the LEI, the IFFC and the SVLK representing each scheme, respectively (Table 3). It is important to note that the cluster is not within the strict boundary. Each actor may cooperate with other parties in other groups if it deemed to be gainful.

Table 3. Cluster of forest certification governance in Indonesia

Key features	Environmental-private governance (FSC)	Hybrid private-public governance (LEI)	Producer-private governance (PEFC/IFCC)	State governance (SVLK)
Rule maker	Environmentalists	Academics	Land owners	Government
State involvement	No	Partially	Limited	Full
Major supporting actor	ENGOs	None	Forest and wood-based Industry	Government
Coverage territory	International	Mainly in Indonesia	International	Mainly in Indonesia
Auditor	Third party	Third party	Third party	Third party
Public examination	Yes	Yes	Yes	Yes
Certificate acceptability	International	National	International	EU and National
Brand image	Strong	Weak	Quite strong	Weak
Industry participation	Voluntary	Voluntary	Voluntary	Mandatory
Influential actor to succeed	Business to business	Business to business	Business to business	Government to government and business to business
Law enforcement	No	No	No	Yes

Except for the LEI, the other three schemes have their supporting actors who promote the schemes' excellence. Due to the lack of capital, human resource and interest affinity, and therefore we argue that the hybrid private-public government will disappear, and the FMUs and wood industries will make alliances with the government to support traditional forest governance. The IFFC could join the SVLK, since the IFFC originally devoted for forest owners and forest managers. Consequently, there will be only two certification schemes competing for authority in the future, namely, the pro-environment and pro-business schemes. The first scheme is represented by the FSC and would be supported by ENGOs, academics and the pro-environmental community, and the LEI, since LEI already signed a joint certification protocol with the FSC. With this collaboration, the FSC has a local partner to introduce its more local-value adapted scheme, and the LEI could maintain its presence in the forest certification field. The increasing number and size of certified forests under FSC indicates the strengthening of this scheme cluster. Conversely, the pro-business scheme consists of SVLK and PEFC will mutually reinforce and possible to make joint recognition for effective certification process, and wider market access. This scheme is supported by the government, landowners, and FMUs. Due to the wider stakeholder involvement and its mandatory nature, this scheme seems to flourish continuously, as indicated by its progress in the year of early implementation.

Through this process, the forest certification regime becomes less complex, since national political processes replace and reduce the international element of international forest regimes, and standard of these schemes may less different (McDermott et al., 2008). Industrial groups, as the users of the certification scheme, have to engage with the SVLK to maintain their business continuity and to enter the market, and only add the FSC to improve their reputation (Wibowo et al., 2015). However, the solidity of this pattern will become more stable under the following conditions. First, if the government of Indonesia promotes the SVLK and its acceptance in other consumer countries, hence industries will need no other certification scheme when exporting their products to such countries. Second, if the government preserves the SVLK as a reputable certification scheme by publishing data having to do with the ways in which the SVLK contributes to the deforestation rate reduction and to the enhancements of society and ecology. Third, if there are fewer or no objections from other stakeholders about the performance of industries that are SVLK-certified, indicating that these companies do business in legal and sustainable ways.

5.2. Obligation and fees as bureaucratic power strategy under SVLK

By making the SVLK mandatory for all FMUs and timber manufacturers, the MoF may gain at least three benefits. First, it could increase its presence in downstream industry circles by financing the first SVLK-certification process for small enterprises (Neraca, 2014). Through this assistance, the MoF would gain power over small enterprises, which in the past have belonged in the domain of the MoI. Second, compelling the enterprises to choose SVLK only or either SVLK and private schemes. The high cost of legality verification and forest certification is the main factor obstructing companies to engage with those processes (Tacconi, 2007; Obidzinski et al., 2014), although it does not always correspond to its benefits. Pratiwi (2015) shows that only around 24% of industry, forest companies and their related associations state that the cost of SVLK-certification is worth its benefits, while only 26% of them acknowledge that they always receive premium prices from their certified products. The gap between the cost and the benefits resulting from certification process forces business actors to be realistic in choosing appropriate certification scheme. This selection strongly depends on their market target and value of the exported products (see Fikru, 2014). If combination of mandatory and voluntary certification costs were higher than the profit, business tended to choose only the SVLK for their products. Third, as consequence of the previous point, the SVLK may be better known and legitimate in domestic and overseas markets. With only around 10% share of Indonesian timber export value goes to Europe (MoF, 2014c), the obligation to put V-legal sign (sign for SVLK-certified products) in all exported timber products will make SVLK better known in other major destination countries like China, Japan, and South Korea.

5.3. Privileging large-scale industries over smallholders

It is widely known that small scale wood industries and tree growers are characterized by limited administration records, production-by-demand orientation, simple management practices, and have less capacity to follow market dynamics (Nurrochmat, 2004). Obidzinski et al. (2014) point out the major challenges to small and medium enterprises (SMEs) in obtaining SVLK-based certificates, namely: business legality status; limited supply of legal timber; small profit margins spread over a long value chain; high certification costs; and small volume and unpredictable harvest times. Under these circumstances it is hard for them to obtain wood legality certificates (S-LK; SVLK-based certificate) even if the trees grown in home yards and their transfers to processing industries are truly legal. The lack of verification bodies and the limited financial support by government are other obstacles for the engagement of small industries in SVLK schemes (Obidzinski et al., 2014). Hence, the policy of implementing SVLK certification in all wood industries is thought to be a simple way to increase export value to European markets by large industries, at the expense of SMEs. On the other hand, the export ban policy for non-certified industry will cut USD 900 million in annual income for SMEs, which amounts to 30% of the total wood export value to non-European markets, excluding pulp and paper (Agroindonesia, 2014). To avoid this loss, the MoT has postponed the implementation of such a policy twice already.

It seems that the MoT places foreign market access and trade performance as the top priority, followed by the survival of large scale industries, wood legality, the subsistence of SMEs, and facility of sales for small tree growers. This policy might limit the circulation of illegal wood, facilitate the supervision of the industry and increase the trust in the SVLK; but it suppresses small and very locally-oriented industries, and creates a disincentive for small tree growers. Small tree growers face legal consequences and may even be found guilty of criminal charges when they harvest and transfer wood from their garden to small wood industries. A more local and adaptive policy that is able to distinguish illegal timber and -incomplete documented timber –like a conformity declaration already being considered– is needed for this kind of business (Kemitraan, 2014), in order to avoid negative social consequences.

6. Conclusions

The Ministry of (Environment and) Forestry was under political pressure from other bureaucracies, and was asked to improve its performance in generating income from forest areas. The Ministry then used its authority to govern forest product circulation and trade in domestic and global markets through the forest certification initiative. This program has at least four goals: to show the public its efforts in reducing illegal logging and the deforestation rate; to restore the forestry sector as the main contributor to state income, as it was in 1990s; to preserve its existence by obliging all FMUs and wood industries to be certified by such a scheme and to become the agency in charge of the Indonesia-EU agreement on FLEGT; and to increase the state sovereignty over forests, by being the institution to decide on the legality of timber. The involvement of the MoF in the forest certification arrangement indicates that private governance is not the final and stable form of this regime, as is argued by some scholars (e.g. Hackett, 2013; Overdevest and Zeitlin, 2014). In addition, the MoF uses its authority to reclaim authority over private actors (see also Giessen et al., 2016). The MoF is likely to succeed, due to the ease of implementation and simple standards of the scheme, which also account for the relatively inexpensive cost of certification assessment fee, in comparison to its competitors (Pratiwi et al., 2015; Maryudi, 2016). Furthermore, the mandatory nature of the SVLK means it has fixed a downstream market, i.e., FMUs and wood industries, whilst upstream it can profit from an educated market that has already been created by private governance. In this case, the state governance seems to be the free rider.

A remaining question needs to answer by both governances is whether their presence is able to reduce forest loss and to sustain forest management. Since 18% of deforested area in Indonesia is open land and the rest is designated for development purposes (Kissinger et al., 2012), we argue that the certification regime will be unable to reduce forest conversions. Most of the forest conversions were planned and allowed by the government, and that is why these, together with other "deforested areas", were not included in the calculation of the deforestation rate (Bellot et al., 2014). In addition, the sustainability criteria of certification regimes do not reflect the fact that small tree growers do clear-cutting when they need cash for unforeseen expenses, like a marriage or educational costs, and that they do not replant their gardens. Hence, legality verification and sustainability certification are disconnected from forest (land) management performed by small tree growers or forest companies. Based on these points, we argue that the certification regime and legality verification are merely instruments to govern timber trade (indicated as transnational business governance by Lesniewska and McDermott, 2014) at the expense of small and traditional tree growers, rather than an effort to secure remaining tropical forests.

References

Literature

Bartley, T. (2007). How foundations shape social movements: The construction of an organizational field and the rise of forest certification. *Social Problems, 54*(3), 229-255.

Bartley, T. (2010). Transnational private regulation in practice: The limits of forest and labor standards certification in Indonesia. *Business and Politics, 12*(3), Article 7. DOI: 10.2202/1469-3569.1321.

Bartley, T. (2014). Transnational governance and the re-centered state: Sustainability or legality?. *Regulation & Governance, 8*(1), 93-109.

Bauer, S., Andresen, S., & Biermann, F. (2012). International bureaucracies. In: Biermann, F., & Pattberg, P. H. (Eds.). *Global environmental governance reconsidered.* MIT Press. p.27-44.

Bell, S., & Hindmoor, A. (2012). Governance without government? The case of the Forest Stewardship Council. *Public Administration, 90*(1), 144-159.

Bellot, F.F., Bertram, M., Navratil, P., Siegert, F. & Dotzauer, H. (2014). *The high-resolution global map of 21st-century forest cover change from the University of Maryland ('Hansen Map') is hugely overestimating deforestation in Indonesia*. Available at: http://forclime.org/images/documents/FORCLIMEOverestimation%20of%20Deforestation.pdf. [Accessed 14 November 2014].

Bernstein, S., & Cashore, B. (1999). World trends and Canadian forest policy: trade, international institutions, consumers and transnational environmentalism. *Forestry Chronicle, 75*(1), 34-38.

Bernstein, S. (2011). Legitimacy in intergovernmental and non-state global governance. *Review of International Political Economy, 18*(1), 17-51.

Blackman, A., Raimondi, A., & Cubbage, F. (2014). Does forest certification in developing countries have environmental benefits? Insights from Mexican Corrective Action Requests. Insights from Mexican Corrective Action Requests (March 31, 2014). *Resources for the Future Discussion Paper*, (14-06).

Boström, M. (2003). How state-dependent is a non-state-driven rule-making project? The case of forest certification in Sweden. *Journal of environmental policy & planning, 5(*2), 165-180.

Brack, D. (2005). Controlling illegal logging and the trade in illegally harvested timber: the EU's Forest Law Enforcement, Governance and Trade Initiative. *Review of European community & international environmental law, 14*(1), 28-38.

Brack, D. (2012). Excluding illegal timber and improving forest governance: The European Union's Forest Law Enforcement, Governance and Trade initiative. In: *High-Value Natural Resources and Peacebuilding,* ed: P. Lujala and S. A. Rustad. London: Earthscan. p. 211-220.

Brack, D. (2014). *Promoting legal and sustainable timber: using public procurement policy*. Chatham House for the Royal Institute of International Affairs.

Burns, S., Yapura, P., Giessen, L. (2016). State actors and international forest certification policy: coalitions behind FSC and PEFC in federal Argentina. *Land Use Policy, 52*, 23-29.

Cai, Z., & Aguilar, F. X. (2013). Meta-analysis of consumer's willingness-to-pay premiums for certified wood products. *Journal of Forest Economics, 19*(1), 15-31.

Cashore, B. (2002). Legitimacy and the privatization of environmental governance: How non– state market–driven (NSMD) governance systems gain rule–making authority. *Governance, 15*(4), 503-529.

Cashore, B., Auld, G., & Newsom, D. (2005). Governing through markets: Forest certification and the emergence of non-state authority. *Journal of Environmental Planning and Management, 48*(4), 619–624.

Cashore, B., Egan, E., Auld, G., & Newsom, D. (2007). Revising theories of nonstate market-driven (NSMD) governance: lessons from the Finnish forest certification experience. *Global Environmental Politics, 7*(1), 1-44.

Derkx, B., & Glasbergen, P. (2014). Elaborating global private meta-governance: An inventory in the realm of voluntary sustainability standards. *Global Environmental Change, 27*, 41-50.

Dharmawan, A. H., Nugroho, B., Kartodihardjo, H., Kolopaking, L. M., & Boer, R. (2012*). SVLK, Jalan Menuju REDD+*. Forest Governance and Multistakeholder Forestry Programme, Ministry of Forestry Indonesia–UK Department for International Development.

Durst, P. B., McKenzie, P. J., Brown, C. L., & Appanah, S. (2006). Challenges facing certification and eco-labelling of forest products in developing countries. *International Forestry Review, 8*(2), 193-200. DOI: 10.1505/ifor.8.2.193.

Espach, R. (2006). When is sustainable forestry sustainable? The forest stewardship council in Argentina and Brazil. *Global Environmental Politics, 6*(2), 55-84.

Fikru, M. G. (2014). International certification in developing countries: The role of internal and external institutional pressure. *Journal of environmental management, 144,* 286-296.

Gale, F., & Haward, M. (2011). *Global commodity governance: state responses to sustainable forest and fisheries certification*. Palgrave Macmillan.

Gan, J., Cashore, B., & Stone, M. W. (2013). Impacts of the Lacey Act Amendment and the Voluntary Partnership Agreements on illegal logging: implications for global forest governance. *Journal of Natural Resources Policy Research, 5*(4), 209-226.

Giessen, L., & Krott, M. (2009). Forestry joining integrated programmes? A question of willingness, ability and opportunities. Allgemeine Forst-und Jagdzeitung, 180(5/6), 94-100.

Giessen L., Burns S., Sahide A. M. K., Wibowo, A. (2016). From governance to government: The key role of state bureaucracies in forest and agricultural certification. *Policy and Society, 35* (1), 71-89.

Gulbrandsen, L. (2003). The evolving forest regime and domestic actors: strategic or normative adaptation?. Environmental Politics, 12(2), 95-114.

Gulbrandsen, L. H. (2004). Overlapping public and private governance: Can forest certification fill the gaps in the global forest regime? Global Environmental Politics, 4(2), 75-99.

Gulbrandsen, L. H. (2014). Dynamic governance interactions: Evolutionary effects of state responses to non-state certification programs. Regulation & Governance, 8, 74–92.

Hackett, R. (2013). From government to governance? Forest certification and crisis displacement in Ontario, Canada. *Journal of Rural Studies, 30*, 120-129.

Halperin, S., & Heath, O. (2012). *Political research: methods and practical skills*. Oxford University Press.

Hawin, M., Nurhayati, I. & Antoni, V. (2010). *Analisis hukum teks Voluntary Partnership Agreement antara Indonesia dan Eropa*. Jakarta: Forest Governance and Multistakeholder Forestry Programme.

Hutan Indonesia. (2013). Mengejar Tenggat Sertifikasi PHPL Mandatory. *Hutan Indonesia, III*, 26-27.

Hysing, E. (2011). Transnational environmental governance: the emergence and effects of the certification of forests and fisheries by Lars H. Gulbrandsen (Cheltenham, UK: Edward Elgar). Environment and Planning. C, *Government and Policy, 29*(3), 567-568.

Kissinger, G., Herold, M., & De Sy, V. (2012). *Drivers of deforestation and forest degradation: a synthesis report for REDD+ policymakers*. Vancouver: Lexeme Consulting.

Krott, M. (1990). *Öffentliche Verwaltung im Umweltschutz. Ergebnisse einer behördenorientierten Policy-Analyse am Beispiel Waldschutz* . Wien: W. Braumüller Verlag.

Krott, M., Bader, A., Schusser, C., Devkota, R., Maryudi, A., Giessen, L., & Aurenhammer, H. (2014). Actor-centred power: The driving force in decentralised community based forest governance. *Forest Policy and Economics, 49*, 34-42.

Lesniewska, F., & McDermott, C. L. (2014). FLEGT VPAs: Laying a pathway to sustainability via legality lessons from Ghana and Indonesia. *Forest policy and economics, 48*, 16-23.

Malets, O. (2015). When transnational standards hit the ground: Domestic regulations, compliance assessment and forest certification in Russia. *Journal of Environmental Policy & Planning, 17*(3), 332-359.

Maletz, O. & Tysiatchniouk M. (2009). The Effect of Expertise on the Quality of Forest Standards Implementation: The Case of FSC Forest Certification in Russia. *Forest Policy and Economics, 11* (5-6): 422-428

Maryudi, A. (2006). Sertifikasi Hutan: Instrumen Baru Dalam Kebijakan Kehutanan. *Rimba Kalimantan, 11*(1), 27-35.

Maryudi, A. (2016). Choosing timber legality verification as a policy instrument to combat illegal logging in Indonesia. *Forest Policy and Economics, 68*, pp.99-104.

Maryudi, A. (2009). *Forest certification for community-based forest management in Indonesia: Does LEI provide a credible option?*. Hayama-mati (Kanagawa-ken): Institute for Global Environmental Strategies, Forest Conservation, Livelihoods, and Rights Project.

McDermott, C. L., Noah, E., & Cashore, B. (2008). Differences that 'matter'? A framework for comparing environmental certification standards and government policies. Journal of Environmental Policy and Planning, 10(1), 47-70.

McDermott, C. L. (2014). REDDuced: From sustainability to legality to units of carbon—The search for common interests in international forest governance. Environmental Science and Policy, 35, 12-19.

McDermott, C. L., Irland, L. C., & Pacheco, P. (2015). Forest certification and legality initiatives in the Brazilian Amazon: Lessons for effective and equitable forest governance. Forest Policy and Economics, 50, 134-142.

Muhtaman, D. R., & Prasetyo, F. A. (2006). Forest certification in Indonesia. In: Cashore, B. et al. (eds). *Confronting sustainability: Forest certification in developing and transitioning countries.* Yale University Faculty of Environmental Studies Publication Series, 33-68.

Nurrochmat, D. R. (2004). Social forestry to alleviate poverty: concepts and experiences in Indonesia. In: *Poverty Alleviation: Concepts and Experiences: Focused on Indonesian Cases*. Cuvillier. Verlag Göttingen. p. 39-44

Nurrochmat, D. R., Dharmawan, A. H., Obidzinski, K., Dermawan, A., & Erbaugh, J. T. (2016). Contesting national and international forest regimes: Case of timber legality certification for community forests in Central Java, Indonesia. *Forest Policy and Economics, 68*, 54-64.

Obidzinski, K., Dermawan, A., Andrianto, A., Komarudin, H., & Hernawan, D. (2014). The timber legality verification system and the voluntary partnership agreement (VPA) in Indonesia: Challenges for the small-scale forestry sector. *Forest Policy and Economics, 48*, 24-32.

Overdevest, C., & Zeitlin, J. (2014). Assembling an experimentalist regime: Transnational governance interactions in the forest sector. *Regulation & governance, 8*, 22-48.

Overdevest, C., & Zeitlin, J. (2014). Assembling an experimentalist regime: Transnational governance interactions in the forest sector. *Regulation & governance, 8*, 22-48.

Pattberg, P. H. (2005). The Forest Stewardship Council: Risk and potential of private forest governance. *The Journal of Environment & Development, 14*(3), 356-374.

Pattberg, P. H. (2007). *Private institutions and global governance: the new politics of environmental sustainability*. Edward Elgar Pub.

Peters, G. (2002). *Politics of Bureaucracy*. Routledge.

Pohnan, E., & Stone, M. (2013). *Can legality verification combat illegal logging in Indonesia? Strategic insights for policy makers and advocates. Issues and Options Brief.* IUFRO Task Force on Forest Governance, 16 November 2013.

Potts, J., van Der Meer, J., & Daitchman, J. (2010). *The state of sustainability initiatives review 2010: Sustainability and transparency.*

Pratiwi, S. (2015). *Comparing forest certification and legality verification system and explaining stakeholder preferences in Indonesia*. Master Thesis. University of Goettingen.

Pratiwi, S., Wibowo, A., & Giessen, L. (2015). Third-Party Certification of Forest Management In Indonesia: Analysing Stakeholders' Recognition and Preferences. *Jurnal Manajemen Hutan Tropika, 21*(2), 65-75.

Raharjo, D. (2013). *Laporan Akhir Multistakeholder Forestry Programme (MFP2)*. Ministry of Forestry of Republic of Indonesia, KEHATI, UKAID. Jakarta. pp. 109.

Sahide, A., Burns, S., Wibowo, A., Nurrochmat, D.R., Giessen, L. (2015). Towards state hegemony over agricultural certification: from voluntary private to mandatory state regimes on palm oil in Indonesia. *Jurnal Manajemen Hutan Tropika, 21* (3), 162-171.

Sikor, T. (Ed.). (2013). *Public and private in natural resource governance: a false dichotomy?.* Routledge.

Solutions, S. G. T. (2003). *Establishing the foundation for sustainable forest management in Africa: Legal origin of timber as a step towards sustainable forest management*. SGS, Geneva. Available at: http://siteresources.worldbank.org/EXTFORESTS/Resources/985784-1217874560960/AFRLegalOrigin.pdf [Accessed 10 January 2015].

Tacconi, L. (2007). Verification and certification of forest products and illegal logging in Indonesia. In: Tacconi, L. (Editor). *Illegal logging: law enforcement, livelihoods and the timber trade. London*: Earthscan. p. 251-274.

Tosun, J. (2012). Forest certification in Europe: Exploring the determinants of cross-country variation. *Marmara Journal of European Studies, 20*(1), 177-197.

UNECE (United Nations. Economic Commission for Europe)/FAO (Food and Agriculture Organization (2013). Forest Products Annual Market Review, 2012-2013. New York and Geneva, United Nations, Geneva Timber and Forest Study Paper, 33.

van Heeswijk, L., & Turnhout, E. (2013). The discursive structure of FLEGT (Forest Law Enforcement, Governance and Trade): The negotiation and interpretation of legality in the EU and Indonesia. *Forest Policy and Economics, 32*, 6-13.

van Kooten, G. C., Nelson, H. W., & Vertinsky, I. (2005). Certification of sustainable forest management practices: a global perspective on why countries certify. *Forest Policy and Economics, 7*(6), 857-867.

van Schaik, L., & Schunz, S. (2012). Explaining EU Activism and Impact in Global Climate Politics: Is the Union a Norm-or Interest-Driven Actor?. JCMS: *Journal of Common Market Studies, 50*(1), 169-186.

Wibowo A., Sahide M.A., Pratiwi S., Dharmawan B. & Giessen L. (2015). Ragam Skema Sertifikasi Hutan Global dan Opsi Transformasinya di Indonesia. *Risalah Kebijakan Pertanian dan Lingkungan. 2* (1), 1-8.

Wibowo, A., & Giessen, L. (2015). Actor Positions on Primary and Secondary International Forest-related Issues Relevant in Indonesia. *Journal of Sustainable Development, 8*(3), 10-27.

Wiersum, K. F., & Elands, B. H. (2013). Opinions on legality principles considered in the FLEGT/VPA policy in Ghana and Indonesia. *Forest Policy and Economics, 32*, 14-22.

Empirical materials and regulations cited

Agrofarm (2013). *Tidak ada insentif, SVLK bebani APHI*. Available at: http://www.agrofarm.co.id /m/kehutanan/121/tidak-ada-insentif-svlk-bebani-aphi/. [Accessed 14 January 2015].

Agroindonesia (2013). *Komitmen Indonesia tetap terjaga*. Available at: http://agroindonesia.co.id /2013/03/19/komitmen-indonesia-tetap-terjaga [Accessed 14 January 2015].

Agroindonesia (2014). *Cukup selipkan satu pasal*. http://agroindonesia.co.id/2014/01/cukup-selipkan-satu-pasal/ [Accessed 14 January 2015].

Agroobserver (2015). ISWA: *Pemerintah diharap jangan hambat industri barecore dari sengon*. Available at: http://agroobserver.com/iswa-pemerintah-diharap-jangan-hambat-industri - barecore-dari-sengon/ [Accessed 14 January 2015].

AMKRI (Asosiasi Mebel dan Kerajinan Indonesia) (2014). *Menteri Perindustrian Saleh Husin setujui penundaan pemberlakuan sistem verifikasi legalitas kayu (SVLK)*. Available at: http://www.amkri.org/berita/menteri-perindustrian-saleh-husin-setujui-penundaan-pemberlakuan-sistem-verifikasi-legalitas-kayu-(svlk)--337.php [Accessed 14 January 2015].

Antara (2015). *WWF-Asmindo bantu industri kecil kehutanan peroleh SVLK*. Available at: http://www.antarajatim.com/lihat3/berita/149907/wwf-asmindo-bantu-industri-kecil-kehutanan-peroleh-svlk [Accessed 20 February 2015].

Anti Forest-Mafia Coalition (2014). *SVLK flawed: An independent evaluation of Indonesia's timber legality certification system*. Available at: http://eyesontheforest.or.id/attach/Anti%20Forest%20Mafia%20Coalition%20%2818Mar14%29%20SVLK%20Flawed%20FINAL.pdf [Accessed 14 January 2015].

EU FLEGT Facility (2014). *VPA countries*. Available at: http://www.euflegt.efi.int/vpa-countries [Accessed 14 June 2014].

FSC (Forest Stewardship Council) (2008, 2009, 2011, 2014). Global FSC certificates: type and distribution. Available at: https://ic.fsc.org/facts-figures.19.html [Accessed 14 Mach 2015].

FSC (Forest Stewardship Council) (2010). FSC Directive on FSC Forest Management Evaluations. Available at: https://www.scsglobalservices.com/files/standards/FSC-DIR-20-007_EN.pdf. [Accessed 14 June 2014].

FSC (Forest Stewardship Council) & PEFC (Programme for the Endorsement of Forest Certification) (2013). Joint statement by the Forest Stewardship Council (FSC) and the Programme for the Endorsement of Forest Certification (PEFC) recommending ISO member to vote against the new work item proposal Chain of Custody of Forest Based Products-Requirements. Available at: http://www.pefc.org/images/documents/external/Joint_StatementFSC_PEFC_July_8_2013.pdf [Accessed 12 June 2014].

FSC (Forest Stewardship Council) (2014). FSC harmonised certification bodies' Forest Stewardship Standard for the Republic of Indonesia. Available at: http://www.scsglobalservices.com/files/standards/fsc-std-idn-01-01-2013_indonesianatural_plantations_and_slimf.pdf [Accessed 14 December 2014].

G8 Action Programme on Forests (2002). G8 action programme on forests—Final report. www.illegal-logging.info/item_single.php?it_id=38&it=document [Accessed 14 January 2015].

Greenpeace (2014). *Weaker certification schemes. Other forest industry driven certification schemes fail to meet basic performance indicators*. Available at: http://www.greenpeace.org/international/en/campaigns/forests/solutions/alternatives-to-forest-destruc/Weaker-Certification-Schemes/ [Accessed 14 December 2014].

Kemitraan (2014). *Terobosan baru dalam SVLK: conformity declaration untuk hutan rakyat dan industri kecil menengah*. Available at: http://www.kemitraan.or.id/sites/default/files/Press%20Release%20Self%20Declare%20SVLK%20Kemitraan-160514final.pdf [Accessed 14 March 2015].

Kompas(2014). *Ini aturan baru legalitas kayu*. Available at: http://bisniskeuangan.kompas.com/read/2014/12/30/090900326/Ini.Aturan.Baru.Legalitas.Kayu [Accessed 14 January 2015].

LEI (Lembaga Ekolabel Indonesia) (2013). *FMU and Manufacture LEI Certified*. Updated: July 2013. Available at: http://www.lei.or.id/files/FMU%20&%20Manufacture%20certified%20LEI July%202013.pdf [Accessed 14 November 2014].

LEI (Lembaga Ekolabel Indonesia) (2014). *Laporan Kegiatan Lembaga Ekolabel Indonesia 2009-2014*. LEI.

MFP (Multistakeholder Forestry Programme). (2014). SVLK: Meningkatkan perdagangan produk kayu legal dari Indonesia. *Infobrief*, December 2014. Available at: http://www.mfp.or.id/attachments/article/89/SVLK%20in%20brief_IND_FA.pdf [Accessed 14 January 2015].

MoEF (Ministry of Environmental and Forestry) (2014a). *Ministry Regulation No P. 95/Menhut-II/2014 on Ministry Regulation No P. 43/Menhut-II/2014 on performance appraisal of sustainable production forest management and timber legality verification.*

MoEF (Ministry of Environmental and Forestry) (2014b). *Circular letter from Director General of Forest Enterprises No: SE.14/VI-BPPHH/2014 on concerning obligations fulfillment related to timber legality verification system (SVLK).*

MoF (Ministry of Forestry) (2009). *Ministry Regulation No P. 38/Menhut-II/2009 on Performance appraisal standard and guidelines of sustainable production forest management and timber legality verification.*

MoF (Ministry of Forestry) (2011). *Ministry Regulation No P. 68/Menhut-II/2011 on First amendment of Ministry Regulation No P. 38/Menhut-II/2009 on performance appraisal standard and guidelines of sustainable production forest management and timber legality verification.*

MoF (Ministry of Forestry) (2012). *Ministry Regulation No P. 45/Menhut-II/2012 on Second amendment of Ministry Regulation No P. 38/Menhut-II/2009 on performance appraisal standard and guidelines of sustainable production forest management and timber legality verification.*

MoF (Ministry of Forestry) (2013). *Ministry Regulation No P. 42/Menhut-II/2013 on Third amendment of Ministry Regulation No P. 38/Menhut-II/2009 on performance appraisal standard and guidelines of sustainable production forest management and timber legality verification.*

MoF (Ministry of Forestry) (2014a). *Ministry Regulation No P. 43/Menhut-II/2014 on performance appraisal of sustainable production forest management and timber legality verification.*

MoF (Ministry of Forestry) (2014b). *Circular letter from Director General of Forest Enterprises No: SE.8/VI-BPPHH/2014 on concerning obligations fulfillment related to timber legality verification system (SVLK).*

MoF (Ministry of Forestry) (2014c). *SVLK development, Commitment to sustainable forests for sustainable trade.*

MoF (Ministry of Forestry) (2013b). *Keunggulan SVLK sebagai penguat produk nasional*. Seminar Kehutanan Indonesia, Jakarta 16 Desember 2013. p.42. (presentation file).

MoF (Ministry of Forestry) (2014). *Data PHPL*. Available at: http://www.dephut.go.id/index.php/common/phpl [Accessed 14 March 2015].

MoT (Ministry of Trade) (2012). *Ministry Regulation No. 64/M-DAG/PER/10/2012 on Export provision on forest industrial products.*

MoT (Ministry of Trade) (2013a). *Ministry Regulation No. 81/M-DAG/PER/12/2013 on Amendment on Ministry Regulation No. 64/M-DAG/PER/10/2012 on export provision on forest industrial products.*

MoT (Ministry of Trade) (2014). *Diplomasi perdagangan Wamendag di London: SVLK berdampak positif bagi ekspor kayu Indonesia*. Available at: http://www.kemendag.go.id/id/news/2014/06/04/diplomasi-perdagangan-wamendag-di-london-svlk-berdampak-positif-bagi-ekspor-kayu-indonesia [Accessed 14 December 2014].

Neraca (2014). *Usaha kecil akan dibantu sertifikasi legalitas kayu*. Available at: http://www.neraca.co.id/industri/43685/Usaha-Kecil-Akan-Dibantu-Sertifikasi-Legalitas-Kayu [Accessed 14 January 2015].

The Jakarta Post. (2014a). *Indonesia aims long-term timber export growth in EU*. Available at: http://www.thejakartapost.com/news/2014/03/01/indonesia-aims-long-term-timber-export-growth-eu.html [Accessed 1 January 2015].

The Jakarta Post. (2014b). *APKI wants more access to local market*. Available at: http://m.thejakartapost.com/news/2014/07/19/apki-wants-more-access-local-market.html [Accessed 14 January 2015].

10

Social Forestry - why and for whom? A comparison of policies in Vietnam and Indonesia

Moira Moeliono[1, *,], Pham Thu Thuy[2], Indah Waty Bong[3], Grace Yee Wong[4] and Maria Brockhaus[5]**

[1] Center for International Forestry Research (CIFOR). Jalan CIFOR, Situ Gede, Sindang Barang, Bogor (Barat) 16115, Indonesia. Email: M.Moeliono@cgiar.org

[2] CIFOR. 172E Quan Thanh Street, Hanoi, 10000. Vietnam. Email: t.pham@cgiar.org

[3] CIFOR. Jalan CIFOR, Situ Gede, Sindang Barang, Bogor (Barat) 16115, Indonesia. Email: I.Waty@cgiar.org

[4] Stockholm Resilience Centre. Stockholm University. SE-10691 Stockholm. Email: grace.wong@su.se

[5] Department of Forest Sciences, University of Helsinki . PO Box 27 Helsinki 00014, Finland. Email: maria.brockhaus@helsinki.fi

* Corresponding author

** With contributions from Mai Yen and Anggiana

Abstract: Community forestry or social forestry (henceforth referred collectively as SF) programs have become new modes of forest management empowering local managers and hence, allowing integration of diverse local practices and support of local livelihoods. Implementation of these initiatives, however, face multiple challenges. State-prescribed community programs, for example, will remain isolated efforts if changes in the overall economic and social governance frameworks, including the devolution of rights to local users is lacking. Financial sustainability of these measures remains often uncertain and equity issues inherent to groups and communities formed for SF, can be exacerbated. In this article, we pose the question: Whose interests do SF policies serve? The effectiveness of SF would depend on the motivations and aims for a decentralization of forest governance to the community. In order to understand the underlying motivations behind the governments' push for SF, we examine national policies in Vietnam and Indonesia, changes in their policies over time and the shift in discourses influencing how SF has evolved. Vietnam and Indonesia are at different sides of the spectrum in democratic ambitions and forest abundance, and present an intriguing comparison in the recent regional push towards SF in Southeast Asia. We discuss the different interpretations of SF in these two countries and how SF programs are implemented. Our results show that governments, influenced by global discourse, are attempting to regulate SF through formal definitions and regulations. Communities on the other hand, might resist by adopting, adapting or rejecting formal schemes. In this tension, SF, adopted to serve the interest of local people, in practice has not fulfilled its promise.

Keywords: Community Forestry; Social Forestry; empowerment; rights; Indonesia; Vietnam

---

1. Introduction

Throughout Southeast Asia, some 140 million people are dependent on forests for their livelihoods and have developed their own systems of managing resources based on traditional knowledge, practices, rules and beliefs for generations ('customary use') (Chao, 2012). The Vietnam Forest Administration (VNFOREST) estimated that in 2008, around 247,030 ha of forestland were managed in a traditional way without legal title in Vietnam (Nguyen, 2009; Sikor and Nguyen, 2011). Of this, 175,395 ha is forest and 71,634 ha is bare land. In Indonesia, 1,500 villages with an area of more than 11 million ha are located within the state forest area, supporting 80 to 95 million people (Ministry of Forestry and Bureau of Statistics, 2009) and many more live in areas bordering forest.

These millions of people developed forest practices according to local conditions and culture, supported by local governance structures that have been effective in maintaining forest and providing local livelihoods. Yet these practices and structures are often invisible to the State, beyond state control and or simply ignored (Scott, 1998; Soriaga and Walpole, 2006). At the same time, people may choose to be at the forest margins (or the periphery) to avoid State control such as the administration of taxes or land control (Scott, 2009). When States claimed rights and control over the forest, people in remote forested areas continued practicing traditional ways of forest and land use (Arnold, 2001), while at the same time incorporating and adapting with the overall political and economic system of the State (Scott, 2009; Herriman and Winarnita, 2016). Today, people in these areas are also the main target for the latest drive of formal state structured social forestry development.

Community Forestry or Social Forestry emerged in the late 1970s, when concern over increasing deforestation rates led to a questioning of state control over forest resources and its capacity to effectively protect and manage forests sustainably. Social or Community Forestry was seen as an alternative to solving forest management problems (Gilmour, 2016), and forest conflicts (Purnomo and Anand, 2014). Over time, ideas of democracy and justice (Brosius et al. 1998; Larson, 2005), coupled with neo-liberal ideas (McCarthy, 2005) and increasing evidence that traditional practices can achieve positive outcomes for conservation (Chomba et al., 2015) shaped how community forestry and social forestry programs developed in different countries.

Although definitions vary, here we use both Community Forestry and Social Forestry, denoted as SF, to mean "any situation which intimately involves local people in a forestry activity" (Arnold, 1992). We follow an accepted understanding that SF includes both: (1) the use of forest resources by local people, on an individual or household basis, for consumption and sale; and (2) the *community management* of forests, which refers to a collaborative enterprise conducted by a group of local people who manage forests either independently or with outside support for the production of resources for consumption and sale (RECOFTC, ASFN and SDC, 2010).

Vietnam and Indonesia are two countries experimenting with decentralization, including devolution of forestland to communities through SF as part of the state regulatory framework. As part of the CIFOR project ASEAN-Swiss Partnership on Social Forestry and Climate Change (ASFCC), the two countries provide an interesting comparison given their rather different histories of democracy and levels of forest abundance and pressures. Implementation and outcomes, however, have been disappointing in the sense that devolution has not included the full suite of rights to accompany the responsibilities of forest management (Maryudi, 2012; Maryudi et al., 2012; RECOFTC 2014; Chomba et al., 2014; Chomba et al., 2015; Gilmour, 2016; Schusser et al., 2015).

We ask the question: whose interests do SF serve? Understanding the effectiveness of SF would first require clarity on the underlying motivations of governments in adopting SF, which would then allow for a proper assessment of how SF policies are implemented. In this article, we apply the 4I (institutions, interests, ideas, and information framework) by Brockhaus and Angelsen (2012) to analyze the underlying power relations, economic incentives and discursive practices. We examine and compare the national policies in Vietnam and Indonesia, changes in policies over time and shifts in discourses to see how SF has evolved, been interpreted and implemented in these two countries.

2. Social Forestry discourse and evolution

Over the more than 40 years of social/community forestry development, different interpretations of SF have emerged, often influenced by different global discourses and often not well adjusted to the local contexts. Understandings of, and perspectives on SF evolved from seeing local people as the problem in driving deforestation and degradation to local people as the solution, from local people being 'allowed' to participate in forest management towards a local governance approach where the management of forests are fully devolved to local communities. Though State motivations for SF might differ, and ideas of justice are rather absent in these, the spirit of SF has been one of devolution and democratization.

Though different discourses emerged at different points of time, core elements in these discourses with regard to problem definition, identified appropriate policy responses and overall objectives of SF remain surprisingly constant over time. One dominant narrative since the 1970s is SF as a solution for poverty, which emerged at the same time as the shift in development discourse where rural people were considered 'poor' and development was aimed to fulfill their 'basic needs' (Pollock, 1973 Streeten, 1981; Todaro and Smith, 2012). Discourses on SF further include ideas such as devolution and restoration of rights; better governance; development of entrepreneurship (RECOFTC, ASFN and SDC, 2010; Sam and Shepard, 2011; de Jong, 2012); empowerment (Maryudi, 2012; Maryudi et al., 2012; RECOFTC, 2014); a mechanism for reducing the fiscal burden of the state; or improving management of natural resources through applying local knowledge (Meinzen-Dick et al., 2008; Arnold, 2001). This mix of discourses and rationales has been adopted by most governments of Southeast Asia and applied in various forms and using different terms: community forestry, participatory forestry, village forestry, community based forest management or social forestry.

Table 1. Discourse on community forestry according to De Jong (2012)

Discourse	Description
Forest resource scarcity CF discourse	- Initiated by FAO - Focus: the perceived danger of forest resource scarcity resulting from unsustainable use would lead to a decline in rural welfare. Limited local resources seen as a constraint in fulfilling basic needs. - Implication: shift focus within development concept/agenda toward rural population and agriculture and to meeting the 'basic needs' of rural dwellers. Projects are aimed to reforestation and afforestation.
Forest conservation CF discourse	- Initiated by conservation NGOs - Focus: Deforestation as a worldwide issue needs to be halted, and 'community forest management through sustained harvesting and sales of NTFP is a more profitable use of tropical forests and at the same time contributes to tropical forest conservation'. - Implication: conservation-development projects working with communities to identify marketable NTFPs and establish the related value chain.
Community enterprise CF discourse	- Development cooperation, forestry agencies and rural development NGOs. - Focus: CF as an abstract objective of communities who manage forest estates, mostly to produce forest commodities. - Implication: promotion of CF enterprise following the western enterprise models with their related hierarchy and business management approach. Forest management is manifested in forest management plans and annual operation plans that are prepared according to legally prescribed norms.

While SF is diverse, governments define three main objectives: 1) to alleviate poverty of forest dependent people, 2) to enable communities to have secure access to, and ownership of, the resources and their benefits, through empowerment and building of capacity for forest management, and 3) to improve the condition of forests (Blaikie, 2006; Maryudi et al., 2012; Chomba et al., 2014; Gilmour, 2016). Reconciling these multiple objectives of conservation, livelihoods, and governance expected from SF can be daunting. In many cases, conservation goals were attained at the expense of local empowerment, showing that the three objectives might not necessarily mutually reinforce one another (Kellert et al., 2000; Arnold, 2001; Ribot et al., 2010; Chomba et al., 2015).

Empowerment appears to be the most challenging. Empowerment implies that local forest users and/or forest dependent communities have autonomy in setting objectives, local control in forest management and utilization, and ownership of the benefits of the forest (Kellert et al., 2000; Blaikie, 2006). It therefore also implies the right to control the use of forestland and resources within their jurisdictions. Yet, empowerment or devolution of power is constrained by conflicts between state and communities to gain control over forestland and resources. Adiwibowo et al. (2016) uses the term 'contested devolution' in which access and control over the forests and ensuing benefits are reformulated within the interaction and negotiation process among parties concerned as they are implemented. The adoption of neo-liberalism (McCarthy, 2005) further drives the discourse that in order to alleviate poverty, SF needs to be managed as a business enterprise (De Jong 2012), thereby reducing forests and forest resources to a commodity. During processes of devolution, negotiation and change, local elites have frequently managed to grab control over benefits at the expense of broader community interests (Maryudi et al., 2012; Schusser et al., 2015).

There appears to be a disconnect between the "spirit" of SF and the policies – is SF really meant to be a devolution of rights for local empowerment and management of forests? Or, as Maryudi (2012) claims, a way for the State and or timber companies to regain control? Whose interests do SF policies in Vietnam and Indonesia serve? To study these questions, we adopt the 4I- Institutions, Ideas, Interest and Information- framework (Brockhaus and Angelsen, 2012), which aims to explain how these factors interact and contribute or actually hinder major policy changes across multiple levels of governance, with a multitude of actors and often conflicting politico-economic interests and objectives. The first I refers to the institutions within a particular policy arena, described as the norms, regulations, and formal and informal arrangements which have shaped the policies. Stickiness of particular rules that aim to govern land and land use practices and persistence of existing arrangements, despite new and emerging evidence can often be attributed to those actors with most power and an interest in maintaining the status quo. A second I is related to interests, and while Brockhaus and Angelsen (2012) refer to this I as material, we would like to stress here also non-material interests, shaped by culture and tradition which are also highly relevant to SF. This links to the third I, Ideas, which refers to ideologies, political beliefs, and mental models, and in fact builds on the other two I's by highlighting the particular role of discourses to frame and shape policy actors' perceptions of policy problems and what is perceived as a solution. Ideas or ideology are important as what actors believe might be just as important as what they want (Vanberg and Buchanan, 1989 quoted in Campbell, 2002). Discourses and particular ideas of actors are also linked to the information available, the fourth I in the framework, and how and by whom facts are selected, interpreted and reinterpreted.

Applying these 4 I's allows us to capture the political economy context of social forestry in South East Asia, which often gets overlooked in literature. With this framework we can explain whose interests and ideas in an existing institutional context are reflected in final decisions over design and implementation of SF and what this would mean for the possibility of change in the wider policy arena.

3. Methods

In our analysis we focus mainly on three of the 4Is, institutions, interests and discourses (ideas) and we operationalized the 4 Is framework in our study through a mix of methods, in particular policy content analysis, literature review, semi-structured interviews between 2012 and 2016, and participatory observation.

In analyzing the main current social forestry policy documents in Vietnam and Indonesia we focused on how SF is defined and regulated, by whom. This gave us insights in the institutions-I and, by looking at policy changes over time, provided information on diverse interests and underlying power relations. Linking this to literature, we analysed how broader discourses around forestry have influenced and driven diverse forms of social and community forestry in the two countries. We then explored to what extent Vietnam and Indonesia have reconciled the interests of the government in terms of control with a rights based approach through SF, or in other words, how institutions deal with the contested devolution.

To provide more grounded discussions, we complemented the review with our personal observations in both following and engaging in the activities of social forestry working groups tasked with developing national policies in Vietnam and Indonesia.

Finally, data was also gathered from informal interviews with key respondents and direct observations during field visits to Son La and Nghe An provinces in Vietnam and West Kalimantan province in Indonesia in 2012 and 2016 as part of the CIFOR ASFCC project.

4. Results

4.1. Pathways of CF policy developments

4.1.1. Vietnam

SF practices were common in Vietnam until the 1960s. After the central state took control of forests through collectivization campaigns, local systems were replaced by cooperatives (Sunderlin and Huynh 2005) and state forest enterprises (SFEs). In 1991, reforms towards a more market based economy included realization that the SFE system was not effective and was failing financially. As a result, the strong state control of forests was rescinded and approaches for privatization of forestry were piloted, mainly though donor funded projects (Wode and Huy, 2009). Villages and or households were designated to replace SFEs as the managers of natural forests (Sikor and Apel 1998). To mobilize people in forest protection and reforestation actions, the Vietnamese Government adopted a series of policies to promote community participation in forestry (Sikor and Nguyen, 2007). Vietnam implemented national programs to (i) prioritize forest land allocation to households and household groups, particularly to the poor and to ethnic minorities whose livelihoods are closely linked to traditional forest management; (ii) decentralize forest management; and (iii) develop pro-poor mechanisms targeting groups involved in innovative forest management solutions. However, it was only with the Law on Forest Protection and Development in 2004, that the concept of 'Community Forest Management (CFM)' was officially recognized. CFM was defined as "any managerial arrangement in which local people share collective responsibility and benefits from managing natural forests, inside their community boundaries, for which they have long-term customary and/or legal rights of entitlement" (Wode and Huy, 2009; RECOFTC, 2014). SF in Vietnam was also closely linked to the government's Forest Land Allocation Program (FLA) (Sunderlin, 2006).

SF in Vietnam can be grouped into two main types. The first is village forest management (VFM), where forest is managed collectively by a forest management group to which all households in a village belong. The second is forest management by groups of households, which is formed by a small number of households – a subset of a village. The HHG model is applied where village population is dispersed in different settlements. Each HHG often manages a patch of forest near their homestead and is characterized by small groups of not more than 20 households and who are usually fairly homogenous in terms of ethnic composition and/or interest. The VFM model, on the other hand involves the whole village population.

SF in Vietnam can be characterized as a practical approach in linking poverty alleviation and forest management goals. While there appears to be some tolerance for inclusiveness and diversity in how SF is implemented, it does reinforce notions of the community or village as homogeneous social structures.

4.1.2. Indonesia

In Indonesia, the state designates forest area to be managed by the state. Failure of effective state management of forests led to the introduction of Social Forestry in the late 1970s and became a government policy in the late 1980s (Lindayati, 2002). SF was defined as a system of management of forests (on either state forest or private forestland) that involves local communities with the goal to improve their wellbeing and realize sustainable forestry (Hakim et al., 2010). Under this definition, SF was not fully meant as a policy for decentralization of rights but rather, was perceived more as a "development" program for villages located in forest areas, and its implementation became the responsibility of the forest corporations holding official exploitation permits in forest concessions. Over the subsequent two decades, SF remained an insignificant part of forestry practices, adopted on small scales by concessionaires and the state forest company primarily as an approach to solve local tenure conflicts (Lindayati, 2002). The forest management practices of local people who lived within or around state designated forest land remained largely invisible.

In the absence of dedicated higher regulations on SF, the ministry developed various community forestry schemes through ministerial decrees. Between 1991 and 1998, several such decrees tried to regulate SF in concessions (HPH-Bina Desa Hutan), in special use zones (KdTK) and in the general forest estate (HKM). The new forestry law of 1999 included provisions for the role of local communities in management of forests (e.g. articles 68-70 and article 34) throughout the forest estate.

The implementing regulation of Law 41, 1999 (Government Regulation (GR) 34, 2002 later replaced by GR 6, 2007, laid out different formal schemes of SF, i.e. community forest, village forest, people's plantation forest and collaborative management, with specific access and use rules. Although explicitly stated to aim for community empowerment, all these formal legal schemes are based on permits to be requested by local communities and granted by the state. Interest in SF since then has waxed and waned according to government priorities but also according to the dominant discourse. In 2002, the empowerment discourse moved SF as an umbrella program but when ministers changed, other priorities took over. SF progressed regardless, but at a very slow pace with only few permits requested and granted.

The global movement for indigenous rights, linked closely with issues of environmental change and sustainable development, gained increasing momentum throughout the 2000s driven by the establishment of AMAN (Alliance of Indigenous Communities of Indonesia) in 1999 (Ardiansyah et al., 2015). The Midterm National Development Plan 2010-2014 set a target to devolve 7.9 million hectares of state forest land to local communities (Praputra et al., 2015). In 2014, the new government, continued this policy commitment to SF by promising to assign 12.7 million hectares to local community management by 2019 (Dirjen PSKL, 2015) and has also framed SF as part of broader initiatives for agrarian reform (Presidential regulation 88, 2017). The stated objectives remain focused on poverty alleviation, envisioned as a lack of access to land. However, it is the pressure of land conflicts, demands for recognition of rights and agrarian reform that has given SF its contemporary urgency in Indonesia.

4.2. How SF is framed in policies

4.2.1. Vietnam

Although the Forest Land Allocation (FLA) is not a SF program per definition, SF was initially framed through FLA policies with the stated objectives to increase forest cover, improved forest quality, and contribute to hunger eradication and poverty reduction in impoverished upland areas (To and Tran, 2014). The two models mentioned above were the offshoot of FLA which was implemented since 1983 and later strengthened by the Land Law of 1993 (Decree No. 02/CP on forest land allocation), and the revised Land Law of 2003. The Land Law confirms state ownership over forests and defines the rights of land users in terms of land use, transfer, concession, lease, mortgage and contribution of capital on the basis of land value. The law further provides use rights for up to 30 ha of production and protection forests for up to 50 years per user/community and was intended to stimulate private investment (Sunderlin, 2006; Wode and Huy, 2009).

FLA itself, is however, not exclusively about allocating land to households and individuals. To and Tran (2014) distinguishes three main forms of FLA: (i) forest land allocation to state organizations, mainly forest companies; (ii) forest land allocation to households, individuals, and communities; and (iii) contract-based allocation of forest land to households and individuals either by forest companies or the state.

There have been many assessments of the impacts of forest and land allocation for households and groups of households with varied results. Helvetas Vietnam (2002) claimed that FLA resulted in improved livelihoods of poor farmers. Yet it is also claimed that households and communities did not gain any actual control over local forests: they still have to seek permission from relevant agencies to harvest trees and use forestland, and they often have limited understanding of their rights and duties (Nguyen et al., 2008). To and Tran (2014) observed the improvement of forest quality but skewed improvements in livelihood where some people were able to disproportionally capture most benefits. Wode and Huy (2009) mentioned the problems with mapping and limited participation resulting in uncertainties of the allocations and disproportionate sharing of benefits.

In 2004, the Vietnam Law on Forest Protection and Development legalized the allocation of natural forests for communities to manage, and thus formally recognizing SF, thereby recognizing indigenous norms and systems of organization (Sikor and Apel, 1998), and encouraging local people to increase the revenue base for community forestry regimes (Sunderlin and Huynh, 2005). Meanwhile, the Vietnam Forestry Strategy 2006- 2020 sets out a clear priority to allocate forests for communities to protect and to benefit from the forests. By the end of 2009, local communities held tenure rights in up to 26% of the total forest area in the country, either as individual households, household groups, or village collectives (Sikor and Nguyen, 2011).

Table 2. SF models in Vietnam

Management regime	Policy target and area implemented	Notes	Relevant legislation
Traditional and customary forests managed by local communities	MARD assumed that this area is relatively small estimated as 247,030 ha	Self sufficiency	2003 Land Law: recognizing and regulating the legal status of village communities in land tenure. 2004 Forest Protection and Development Law: recognizing legal status of village communities in forest management. Decree 29/1998/ND-CP in May 1998, later amended to Decree 79/2003/ND-CP in July 2003: on grassroots democracy
Forests and Forest land that the State allocates to communities to manage	524,477 hectares according to MARD (2013)	Must be managed as forest; small scale wood production for industry	Decree No. 02/1994/ND-CP: regulating forest land allocation to organizations, households and individuals for long term and stable use on forestry purpose. Decree No. 01/1995/ND-CP: on land allocation for use in agricultural, forestry production, aquatic product rearing within state-owned enterprises. Decree No. 01/CP/1995: on contracting agricultural land, production forest and surface water for aquaculture production to organizations. Decision 799/QĐ-TTg dated 27/6/2012 on approval of the National REDD+ Action Program
Group of households that collectively manage area of forests that are allocated to their individual HHs	No figure available		Decision No. 187/1999/QĐ-CP: on renovation of organizational structure and management mechanism of SFE directing SFEs to give forest lands back to the districts, so that they could be further allocated to households. Decree No. 163/1999/ND-CP: on assigning and leasing forestland.

Vietnam has also linked different payment for environmental services, such as the PFES (Payment for Forest Environmental Services) and REDD+ (Reducing Emissions from Deforestation and forest Degradation and enhancement of carbon stocks) to forms of community-managed forests. The PFES program has been implemented nationally since Decision 99 in 2010. Article eight in this Decision states that village communities, who are allocated forestland for long-term usage, are the beneficiaries of forest payment. As an incentive to encourage communities to get involved in forest protection and management, the PFES scheme compensates communities for forest protection activities. Moreover, the implementation of PFES acts as a catalyst to move forward forest allocation to communities.

In 2012, the Vietnamese Government issued Decision 799/QĐ-TTg on 27/6/2012, which approved the REDD+ National Action Plan. According to this, communities are at the forefront of forest protection and stand to benefit from REDD+ payments. Both PFES and REDD+ are designed as incentives for community forestry management and protection.

Forestry regulations are also often implemented in different ways in different provinces or regions. Some provinces are more flexible and progressive than others and can set up trial sites for community forestry before the central government legally acknowledges this form of forest management (Pham et al., 2012). SF practices might therefore not fit neatly into the government schemes as shown in Table 2 for Vietnam. Indeed, in practice people often ignore government programs or adapt the programs in accordance to local context.

4.2.2. Indonesia

Law 41, 1999 on Forestry and its derivatives (in particular government regulation (GR) no 6, 2007 and its revision (GR) no 3, 2008, provided an early elaboration on how local communities can gain access to state forest, i.e. through permits to manage community forests (HKm), village forests (HD) or small scale plantations (HTR) in accordance to the forest zoning, i.e. production and protected forest (See Table 3). A latest iteration of implementing regulations is Ministerial Decree P 83/2016.

Table 3: SF schemes according to official policies in Indonesia

Management regime	Policy target and area implemented	Notes	Relevant legislation
Groups with management rights to production or protection forests (community forests or HKM)	Target to 2020 is 12.7 ha for all schemes. In 2016, 55000 ha was approved but permits were issued for 2465 ha		Permenhut P.52/2011 replaced by P.83/2016
Villages with management rights to a production or protection forests (Village forests or HD)	Target to 2020 is 12.7 ha for all schemes. In 2016, 114,496 ha was approved and 73,190 ha was issued permits		Permenhut P.53/2011. Now replaced by P.83/2016
Individuals or groups managing plantations in the forest zone (HTR)	Target to 2020 is 12.7 ha for all schemes. In 2016, 22,853 ha was approved and 14.131 ha was issued permits	This can be done in partnership with private or state owned companies	Permenhut P.55/2011. Now replaced by P.83/2016
Individuals or groups managing tree plantations on privately owned land		Although mentioned in the laws, private forest is often outside state control	Now replaced by P.83/2016
Customary communities managing customary forest (HA)			Constitutional Court Decree (MK) 35/PPU-X/2012 and P.83/2016
Forest in all zones managed in partnership between communities and (legal forest managers)	In 2016, 24,468 ha was allocated for collaborative management	The forest managers are state agencies but can be companies.	P.83/2016

Source: KEMENLHK (Ministry of Environment and Forestry), 2016

Collaborative management in conservation areas is regulated by yet another decree (P.49/2008). Outside the State Forest land, forests on privately owned land are managed by the rights holders as private forest (Law 41, 1999) or customary forest (MK 35, 2012) although no further special regulations currently exist.

In general government schemes allow only rights of use linked to responsibility to maintain the forest. The different rights are described in Table 4 below.

Table 4: Extent of rights in government SF schemes:

Scheme	Extent of rights	Time limit	Purpose	Discourses/ Objectives as stated in draft regulation
HKM (community forest)	Rights to manage and use NTFPs allocated to groups	35 years extendable	For harvesting timber an additional permit is required	To empower
HD (Village forest)	Rights to manage an use the forest (NTFPs and or environmental services) allocated to villages	35 years extendable	Often managed by the Village Enterprise even when mainly managed for ecosystem services	To improve welfare/well being for the village
HTR(People's plantation forest)	Use rights to land in production forest to manage and use timber planted. Allocated to individuals or cooperatives	35 years extendable	these small scale timber plantations are often managed in collaboration with a large corporation	To improve the potential and quality of production forest
Partnership	Rights to collaborate in activities related to management of forest		Limited to 2 Ha per head of HH or 5 Ha per HH	Empowerment of local communities by forest managers
HA (Customary forests)	Full ownership by customary communities	No time limit	The follow-up regulations detailing the scopes of the rights and right holders of the HA are yet to be issued by the local government.	

SF is the legal mandate of the MoEF and therefore primarily regulated by forestry laws, regulations and ministerial decrees. However, as a category of land use, SF is also bound by the law on spatial planning, the land law, the law on regional government and the law on villages, all mandated by different authorities. Further, SF should defer to the national and regional development plans, and the use of land for 'strategic' purposes such as mining and roads. At local level, this plethora of laws and regulations leads to contradictory and uneven implementation.

4.3. SF within decentralization policies

Vietnam and Indonesia are two countries that have experimented broadly with decentralization processes, including through SF programs. The Vietnam SF policy, for example, is closely linked to Forest Land Allocation (FLA) but has also been driven by decentralization processes such as Decree 29/1998 and its amendment Decree 79/2003 on Grassroots democracy aimed to enhance wider public and social participation of different non-state actors group including communities. In Indonesia, administrative decentralization policies have gone further than Vietnam, although actual devolution of management rights of forest to local communities has been slow. Targets are ambitious, 4 million ha of the 15.2 million ha forest land in Vietnam (RECOFTC, 2014) and 12.7 million ha of the 143 million total forest land in Indonesia are to be devolved to local people by 2020 (Dirjen PSKL. 2015).

Devolution of forestland and resources from the state to local people is assumed to result in improved forest as well as improved livelihood. Implicit is empowerment of people. As the Vietnam policy states: "land allocation makes forests officially owned, attaching rights to responsibilities, and offering favorable conditions for local people to protect forest, who find it secure to manage, invest, and develop and allocate forest." (MARD's master plan for forest lease and allocation, 2007: 5 quoted in To and Tran, 2014).

In practice, both countries are very centralized with regard to forest governance (Brockhaus and Di Gregorio, 2014). In Indonesia, SF is simply considered a permit to manage forests with implementation rigidly regulated and or tied within a web of complex regulations around forest products harvesting, transporting and selling (Ministerial Decree P.83, 2016). As found elsewhere, the rural poor under well intended SF policies must compete on "an uneven playing field of ... social inequities and economic hurdles" (Larson and Ribot, 2007). Indeed, it might seem that the official SF policy is an attempt to re-centralize, to re-assert control over forest resources by the government (Maryudi, 2012; Schusser et al., 2015; Adiwibowo et al., 2016). Simply put, state-led land and forest reform strategies often do not succeed because "bureaucratic modalities cannot accommodate the varying meanings of land, plural notions of property, and diverse political-economic contexts" (Sikor and Müller, 2009).

As mentioned earlier, local people are also subject to policies from other sectors, which are often contradictory or inconsistent, and add to the complexity of SF. In the Vietnam Forest Protection and Development Law of 2004, for example, communities are recognized as legal entities and allowed to enter forest protection contracts. Yet, in the Land Law and Civil Code 2005, communities are not allowed to enter any legal contract, unless registered in the commune as a group of households (civil code) or as cooperatives (circular 07/TTLT MARD and MONRE).

In Indonesia, the new law on village governance (Law 6, 2014) gives villages autonomy to manage their assets, including village owned forests. Yet the Forestry Law sets state authority over all forests at national level. Furthermore, it remains unclear to what extent state forest within the village area is a village asset. On the other hand, in practice, rights of a village to the exclusive use of major forest products from forests in its vicinity have always been de facto recognized.

In the context of overall forest governance, the plethora and complexity of laws and regulations pertaining to local land use that appear unrelated and contradictory is a main issue. Because of this complexity, local communities have difficulty in following the process and the local government tends to treat SF as an administrative issue, an exercise of ticking the boxes, reducing SF permits to merely confirm de facto use of forest land by local people without real attempts to improve governance of forests or empower local people. Under lack of clarity, forest administrators tend to merely follow the letter of the law and forget the spirit of the law.

5. Discussion

In the following sub-sections, we discuss the (in)effectiveness and adequacy of existing SF policies as an instrument for recognizing community rights, empowering local forest communities, reducing poverty, and for improving forest governance based on the literature review and complemented with findings from our field research.

5.1. The spirit of SF laws: competing or complementary values and discourses?

Like general laws, SF laws and regulations are shaped by the political economy of a nation as well as global discourse on environmental and rights issues (Brosius et al 1998; Li, 2002). Although on the one hand, SF emerged out of global ideals of democracy and local rights (Chomba et al., 2014), as McCarthy (2005) claims, SF has also developed from the confluence of environmental governance, neoliberal policy agendas and responsiveness to contingent historical and geographical factors.

These social political contexts at different times gave rise to different discourses which influenced how SF was perceived. The perceived failure of the forest industry model of the state on the one hand, and the increasing rate of deforestation and forest degradation on the other hand (Gilmour, 2016) led to a discourse of resource scarcity. The belief that community control over forests is or can be more ecologically friendly with properly crafted rules and norms governing use led to SF for tropical forest conservation. The rising dominance of the neo liberal economy in the 1990s, gave rise to the discourse of community forestry enterprise (de Jong, 2012), assuming that communities who are able to profit substantively from the forests will be inclined to manage them sustainably.

While these SF discourses are somewhat sequential (de Jong, 2012), they are not exclusive and overlap and mix with parallel discourses. For example, the social justice discourses driving participatory development and 'Bottom up Planning' approaches of the early 1980s and the movement for rights of indigenous and other local communities to exercise control over their traditional natural resources have contributed to the development of SF as a rights-based approach.

The discussion above highlights the influence of discourses and values and interests of the national institutions on the objectives and regulations on SF. However, all regulations and policy, highlight that SF intends that forest should be for the people by the people. Yet, institutional stickiness and attachment to power has made full devolution of forests an elusive dream. Time and again, powerful actors remain reluctant to give up management power and the benefits to be gained (Edmund and Wollenberg, 2003; Dahal and Capistrano, 2006; Sunderlin, 2006; Schusser et al., 2015; Adiwibowo, 2016).

There remains a tension between governments (and powerful non-state actors) trying to exert control over forestland and local people claiming or reclaiming rights over resources they consider their heritage. There are tensions between processes of recentralization and decentralization, and between processes of gaining access and exercising access. An example is how in Kalimantan, the decentralization in 1999, allowed local communities the right to log their forests. This led to a logging boom benefiting local people but also leading to some environmental damage. In 2002, this right was re-centralized to the province and communities were no longer allowed logging rights. In Vietnam, there are cases where large, often state owned companies took over the land already allocated to individual households. Although the grassroots democracy decree promises protection of local rights, State interest takes precedence in practice. Indeed a contested devolution.

5.2. SF as a rights based approach?

At the same time, in both Vietnam and Indonesia, SF has become one approach to solve conflicts and to recognize rights of indigenous or local communities. Yet, rights are allocated rather than recognized, and limited to use and management only rather than ownership or control. The constitution of Vietnam states that all land belongs to the state and citizens have only use rights. The government of Indonesia does not claim to *'own'(memiliki)* all land but instead uses the word *'control'(menguasai)* thereby claiming the right of disposal and thus the right to allocate land or 'permit' others to use the land.

In the context of SF, forest management is considered a communal effort and land is allocated to groups or communities. However, traditionally, while the rights to a forest or landscape might be communal, there is often a nested mix of communal and individual rights, e.g. in Kalimantan the *tembawang* agro-forest is a communal property with individual rights to certain trees (Peluso 1996), and in Sulawesi, forest farmers groups were formed to gain a CF permit but in practice, little patches of forest are used and managed individually. This is in contrast to some cases in Vietnam where forestland is allocated to individual households while the community might in fact rather manage it communally (Sikor and Nguyen, 2011). The disconnect between rights allocation and household and community forest use strategies could disrupt local governance structures and how benefits are captured. It is therefore necessary to better understand the interlinkages between communal and individual rights and how communities operate in relation to forest as part of SF policy design if objectives of empowerment and devolution of rights are to be taken seriously.

In both Vietnam and Indonesia, legal rights of use to forestland and resources come with restrictions: they are usually time bound, allow only certain uses, and with only part of the bundle of rights (e.g. no alienation or conversion to other uses) are transferable. These rights also come with associated responsibilities: to manage the forest in accordance with technical requirements often for forest protection and rehabilitation rather than production for local livelihood needs, and includes submitting management plans and budgets often beyond the ability of local people.

Local communities might therefore reject these programs, either because they are not interested in managing forest or because it does not recognize their historical rights, as occurred in one community forest area in South Sulawesi (Moeliono et al., 2015b). Instead, it reinforces what is 'given' by the state and therefore, can also be revoked by the state.

Although, SF is positioned as a solution within the rights based approach, it is also recognized that, even when communities desire SF, tenure rights alone do not guarantee its success (Sikor and Nguyen, 2011; Larson et al., 2010). Secure tenure leads to desirable environmental, economic, political, and cultural outcomes only if local communities can realize the rights given to them in legislation. For example, tenure transfers have little meaning if forest regulations and logging bans severely restrict the concrete rights accorded to people to use and extract timber and other resources. Transfers also possess little value if they emphasize protection obligations over rights to forest use and management. In addition, legal tenure rights often do not translate into real rights on the ground if local authorities and communities do not have the capacity to implement legal stipulations (Larson et al 2010; Barry et al., 2010).

At the same time, focus on tenure has been at the expense of the more technical matters of good forest management (Moeliono et al., 2015b). Often communities are given areas that are less productive or poor quality forest, or are expected to restore degraded land without adequate technical input (Sunderlin, 2006; Wode and Huy, 2009; Le, 2010; Pulhin et al., 2010). As Dressler et al (2010) also found, governments tend to consider their task done when policies are made, communities informed, land allocated or permits issued. There is little effort to help communities understand their rights and responsibilities regarding the allocated forests. In most instances, local people accept their responsibility to protect or manage the allocated forest, but have little understanding of the related responsibilities of others, particularly state actors, to support them in realizing their endowed rights (Maryudi et al., 2012)

Furthermore, the institutional aspects of governance are often neglected. Permits or rights were at times allocated to groups where such groups are not well established nor have a communal identity (Indonesia) or to individuals where traditional systems are communal (Vietnam). Rather than strengthening governance, the introduced schemes can cause disruption and conflict and lead to disempowerment (Blaikie, 2006; Moeliono et al., 2015b; Maryudi et al., 2016).

5.3. SF as a new mode of governance?

SF might also be considered as a new mode of governance, where the failure of the state to manage forests sustainably and or solve forest land conflicts peacefully, calls for redistribution of power and decision making over the use of forest with the expectation that this would lead to more sustainable forest management. Although as in Indonesia, while the law mandates community empowerment, it is unclear as to what it implies and who should do it. 'Empowerment' is thus simply translated as participation or involvement of local community in forestry activities. Some argue that participation does not always empower as often only the elites in a community are able to participate. In that case, participation might even be disempowering (Agrawal, 2001, Cooke and Kothari, 2001)

Community empowerment can be perceived from two aspects: firstly, it relates to the ability of an individual or group to exercise agency, to make decisions, and take ownership of their lives. The second aspect relates to institutional structure, both state and non-state, that are more responsive and accountable to people and thus enables agency. Empowerment is thus located at the intersection of asset-based agency and institution-based opportunity with transformative processes of democracy and transfer of powers, respectively (Chomba et al., 2014). However, in the practice of SF, empowerment is understood as the state providing access to forestland and resources, thereby simply assuming that access will automatically lead to better management of forest and improved livelihood.

There is no acknowledgement of the ability of communities to manage forests traditionally. There is no recognition that a "community" comprises a complex set of actors, with different social, economic, and political characteristics such as wealth, gender, age, ethnicity, and castes (Agrawal and Gibson 1999), with different access to resources and power, and different interests in, and claims over, various environmental goods and services (Leach et al., 1999, Nygren, 2000; Chomba et al., 2016).

Power relations are often the most resistant part of institutions to change (Wilshusen, 2009; Sneddon and Fox 2007; Wong, 2010), perpetuated through political ties, family assets and land holdings, family networks, and religious affiliations (Dasgupta and Beard 2007; Lund and Saito-Jensen 2013; Schusser et al., 2016). In the processes around SF, the State usually deals with the village institution, and this institution is often also an apparatus of the State. Indeed, as we observed in rural settings in Vietnam and Indonesia, local institutions led by a village head is often the only bridge of communication between the community and external interventions. The village head is usually nominated by the state and although might be voted for local people, but they still need to get formal approval from the party to be formally recognized as the village head. The power of access and information allows the local elites to "exert disproportionate influence over a collective action process" (Beard and Phakpian, 2009). Elites tend to position themselves as decision makers and promote their own agenda (Thoms, 2008). Many assessments on SF include reports on how elites in a community are able to capture most of the benefits (To and Tran, 2014; Maryudi et al., 2012). While these examples all refer to power relations at the local level, when placing SF into a wider political economy framework, the lack of changes in power relations beyond the local becomes very visible: the established power relations between state- and non-state actors; state, private sector and civil society; and national and local level power relations all have changed so far very little over the course of the different waves of SF, if at all.

SF interventions have contributed to inequality and elite capture, especially where local elites dominate the village enterprises (often a State or State-initiated institution) assigned to manage village forests in Indonesia (Moeliono et al., 2015a). This same problem can be exacerbated when future benefits tied to community forestry practices (such as PES or REDD+) come into play. Studies from West Kalimantan, suggests that the implementation of government development projects or REDD+ without intimate knowledge of the political landscape of resource struggle can easily lead to benefits appropriated by local elites (Kallio et al., 2016; Eilenberg, 2015)

Policy documents often imply that local communities are weak and incapable of managing forest sustainably without prescriptive rules. Yet proponents of SF often mention the weakness of the state and the inability of government institutions to manage the forest. Thus on the one hand, the state ignores the different local institutions and their different interests and on the other, an often mentioned constraint is the lack of government institutions responsible for SF.

In Vietnam, there is no official government institution that is responsible for community forest management at both central and local levels. High ranking policy makers in MARD are not convinced about the important role of SF due to small areas of only 3.76% of forest that are currently managed by communities, and the mixed results of pilot programs which shows SF is not always effective[1]. However, this figure only refers to land allocated to communities for communal management and overlooks the 25% of total forests that are under community management (RECOFTC, 2014) allocated to households, as a large proportion of individual land is managed at communal level.

In Indonesia, the new government of Joko Widodo merged the Ministry of Forestry with the Ministry of Environment in 2015, and formed a special Directorate General to manage SF. The Directorate General is expected achieve the target of allocating 12.7 million hectares for SF by 2020 (RPJM). However, SF is challenged by a small budget, (about half of what is allocated for the forest production sector), and this is also the case in Vietnam.

The SF assemblage. We have discussed how a national-led reform process does not translate well at the local level. Indeed, community forest management is what Li (2009) terms an 'assemblage', where competing and overlapping responsibilities, discourses and interests among provincial government, district government, the various Technical Implementing Units of the Ministry (e.g. SF, Watershed Management, Park management), different non-government actors and local communities have made the cohesive governance and management of forest so difficult. Changes at national level and the establishment of new agencies to manage SF, as occurred in Indonesia, have led to confusion and insecurity. Unless mandates and roles are clarified, a coordinating mechanism is established and funding allocated accordingly, targets will likely not be achieved.

---

[1] Interview results, 2014

Despite the decentralization processes associated with SF in Vietnam and Indonesia, the State has remained a dominant presence. The governmentality of SF can be seen in two ways: firstly, SF is perhaps not so much meant to seriously transfer authority to communities, but to maintain control over forests in a different manner, for example, through complex permitting applications and reporting obligations. Decentralization also offers an opportunity for States to transfer their unresolved forest problems and financial deficits, while maintaining central control over valuable assets and resources. This is certainly the case for Vietnam in terms of state failure and budget deficits in forest management, and Indonesia in terms of appeasing local conflicts. Hence, what seems empowerment of and increased autonomy for and communities at first glance, is actually driven by the central state to dispose of its wicked problems (Arts, 2014).

Secondly, SF is not about increasing decision-making power for ordinary people, but is really about shaping of the community as environmentally responsible subjects and the creation of mutual consent around local forestry problems and objectives (Agarwal 2005). Given these two perspectives and state failure to transform SF as a diverse and complex system of local and nested forms of governance, we should be "cautious about accepting too optimistic or too naïve governance claims of devolution, autonomy and empowerment." (Arts 2014).

6. Conclusions

Although based on different contexts and for different reasons, SF was adopted by the governments of Vietnam and Indonesia as a government program to achieve the multiple objectives of improving livelihoods, empowering communities and improving forest governance. The governments regulate SF by legislating multiple policies and regulations that have had at times contradictory effects to their stated objectives. Based on a literature review and data from the field, we find that conflicts over resources, incomplete allocation of rights, misalignment between formal SF rules and local governance systems, and actors involved have rendered SF ineffective in general.

Several interpretations can be identified. First is that the government itself has no serious interest in the implementation of SF as it not only would lose control over the resources, but also must undergo transformational change. Old habits die hard and institutional stickiness and vested interests is a common problem (Brockhaus et al., 2014). Secondly, either there is a lack of trust, or communities are indeed unwilling to manage forests on these terms. Local people do not hesitate to change their livelihood system if they see economic opportunities (Feintrenie et al., 2010) but incentives need to be clear.

Thirdly, empowerment needs more than SF, it would need respect for the local people and structural reform to provide clear rights and responsibilities allowing local people autonomy to exercise power. It requires providing understanding and skills that allows local people to govern resources. Equally important, it also requires an understanding that the interlinkages between the community and forest are not static, but an adaptive and dynamic practice of governance to changing development, movements of people, market, policy and environmental processes. As the experience from Vietnam and Indonesia suggest, SF in its current forms fail to meet up to these requirements. In line with other authors (Adiwibowo, 2016; Maryudi, 2012) it may be that SF is indeed not fully serving the interests of the people, but perhaps that of the State.

Acknowledgement

This study was conducted under the ASEAN-Swiss Partnership on Social Forestry and Climate Change (ASFCC), funded by the Swiss Agency for Development and Cooperation (SDC) – Switzerland F-07476.02.04 (Social Forestry, Swidden Systems and Social Networks: Lessons for Designing REDD+ in Southeast Asia, Contract n°810245657).

The research was led by the Centre of International Forestry Research (CIFOR) in collaboration with National University of Laos (NUOL), the Vietnamese Academy of Forest Sciences (VAFS), and the University of Tanjungpura (UNTAN). We are grateful to the ASEAN Working Group on Social Forestry (AWG-SF), and our local partners and informants for their instrumental support, advice and patience during the course of this research. We especially thank the respondents for their valuable time and rich insights on which this paper is based. We also thank the anonymous reviewers for their constructive and very helpful contributions to this paper.

Additional funding was provided by the Norwegian Agency for Development Cooperation; the Australian Agency for International Development; the European Union; the International Climate Initiative (IKI) of the German Federal Ministry for the Environment, Nature Conservation, Building and Nuclear Safety (BMUB); the United Kingdom Department for International Development; and the CGIAR Research Program on Forests, Trees and Agroforestry.

References

Adiwibowo, S., Shohibuddin, M., & Kartodihardjo, H. (2016). Contested devolution: the political ecology of community-based forest management in Indonesia.

Agrawal, A. (2005). *Environmentality: Technologies of Government and the Making of Subjects*. Duke University Press, Durham and London.

Agrawal, B. (2001). Participatory exclusions, community forestry, and gender: An analysis for South Asia and a conceptual framework. *World Development* 29(10):1623–1648.

Agrawal, A. and Gibson, C.C. (1999). Enchantment and disenchantment: the role of community in natural resource conservation. *World development* 27(4):629-649.

Ardiansyah, F., Marthen, A. A., & Amalia, N. (2015). Forest and land-use governance in a decentralized Indonesia: A legal and policy review. Occasional Paper 132. Bogor, Indonesia: CIFOR.

Arnold, J.E.M. (1992). *Community Forestry: Ten Years in Review*. Rome, Italy: Food and Agriculture Organizations of the United Nations. http://www.fao.org/docrep/u5610e/u5610e00.HTM

Arnold, J.E.M. (2001). *Forests and people: 25 years of community forestry*. Rome, Italy: Food and Agriculture Organizations of the United Nations. http://www.fao.org/docrep/012/y2661e/y2661e00.pdf

Arts, B. (2014). Assessing forest governance from a 'Triple G' perspective: Government, governance, governmentality. *Journal of Forest Policy and Economics* 49:17-22

Barry, D., Larson, A.M., & Colfer, C.J.P. (2010). Forest tenure reform: an orphan with only uncles. *In* Larson AM, Barry D, Dahal GR and Colfer CJP, ed. *Forests for people: community rights and forest tenure reform*. London, UK: Earthscan. 19-39.

Beard, V., & Phakphian, S. (2009). Community-based Planning in Chiang Mai, Thailand: Social Capital, Collective Action and Elite Capture. Paper presented at Dialogical Conference 'Social Capital and Civic Engagement in Asia', 7-10 May 2009. University of Toronto.

Blaikie, P. (2006): Is Small Really Beautiful? Community-based Natural Resource Management in Malawi and Botswana. *World Development* 34(11): 1942-1957.

Brockhaus, M., & Di Gregorio, M. (2014). National REDD+ policy networks: from cooperation to conflict. *Ecology and Society 19*(4):14.

Brockhaus, M., Di Gregorio, M., and Carmenta, R. (2014). REDD+ policy networks: exploring actors and power structures in an emerging policy domain. *Ecology and Society* 19(4):29.

Brockhaus, M., & Angelsen, A., (2012). Seeing REDD+ through 4Is: a political economy framework. Analysing REDD+: challenges and choices. Center for International Forestry Research, Bogor, Indonesia, pp.15-30.

Brosius, J.P, Tsing, A.L., & Zerner C. (1998). Representing communities: Histories and politics of community-based natural resource management. *Society & Natural Resources* 11(2):157-168.

Campbell, J.L. (2002). Ideas, Politics and Public Policy. Annual Review of Sociology 28: 21-38.

Chao, S. (2012). *Forest Peoples: Numbers across the world.* UK: Forest Peoples Program. http://www.forestpeoples.org/sites/fpp/files/publication/2012/05/forest-peoples-numbers-across-world-final_0.pdf

Chomba, S., Kariuki, J., Lund, J.F., & Sinclair, F. (2016). Roots of inequity: How the implementation of REDD+ reinforces past injustices. *Land Use Policy* 50:202-213.

Chomba, S., Nathan, I., Minang, P.A., & Sinclair, F. 2015. Illusions of empowerment? Questioning policy and practice of community forestry in Kenya. *Ecology and Society 20*(3):2.

Chomba, S., Treue, T., & Sinclair, F. (2014). The political economy of forest entitlements: can community based forest management reduce vulnerability at the forest margin? *Forest Policy and Economics* 58: 37-46.

Cooke, B., & Kothari, U. (2001). *Participation: The new tyranny?* New York: Zed Books.

Dahal G.R., and Capistrano, D. (2006). Forest governance and institutional structure: an ignored dimension of community based forest management in the Philippines . International Forestry Review 8 (4) :377-394. ISSN: 1465-5489

Dasgupta, A., & Beard, V.A. (2007). Community driven development, collective action and elite capture in Indonesia. *Development and change 38*(2): 229-249.

De Jong, W. (2012). Discourses of community forestry. *In* Arts B, van Bommel S, Ros-Tonen M and Verschoor G, ed. *Forest-people interfaces: from local creativity to global concern,* pp. 107-120. Netherland: Wageningen Academic Publishers.

[Dirjen PSKL] Direktorat Jenderal Perhutanan Sosial dan Kemitraan Lingkungan. (2015). *Rencana Strategis 2015-2019.* Jakarta, Indonesia: Kementrian Lingkungan Hidup dan Kehutanan.

Dressler, W., Büscher B., Schoon, M., Brockington, D.A.N., Hayes, T., Kull, C.A., McCharthy, J., & Shrestha, K. (2010). From hope to crisis and back again? A critical history of the global CBNRM narrative. *Environmental conservation 37*(1):5-15.

Edmunds, D. & Wollenberg, E. (2003). Whose Devolution is it Anyway? Divergent Constructs, Interests and Capacities between the Poorest Forest Users and States. Local forest management: the impacts of devolution policies. Edmunds, D., Wollenberg, E. (eds.) 150-165

Eilenberg, M. (2015). Shades of green and REDD: Local and global contestations over the value of forest versus plantation development on the Indonesian forest frontier. *Asia Pacific Viewpoint* 56(1), pp.48-61.

Feintrenie, L., Schwarze, S., and Levang, P. 2010. Are local people conservationists? Analysis of transition dynamics from agroforests to monoculture plantations in Indonesia. Ecology and Society 15(4).

Gilmour, D. (2016). *Forty years of community-based forestry: A review of its extent and effectiveness.* Rome, Italy: Food and Agriculture Organizations of the United Nations. http://agris.fao.org/agris-search/search.do?recordID=XF2017000018

Hakim, I., Irawanti, S., Murniati, Sumarhani, Widiarti, A., Effendi, R., Muslich, M., & Rulliaty, S. (2010). *Social Forestry Menuju Restorasi Pembangunan Kehutanan Berkelanjutan*. Bogor, Indonesia: Pusat Penelitian dan Pengembangan Perubahan Iklim dan Kebijakan.

Helvetas Vietnam. (2002). *Vietnam* experience & possible contribution to SDC development goal: Sustainable livelihoods & *poverty reduction in the upland.* Hanoi, Vietnam: Helvetas Vietnam.

Herriman, N., & Winarnita, M. (2016). Seeking the State: Appropriating Bureaucratic Symbolism and Wealth in the Margins of Southeast Asia. Oceania (2016) DOI:10.1002/ocea.5122

Kallio, M.H, Moeliono, M., Maharani, C., Brockhaus, M., Hogarth, N.J., Daeli, W., Tauhid, W., & Wong, G. (2016). Information exchange in swidden communities of West Kalimantan: lessons for designing REDD+. *International Forestry Review, 18*(2): 203-217.

Kellert, S.R., Mehta, J.N., Ebbin, S.A., & Lichtenfeld, L.L. (2000). Community natural resource management: promise, rhetoric, and reality. *Society & Natural Resources, 13*(8): 705-715.

[KEMENLHK] Kementerian Lingkungan Hidup dan Kehutanan. (2016). *Capaian Perhutanan Sosial Tahun 2016.* Jakarta, Indonesia: KEMENLHK. Accessed 21 August 2017. http://www.menlhk.go.id/tinymcpuk/gambar/file/3.%20CAPAIAN%20PS%20TH%202016.pdf

Larson, A.M. (2005). Democratic decentralization in the forestry sector: Lessons learned from Africa, Asia and Latin America. *In* Colfer CJP and Capistrano D, ed. *The Politics of Decentralization: Forests, Power and People*. London, UK: Earthscan.

Larson, A.M., & Ribot, J.C. 2007. The poverty of forestry policy: double standards on an uneven playing field. *Sustainability Science* 2(2):189-204.

Larson, A.M., Barry, D., & Dahal, G.R. (2010). New rights for forest-based communities? Understanding processes of forest tenure reform. *International Forestry Review*, 12(1): 78-96.

Le, K.C. (2010). Vietnam national forest programme process evaluation report (2nd draft). Available at: http://vietnamforestry.org.vn/NewsFolder/NFP%20Assessment%20Report_EN.pdf [2 February 2012].

Leach, M., Mearns, R., & Scoones, I. (1999). Environmental entitlements: dynamics and institutions in community-based natural resource management. *World development 27*(2):225-247.

Li, T.M. (2002). Engaging simplifications: community-based resource management, market processes and state agendas in upland Southeast Asia. *World development* 30.2 (2002): 265-283.

Li, T.M. (2007). Practices of assemblage and community forest management. *Economy and society* 36(2):263-293.

Lindayati, R., (2002). Ideas and institutions in social forestry policy. Which Way Forward? People, Forests, and Policymaking in Indonesia, pp.36-59

Lund, J.F., & Saito-Jensen, M. (2013). Revisiting the issue of elite capture of participatory initiatives. *World development* 46:104-112.

[MARD] Ministry of Agriculture and Rural Development. (2007). *Project of forest allocation and lease for the period of 2007–2010*. [Attached to Decision No. 2740 /QĐ- BNN-KL.] Hanoi, Vietnam: MARD.

Maryudi, A., Devkota, R.R., Schusser, C., Yufanyi, C., Salla, M., Aurenhammer, H., Rotchanaphatharawit, R., & Krott, M. (2012). Back to basics: considerations in evaluating the outcomes of community forestry. *Forest Policy and Economics* 14(1):1-5.

Maryudi, A. (2012). Restoring state control over forest resources through administrative procedures: Evidence from a community forestry programme in Central Java, Indonesia. *ASEAS – Austrian Journal of South-East Asian Studies* 5(2): 229-242.

McCarthy, J. (2005). Devolution in the woods: community forestry as hybrid neoliberalism. *Environment and Planning A* 37(6):995-1014.

Meinzen-Dick, R., Di Gregorio, M., & Dohrn, S. (2008). *Decentralization, pro-poor land policies, and democratic governance*. CAPRi Working Paper No. 80. Washington DC, USA: International Food Policy Research Institute (IFPRI).

Ministry of Environment and Forestry, Indonesia. (2016). Dataset Lingkungan Hidup Dan Kehutanan Tahun 2016. http://www.menlhk.go.id/berita-129-dataset-lingkungan-hidup-dan-kehutanan.html

Ministry of Forestry and Bureau of Statistics (DepHut and BPS, 2009). Identifikasi Desa di Dalam dan di Sekitar kawasan Hutan. 2009. (identification of villages in and around forest areas).

Moeliono, M., Mulyana, A., Adnan, H., Manalu, P., Yuliana, L., and Balang. (2015a). *Village forests (Hutan Desa): empowerment, business or burden*. Brief 51. Bogor, Indonesia. World Agroforestry Centre (ICRAF) Southeast Asia Regional Program.

Moeliono, M., Mulyana. A., Adnan, H., Yuliani, E.L., Manalu, P., & Balang. (2015b). *A permit is not enough: community forests (HKM) in Bulukumba*. Brief 49. Bogor, Indonesia: World Agroforestry Centre (ICRAF) Southeast Asia Regional Program.

Nguyen, B.N. 2009. *Report on community forest management in Vietnam: status, problems and solutions*. IUCN Vietnam Strengthening Voices for Better Choices Project

Nguyen, Q.T., Nguyen, B.N., and Nguyen, N.T. (2008). *Forest tenure reform in Vietnam: case studies from the northern upland and central highlands regions*. Hanoi, Vietnam: The Center for People and Forests (RECOFTC) and the Rights and Resources Initiative (RRI).

Nygren, A. (2000). Development discourses and peasant-forest relations: natural resource utilization as social process. *Development and Change* 31(1): 11-34.

Peluso, N.L. 1996. Fruit trees and family trees in an anthropogenic forest: Ethics of access, property zones, and environmental change in Indonesia. *Comparative studies in Society and History* 38(3): 510-548.

Pham, T.T., Moeliono, M., Nguyen, T.H., Nguyen, H.T., and Vu T.H. (2012). *The context of REDD+ in Vietnam: Drivers, agents and institutions*. CIFOR Occasional Paper no. 75. Bogor, Indonesia: Center for International Forestry Research.

Praputra, A.V, Sahide, M.A.K., & Santosa, A. (2015). *Menata Era Baru Kehutanan Masyarakat*. Policy Brief Kehutanan Masyarakat. Bogor, Indonesia: FKKM.

Pollock., & David, H. (1973). The Cruel Choice: A New Concept in the Theory of Development. By Goulet Denis. *The Americas* 29.3 (1973): 395-397.

Pulhin, J.M., Larson, A.M., & Pacheco, P. (2010). Regulations as barriers to community benefits in tenure reform. In Larson AM, Barry D, Dahal GR and Colfer CJP, ed. *Forests for people: community rights and forest tenure reform*, pp. 139-159. London, UK: Earthscan.

Purnomo, E.P., & Anand, P.B. 2014. The Conflict of Forest Tenure and the Emergence of Community Based Forest Management in Indonesia. *Journal of Government and Politics* 5(1).

RECOFTC. (2014). *Current status of social forestry in climate change mitigation and adaptation in the ASEAN region: Situational analysis 2013*. Bangkok, Thailand: RECOFTC – The Center for People and Forests.

RECOFTC, ASFN & SDC. 2010. *The role of social forestry in climate change mitigation and adaptation in the ASEAN region. Assessment 2010*. Bangkok, Thailand: RECOFTC – The Center for People and Forest.

Ribot, J.C, Lund, J.F., and Treue, T. (2010). Democratic decentralization in sub-Saharan Africa: its contribution to forest management, livelihoods, and enfranchisement. *Environmental Conservation* 37(1):35-44.

Sam, T., & Shepard, G. (2011). *Community Forest Management*. Background Paper for The United *Nations* Forum on Forests Secretariat UNFF9: "Forests for People, Livelihoods and Poverty Eradication". UN Headquarters, 24 January to 4 February 2011. New York, USA: UN. http://www.un.org/esa/forests/pdf/publications/CBFM.pdf

Schusser, C., Krott, M., Movuh, M.C.Y., Logmani, J., Devkota, R.R., Maryudi, A., Salla, M., & Bach, N.D. (2015). Powerful stakeholders as drivers of community forestry—Results of an international study. *Forest Policy and Economics, 58:* 92-101.

Scott, J.C. (2009). *The art of not being governed: An anarchist history of upland Southeast Asia*. Yale University Press.

Scott, J.C. (1998). *Seeing Like a State. How Certain Schemes to Improve the Human Condition Have Failed*. Connecticut, USA: Yale University.

Sikor, T., & Apel, U. (1998). *The possibilities for community forestry in Vietnam*. Working paper series No.1 (Asia Forest Network). Berkeley, USA: Center for Southeast Asia Studies, University of California.

Sikor, T., & Nguyen, Q.T. (2007). Why May Forest Devolution not Benefit the Rural Poor? Forest Entitlements in Vietnam's Central Highlands. *World Development* 35(11): 2010-2025.

Sikor, T., & Müller, D. (2009). The limits of state-led land reform: An introduction. *World Development, 37*(8): 1307-1316.

Sikor, T., & Nguyen QT. (2011). *Realizing Forest rights in Vietnam: Addressing Issues in Community Forest Management*. Hanoi, Vietnam: The Center for People and Forests.

Sneddon, C., and Fox, C. (2007). Power, development, and institutional change: Participatory governance in the lower Mekong basin. *World Development* 35(12): 2161-2181.

Soriaga, R., & Walpole, P. (2006). *Forests and Poverty Reduction: Opportunities in Asia-Pacific Region*. Paper Presented during the Asia-Pacific Forestry Commission Pre-Session Workshop, 15 April 2006. Dehradun, India: Asia Forest Network. http://www.asiaforestnetwork.org/pub/pub77.pdf

Streeten, P., Burki, S.J., Haq, U., Hicks, N., & Stewart, F. (1981). First things first: meeting basic human needs in the developing countries.

Sunderlin, W.D. (2006). Poverty alleviation through community forestry in Cambodia, Laos, and Vietnam: An assessment of the potential. *Forest Policy and Economics, 8*(4): 386-396.

Sunderlin, W.D. and Huynh, T.B. (2005). *Poverty alleviation and forests in Vietnam*. Bogor, Indonesia: Center for International Forestry Research.

Thoms, C.A. (2008). Community control of resources and the challenge of improving local livelihoods: A critical examination of community forestry in Nepal. *Geoforum* 39 (3): 1452-1465.

To, X.P., & Tran, H.N. (2014). *Forest Land Allocation in the Context of Forestry Sector Restructuring: Opportunities for Forestry Development and Uplands Livelihood Improvement*. Hue, Viet Nam: Tropenbos International Viet Nam.

Todaro, M.P., Smith, S.C. (2012). Economic development. George Washington University. 2012.

Wilshusen, P.R. (2009). Social process as everyday practice: the micro politics of community-based conservation and development in southeastern Mexico. *Policy Sciences* 42(2): 137-162.

Wode, B., & Huy, B. 2009. *Study on state of the art of community forestry in Vietnam*. Hanoi, Vietnam: GTZ and GFA Consulting Group.

Wong, S. (2010). *Elite capture or capture elites? Lessons from the'counter-elite'and'co-opt-elite'approaches in Bangladesh and Ghana*. Working paper 2010, 82. Helsinki, Finland: World Institute for Development Economics Research.

Vietnam regulations and policies

Decision 799/QĐ-TTg dated 27/6/2012 on approval of the National REDD+ Action Program

Decision No. 187/1999/QĐ-CP: on renovation of organizational structure and management mechanism of SFE directing SFEs to give forest lands back to the districts, so that they could be further allocated to households

Decree No. 163/1999/ND-CP: on assigning and leasing forestland.

Decree 29/1998/ND-CP in May 1998, later amended to Decree 79/2003/ND-CP in July 2003: on grassroots democracy

Decree No. 01/1995/ND-CP: on land allocation for use in agricultural, forestry production, aquatic product rearing within state-owned enterprises.

Decree No. 01/CP/1995: on contracting agricultural land, production forest and surface water for aquacultural production to organizations

Decree No. 02/1994/ND-CP: regulating forest land allocation to organizations, households and individuals for long term and stable use on forestry purpose

Forest Protection and Development Law of 2004: recognizing legal status of village communities in forest management

Land Law of 2003: recognizing and regulating the legal status of village communities in land tenure

Indonesia regulations and policies

Constitutional Court Decree (MK) 35/PUU-X/2012 dated 26/03/2013 about Customary Forest (*Hutan Adat*)

Government Regulation No. 6/2007 on Forest Systems and the Formulation of Forest Management and Use

Government Regulation No. 3/2008 on Amendments to Government Regulation No. 6/2007 on Forest Systems and the Formulation of Forest Management and Use

Ministerial Decree No. P.83/MENLHK/SETJEN/KUM.1/10/2016 on Social Forestry (*Perhutanan Sosial*)

Ministerial Decree No. P.52/MENHUT-II/2011 on third amendments of Ministry of Forestry Decree No. P.37/MENHUT-II/2007 on Community Forestry (*Hutan Kemasyarakatan*)

Ministerial Decree No. P.53/MENHUT-II/2011 on second amendments of Ministry of Forestry Decree No. P.49/MENHUT-II/2008 on Village Forest (*Hutan Desa*)

Ministerial Decree No. P.55/MENHUT II/2011 on procedure to request permits to use non timber products in planted forest (*tentang Tata Cara Pemohonan Izin Usaha Pemanfaatan Hasil Hutan Kayu pada Hutan Tanaman Rakyat dalam Hutan Tanaman*)

Ministerial Decree No. P.49/MENHUT-II/2008 on Village Forest (Hutan Desa)

Presidential regulation no 88, 2017 on solving conflicting claims in forest areas (tentang *Penyelesaian Penguasaan Tanah dalam Kawasan Hutan*)

Law No 41/1999 dated 30/09/1999 on Forestry

Law No.6/2014 on Village (tentang Desa).

11

A synthesis of the implementation ambivalence of REDD+ in Sub-Saharan Africa and Southeast Asia

Divine O. Appiah[1], **Stephanie E. A. Gbeddy**[2,*]

[1] Department of Geography and Rural Development, KNUST, Kumasi, Ghana ; doappiah.cass@knust.edu.gh
[2] Department of Geography and Rural Development, KNUST, Kumasi, Ghana
* Correspondence author: stephaniegbeddy@gmail.com

Abstract: Reducing emission from deforestation and forest degradation and associated benefits (REDD+), has received much attention as one of the most controversial climate change initiatives, especially by forest fringed community actors in sub-Saharan Africa (SSA) and Southeast Asia, (SEA) who are skeptical of the scheme. The object of this paper is to examine the seeming potential benefits and accompanying risks and challenges of REDD+ on the livelihoods among smallholder farmers in SSA and SEA. The paper espouses the sustainability context of REDD+ projects as pro-poor forest management mechanisms; through the provision of alternative livelihood. This is achieved through critical review and critique of scientific articles, project reports and relevant documents on REDD+ interventions from a worldwide, regional to local scale. The paper identifies projects that seem to solidify claims that REDD+ projects are simply a new form of colonialism; which the West is using to take advantage of vulnerable groups in the South. The paper concludes with the need to actively engage sub-Saharan African and Southeast Asian women in climate change mitigation benefit schemes on account of the expedient role women play in agricultural activities (which may involve deforestation and forest land degradation).

Keywords: Community development; Community participation; Gender mainstreaming; REDD-Plus; Sub-Saharan Africa; Southeast Asia

---

1. Introduction

The importance of taking into consideration the benefits that local communities stand to derive from reducing emissions from deforestation and forest degradation and associated forest carbon stock management (REDD+) schemes, have been recognized by the United Nations programme (UNFCCC, 2010). Despite the fact that the REDD scheme was originally focused on reducing carbon emissions by preventing forest loss, the United Nations Framework Convention on Climate Change Conference of Parties (UNFCCC COP 16) in Cancun, promoted co-benefits (or multiple benefits) and safeguards for REDD+, which entails the reduction in emissions from deforestation and forest degradation with the additional task of conservation, sustainable forests management, for the enhancement of forest carbon stocks in developing countries. This placed the two concepts, high on the international climate change agenda (Lee et al., 2011; Leventon et al., 2014).

The Cancun Agreements sought to stimulate the complete and effective participation of key stakeholders including indigenous communities in REDD+ related actions. This is because; well-designed benefit sharing mechanisms can enhance livelihoods of local communities as well as strengthen the rights and welfares of indigenous peoples (Lee et al., 2011). The growth of REDD+ has instigated the notion that conservation programs are fairly capable of delivering win-wins by protecting the environment and enhancing community development to reduce poverty (Lawlor et al. 2013). The poverty-environment trap theory postulates that since the environment is a major contributor to a household's production and income, the destruction of the environment, will further aggravate household's poverty scenarios.

This is because the individual's primary asset would be lost (Lawlor et al., 2013). Globally, there is an estimated 850 million hectares of degraded forests, with deforestation and forest degradation responsible for approximately 17.4% of annual global carbon dioxide emissions (IPCC, 2007a; UNFCCC, 2007b). Factors accounting for deforestation include; agricultural expansion, unsustainable extraction of forestry products such as fuel wood, forest-residential land use conversions, logging activities, among others (Streck, 2010; Population Action International, 2011; Kissinger et al., 2012). Though carbon dioxide emissions from land use change may seem to be little – between 6% and 17% of total carbon dioxide emissions – forests can play a vital role in reducing emissions as well as augmenting the sequestration of carbon in terrestrial reservoirs (Forsell et al., 2016). Thus, interventions aimed at sustainable management of the environment will be beneficial to the rural poor; though this benefit is one targeting poverty stabilization and not necessarily, poverty reduction strategy. However, the introduction of conservation programs such as REDD+ will enhance sustained livelihoods in local populations' property rights and tenure security in their land-based activities (Lawlor et al., 2013).

1.1. REDD+ Mechanism in Perspective

Reducing emissions from deforestation and forest degradation (REDD) refers to forest and land based activities which abates forest carbon emissions by minimizing or preventing forest conversion and destruction (Virgilio et al. 2010). The REDD project was implemented due to the evidence that a large percentage of current annual carbon emissions resulted from the loss of tropical forest (Madeira, 2008). Unlike REDD, REDD+ goes beyond deforestation and forest degradation to encompass additional function of ensuring that there is sustainable forest management, including the enhancement of forest carbon stocks. It was further based on the idea that a developing country may receive compensation if it develops a national strategy to reduce its deforestation rate, in proportion to the amount of carbon emissions that are reduced (UNFCCC, 2010). The UN-REDD, was introduced in 2008 to build on the convening role and technical expertise of the United Nations Development Programme (UNDP), the Food and Agriculture Organization of the United Nations (FAO), as well as the United Nations Environment Programme (UNEP) (UN-REDD, 2016).

Policy and economic measures to address these emissions have the tendencies to augment the ability of forests to sequester and store carbon. The concept of carbon sequestration connotes the processes by which green vegetation, through photosynthesis uptake atmospheric carbon dioxide for storage as carbon in biomass and soils (IPCC, 2007b as cited in Virgilio et al., 2010). Since poverty has been recognized as the primary driver of deforestation, adaptation for and mitigation of climate change has in recent times, been addressed in unison with poverty alleviation programmes (FAO, 2010). A community forestry development process initiated in Cameroon in 2001 introduced an environmentally friendly, low-impact approach to forest exploitation, which subsequently led to the locales generating substantial sums of money, in excess of USD $87,063 to develop their community for income earning engagements (Minang, 2010).

It is on this account that forest carbon emission mitigation programmes such as REDD+, offer the potential for community development and biodiversity conservation aside its main focus, of climate change mitigation (Virgilio et al., 2010). Commitment from government, corporate entities and individuals in response to reducing carbon dioxide emissions has led to the branding of carbon as a priced environmental commodity on the global marketplace (FAO, 2010). Africa's proportion of the carbon market stands at only 2% by 2009 (Institute for Security Studies, 2011). Three flexible mechanisms in the Kyoto Protocol allows countries in the heavy emitter countries, particularly in the global north to obviate the adversities of global warming on the south, by investing into emission reduction and carbon sequestration; especially so when the developed countries are seemingly not prepared to commit themselves to specific emissions reduction targets (Virgilio and Marshall, 2009).

An aspect of the Kyoto Protocol, was the Clean Development Mechanism (CDM), created to help some countries meet their emission reduction goals. This mechanism was implemented to allow for emission reduction in the forest sector through afforestation and reforestation programmes. Another focus was to improve forest management with the exclusion of activities aimed at REDD; due to skepticism about its credibility as a carbon market scheme (Virgilio and Marshall, 2009).

The sustainable quality and successes of REDD+ initiatives are highly dependent on the participation of local communities, especially smallholder farmers that live in forest fringe areas (Scriven and Malhi, 2012). This review identifies and explains the critical relevance of effective stakeholder engagement and participation in REDD+ project development and implementation. It also details the engagements, especially in relation to the disaggregation of the farmer responses and implementation by gender, an attribute that is crucial in comparative natural resources management.

Thus REDD+ serves as a benefit-sharing system which entails institutional approaches, mechanism and legal instruments for funds distribution and other additional benefits from the programmes. This initiative is vital for providing the necessary inducements to forestall deforestation and forest degradation behaviours, ultimately reducing carbon emissions (Brockhaus et al., 2013). There are, however, some challenges of implementation as critics of the scheme tend to focus on the livelihood dimensions of forest fringe communities. Some of the factors that have and hold further potential to affect the implementation success or otherwise include; security of financial base, human attitudinal responses to the scheme across different scales, access to forest land and related tenure arrangements pertaining to the ownership and organization of forest rights and responsibilities in sub-Sahara Africa (SSA) and Southeast Asia (SEA).

By virtue of reducing deforestation and forest degradation, off-reserve smallholder farmers have viewed the scheme as depriving, of their efforts to expand their agricultural land frontiers by clearing new forest lands for cultivation. In the absence of comprehensive and transparent benefits sharing mechanisms, which invariably characterize some of these schemes in sub-Sahara Africa (SSA) and Southeast Asia (SEA), it has become extremely difficult for the potential implementers, that is, the off-forest reserve farmers to overtly accept the scheme for implementation. It essentially imperative to diagnose the various setbacks and as well as the factors that would impede or promote the successes of REDD+ schemes in the sub-regions, as a way of informing the climate change mitigation components of member countries of these sub-regions' intended nationally determined contributions (INDCs), towards the emission reductions, via the implementation of the REDD+ mechanisms.

The objectives of this paper are to elucidate the various community benefit accruals from existing REDD+ programmes in sub-Saharan Africa and Southeast Asia as well as interrogate some of the implementation ambivalence arriving from human attitudinal responses associated with the implementation of the REDD+ schemes in SSA and SEA. Clarifications have been given concerning the concept of benefit sharing (derived from REDD+ implementation which can be associated with the Payment for Environmental Service (PES) scheme, where tangible benefits (income) are enjoyed by the farmers whilst the ecosystem enjoys the intangible benefits, that is, forest carbon emissions. Details have also been given about the gendered dimension of the REDD+ project in SSA and SEA which is based on the idea that majority of subsistence farmers are women and are thus more likely to be affected by climate change.

2. Materials and Methods

2.1. Profile of Sub-Saharan Africa (SSA)

Sub-Saharan Africa (SSA), with the total human population of 926 million, politically encompasses forty-eight (48) countries fully or partially located south of the Sahara (PRB, 2013). The region is endowed with vast amount of natural resources ranging from forestry, minerals, energy, and wildlife amongst others. A huge percentage of the region's total population habituates in rural areas where agriculture is the main livelihood activity. Smallholder agriculture is the dominant form of farm organization in this region (FAO, 2009). It is thus no wonder that agricultural income provides the largest source all incomes and offers employment to about 65% of the region's population.

High population growth in SSA has predisposed the region to certain threats or challenges. Increasing population implies rising percentage of dependents in the population and thus a heavier burden on the working group. This has heightened the need for families to undertake all forms of activities, both legal and illegal, to acquire income for survival.

Such activities include agriculture, illegal and destructive mining and logging practices which has led to large scale clearing of land as well as caused severe damage to other natural resources (Naoto, 2006). According to Naoto (2006), SSA has one of the highest rate of deforestation; an activity which is believed by some to be caused by poverty.

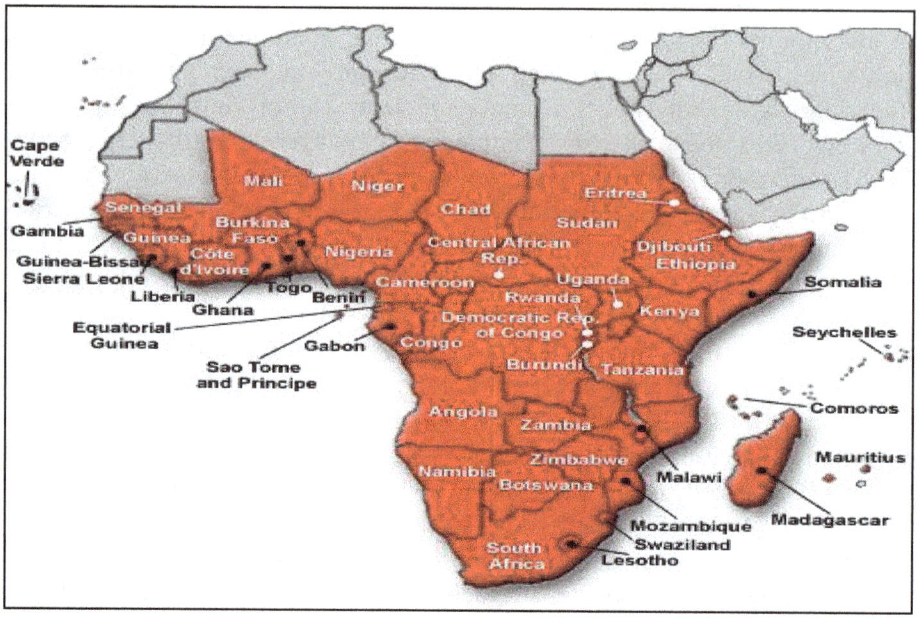

Figure 1: Map of Sub-Saharan Africa

Although Africa has some of the lowest annual greenhouse gas emissions that cause global warming, the impacts of climate change on SSA is very high. This is particularly due to its weak institutions, poor planning and limited finance and information capacities to adapt (WFSE-IUFRO, 2009). Environmental services, such as carbon sequestration, therefore, offer the continent opportunities in the form of the growing carbon markets, by virtue of the continent's rich forest land cover. These forest resources increase the region's potential for REDD+ (aimed at reducing emissions from forests) and its associated benefits (enhanced livelihood for smallholder farmers and protection of biodiversity) (WFSE-IUFRO, 2009). The UN-REDD has played a pivotal role in the development of REDD+ policy framework in many countries in SSA, with countries like Ghana, Cameroon, Tanzania, Benin, Gabon and Zambia (Spratt, 2016; Shanahan et al., 2013).

2.2. Profile of Southeast Asia (SEA)

About 300 million people reside in rural parts of Southeast Asia, with around 70 million relying on forests for their livelihoods, fiber, fuel and nutrition (ASEAN Swiss Partnership on Social Forestry and Climate Change, 2015). Despite the fact that forests cover about 50% (734 million hectares) of Southeast Asia's land area, the region has the highest rate of deforestation and forest degradation in the tropics as a result of destructive activities such as logging (mostly in Papua New Guinea, Indonesia, Myanmar, Malaysia and Cambodia); agriculture - notably oil palm (Wertz-Kanounnikoff and Kongphan-Apirak, 2008; ADB, 2010). Poverty remains prevalent in forested areas due to limited and, or absence of basic services and markets (ADB, 2010). Indonesia has been recognised as one of the top ten emitters of greenhouse gases in the world -with land use change and forestry accounting for 65.5% of its total emissions as at 2013 (USAID, 2017) - largely due to peatland degradation, deforestation and forest fires. The region's vast forest land cover – which serve

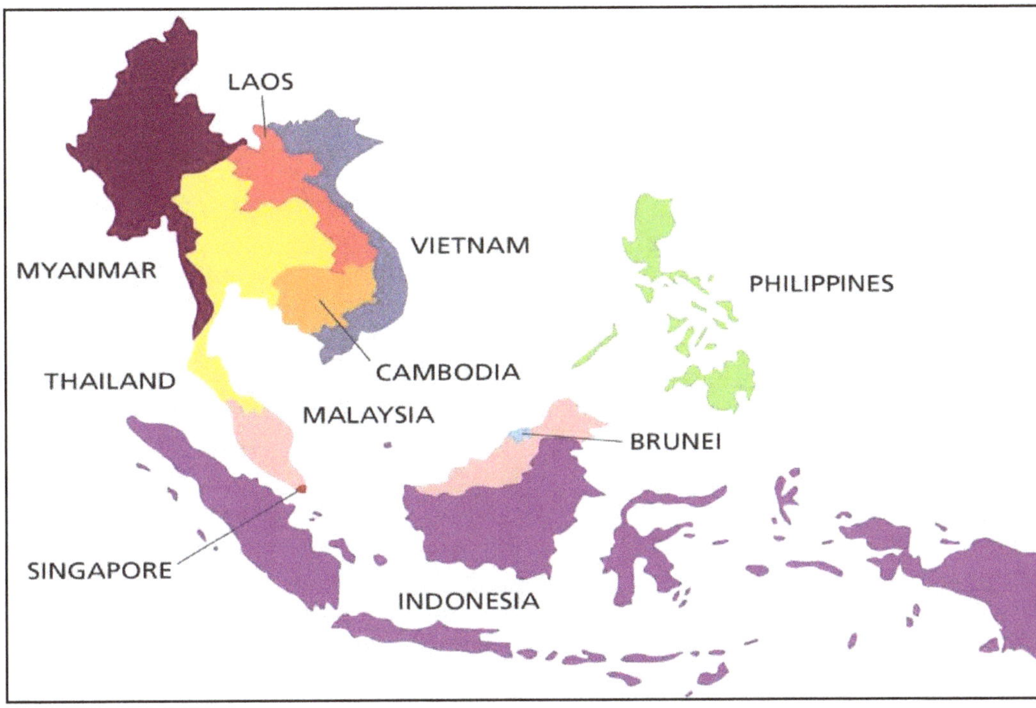

Figure 2: Map of Southeast Asia as significant carbon sinks - offers huge potential to benefit from REDD+ by facilitating 40% of the scheme's ability to reduce carbon dioxide emissions by 2050 (ADB, 2010).

2.3. *Methods*

The conceptual foundation upon which the discussions were based was on three facts: 1 Forests cover about 50% of the land cover of Southeast Asia and Africa is home to 25% of the world's remaining rainforests. 2 Both regions are highly vulnerable to the brunt of climate change. 3 A large percentage of their population rely on the forests for their livelihoods, ultimately increasing the rate of deforestation (ASEAN, 2015; World Bank Africa, 2017). This presents a firm case for the regions to access funding to promote mitigation and adaptation strategies against climate change (Nakhooda et al., 2011; Kula et al., 2013).

Data attained specifically emanated from project report documents, articles, institutional websites as well as government documents on relevant topical issues with bearings on the subject matter under discussion. The selection of the countries with REDD+ projects was predicated on the principles of acceptability, implementation successes and failures as well as the premium put by each country on the REDD+ mechanism as a potential climate change mitigation measure. A combination of keywords related to REDD+ in sub-Saharan African and Southeast Asia were used to retrieve the relevant works, of which include 'climate change' 'climate impacts' 'adaptations', 'climate change mitigation', 'forest management', 'REDD+', 'community development', 'benefits sharing' and 'gender'. The results of the top 20 literature out of the total of 131 reviewed and referenced works, under various thematic and spatial scopes with special reference to Sub-Saharan Africa and Southeast Asia are displayed in table 1.

Analysis and discussions have laid out the tripartite benefits that serve as community development, enhancement of alternative livelihoods poverty reduction, and gendered ramification, from REDD+ project in SSA and Southeast Asia.

Table 1: Top 20 relevant references synthesized as the core of the paper

SN	Thematic area	Author	Scope	Indicators of REDD+ for mitigation successes
1	REDD+ and Gender	Nhantumbo and Chiwona-Karltun (2012)	Global	*'Deliberate choice'* to make payments directly to women was necessary for the success of the REDD+ programme in Amazonas, Brazil.
2	Trends in global CO_2 emissions	Olivier, et al (2013)	Global	REDD+ involves the storage of carbon that would have otherwise been emitted into the atmosphere.
3	National REDD+ Strategies	Asian Development Bank (2010)	Asia	Stakeholder *'partnership'* among countries within regions enablied dialogue on best practices, safeguards and coordination of REDD+.
4	Governance of forests	Agrawal et al. (2008)	Global	*'Participation'* from local communities in the management of local forests enhanced the implementation.
5	Forest and People	Parrotta et al (2012)	Global	*'Proper management actions'* (such as reduced impact logging (RIL)), promotes REDD+ implementation successes.
6	Kariba REDD+ project	Gogo (2014)	Africa	The factor of *'investments in training and education'* of farmers is essential for alternative livelihoods trade-off for REDD+.
7	Gender and REDD+	Setyowati (2012)	Global	Women *'capacity-building'* in literacy, numeracy, advocacy, negotiation, community-organisation and skills inures to the successes of REDD+ projects.
8	REDD	Hall (2014)	Global	REDD+ must not ignore *'local people's links to, and knowledge of forests'*.
9	Intergovernmental Panel on Climate Change	IPCC (2007)	Global	REDD+ raises *'awareness'* and has played a role in the evolution of many national forestry policies.
10	Forest carbon offsets	Kill (2013)	Global	*'Local interest'* is necessary in order to avoid leakage (people going elsewhere to harvest trees or grow crops they could no longer grow in the project area).
11	Drivers of deforestation	Kissinger et al (2012)	Global	REDD+ has promoted *'sustainable practices'* such as agroforestry and afforestation to address demand for fuel wood and to increase carbon stocks.
12	Land tenure and REDD+	Larson et al (2013)	Global	A defining factor of *'land tenure'* is crucial in ensuring the success of REDD+ projects, in developing countries.
13	REDD+ and Gender	Livingston (2015)	Africa	*'Women's groups'* have enabled a spreading boon of economic growth in the Kasigau region.
14	Benefit sharing mechanisms of REDD+ projects	Madeira et al (2012)	Africa	The factor of *'legislation and policy'* instruments are needed to governs land tenure, revenue sharing etc., for smooth REDD+ implementation.
15	Carbon markets and REDD+	Pye (2012)	Southeast Asia	REDD+ has brought to the fore, the need to safeguard the *'rights of indigenous people'* in vulnerable societies.
16	Livelihood implications of REDD+ projects	Mutabazi et al (2014)	Africa	The factor of *'innovation'* in terms of REDD+ project encourages agricultural intensification for enhanced productivity.
17	Climate change	IPCC (2014)	Global	Mitigation strategies should be *'integrated into national development policies'*.
18	Impacts & adaptations of climate change in developing countries.	UNFCCC (2007a)	Global	Developing *'communication strategies'* to make mitigation strategies accessible to all.
19	Climate change mitigation	Virgilio and Marshall (2009)	Global	Providing *'incentives'* for conservation and better forest management is an ideal way of mitigating climate change.
20	REDD+ and Local Democracy	Ribot (2011)	Global	The factor of *'democratic representation of local populations'* at all decision levels on forest governance is a must for REDD+ to be a tool of justice.

3. Results and Contextual Discussions

3.1. REDD+ and community development

One of the main challenges of forest conservation practices is keeping the local populace from undertaking activities that are detrimental to forests but vital for their livelihoods. There seems to be a conflict of interest especially in cases where the people have not been offered sustainable livelihood alternatives which would otherwise restrain them from resource-degrading activities (TFD, 2008). A large proportion of sub-Saharan Africans and Southeast Asians depend heavily on forests for their subsistence, income generation and energy, accounting for the high rates of deforestation in the continent (TFD, 2008; ASFN, 2011).

A study by Madeira et al (2012) indicated that efforts to manage forests sustainably must endeavor to address the root causes of deforestation and forest degradation. According to a report by Funder et al (2009), common causes of deforestation include smallholder agricultural expansion, logging and infrastructure development. Since forests therefore serve as a vital element in the livelihoods of local communities, action must be taken to support a pro-poor forest management mechanism which will consider the needs of the people in its operations.

Many developing countries believe that REDD+ projects have great potential to raise funds for poverty alleviation (Kanninen et al., 2010). Thus, the crucial question is; how can REDD+ benefit the poor? Giving local people's control and utilization power over forest resources deforestation has been the outcome of the over use of the forest resources. Deriving some incentive from benefits sharing of REDD+ is imperative to reduce the deforestation and forest degradation, better and more profitable than cutting down of trees. The REDD+ initiative have potentials to serve as a scheme that provides more benefits both in socioeconomic and ecological sense, than deforestation (Romero et al., 2013). Monies accrued from REDD+ projects are given to participating communities as Payment for Ecosystem Services (PES) in project areas; mostly smallholder farmers to undertake alternative livelihood activities, in a bid to reduce pressure on forests for their sustenance (IUCN, 2011a). This initiative is vital for providing the necessary incentives to change deforestation and forest degradation behaviours, ultimately reducing carbon emissions (Brockhaus et al., 2013; Funder et al., 2009).

Peskett (2011) views benefit sharing as a system that involves the distribution of potentially large financial benefits among all stakeholders linked to minimization of deforestation, degradation and forest regeneration. Aside generating material opportunities for better livelihoods, REDD+ programs have ensured the safety of local population in terms of food and water security. It has also, enhanced land and tree tenure security and improved adaptability to climate change as well as facilitated the enablement of communities to participate in environmental decision making activities (Madeira et al., 2012). Researchers from Sokoine University of Agriculture, Tanzania found that people in the area of Kilosa REDD+ pilot project, preferred alternative forms of compensation such as employment opportunities, infrastructure, and seedlings for woodlots outside protected forest sites, over direct payment (Kajembe et al., 2013).

Emphasis on this was further made by Nabanoga et al. (2012), in their study of poverty impacts of Ongo REDD+ pilot project in Uganda. This study noted that more than 80% of respondents of study villages preferred indirect sources of payment as compared to 60% of respondents who agreed to direct payment. This was based on apprehension that direct compensation may be inadequate. One example of the benefits accrued by REDD+ projects is the Mbire council; one of the communities under the Kariba REDD+ project in Kenya. According to the council's chief executive, about 300 farmers receive inputs for conservation farming each year and about 47 boreholes have been sunk to aid improvements in health systems by delivering clean drinking water (Gogo, 2014). Meanwhile, coffee growers within protected forest areas in Sumberjaya, (a sub-district in Western Sumatra, Indonesia) could receive rewards in the form of land tenure as long as they comply with the necessary cultivation and conservation practices (Wertz-Kanounnikoff and Kongphan-Apirak, 2008).

REDD+ projects would inevitably have an impact on how much land a farmer will be able to access since the projects involves conserving large plots of land (Larrazabal et al., 2012). Farmers who once harvested on that land will lose their jobs and may be forced to clear forests outside the protected site in a bid to cultivate food crops and this threatens the sustainability of the project (Mutabazi et al., 2014; Larrazabal et al., 2012). There are fears that agricultural restriction in lieu of REDD+ projects, could have adverse implications for food security and other products from agriculture (Dooley and Chapman, 2014). In the Southeast Asian region, it is believed that unsustainable practices and land ownership laws will likely serve as roadblocks to agricultural productivity and lead to the malnourishment of about 60 million in the region (OECD, 2017). A study by Pokorny et al (2013) reports that environmental and local development successes of major forest carbon strategies such as REDD+ are in themselves limited.

It is on this basis, that recent attention has been given to agroforestry as a burgeoning aspect of REDD+ projects especially in sub-Saharan Africa where agricultural activities are predominant and a primary agent of deforestation. Minang et al. (2013), reports that about 40% of eleven countries in Africa have at least two agroforestry-based REDD+ strategic options. Agroforestry can be described as a climate-smart agriculture intervention that integrates tree cultivation with agricultural production (Minang, 2010; Dooley and Chapman, 2014). It is a very necessary REDD+ scheme which helps to decrease pressure on forests for agricultural land use and fuel; it improves soil fertility; controls soil erosion as well as mitigate carbon dioxide emissions (Murthy et al., 2013). Though this component (agroforestry) sequesters very minimal amount of carbon as compared to protected forests, it helps to provide additional livelihood benefits, food security and save forests that would have otherwise been cut, especially areas outside the protected REDD+ sites (Kill, 2013). Furthermore, Gockowski and Sonwa (2011) have shown that the integration of cacao with timber tree species in the Guinean rainforest of West and Central Africa could have spared 21,000 km2 of forests; reduced emissions of about 1.4 billion tCO2 and also provided income. Portal Limited is an agroforestry business in Ghana that incorporates potential carbon income to earn source of income, thereby making the project a sustainable source of revenue for the local populace (Metzel, 2015).

3.2 Specific sub-regional cases of REDD+ projects

Although REDD+ projects have had positive socio-economic outcomes, wrongful management of this intervention can have several adverse impacts on the vulnerable group, that is, the poor. Research in this context has highlighted a number of complaints about various REDD+ projects in Sub-Saharan African and Southeast Asia. According to Shames (2013) and Hall (2014), the unstable price or value of carbon on the carbon markets has played a pivotal role in the unsuitability of REDD+ as a source of finance for forest conservation. Several complaints have been issued by farmers concerning the varied nature of payments and many have resented the fact that carbon prices are determined by global commercial structures that see the former (farmers) as pawns. Studies by Lohmann (2006) and PACJA (2012) as cited in (Lang, 2015), have shown that the low value of carbon on the global markets may not be enough to curb deforestation as locales may find it difficult to subsist on such meagre allowances and may be forced to resort to agricultural and logging activities in other forests, in a process called ‘activity shifting’ or more popularly, leakage. A research by the center for international forestry research (CIFOR) and centre de coopération internationale en recherche agronomique pour le développement (CIRAD), revealed that the current prices on the carbon market bring limited additional benefits, which threatens sustainable forest management efforts (Baxter, 2015).

Furthermore, the vulnerability of forests to future climatic events, fire and other degrading activities, creates a situation where receipts of payments to locales have become erratic. This notion is emphasized by a group of people representing the ‘No REDD in Africa Network’; a group which seeks to put an end to ‘carbon colonialism’. According to Nnimmo Bassey, former Executive Director of environmental rights action (ERA), the Nigerian chapter of Friends of the Earth; REDD+ in Africa has emerged as a neocolonialist cliché and a major driver, towards land grabbing (Lang, 2013).

The above opinion can be argued when looking at some current projects, specifically, the much talked about N’hambita Community REDD+ Carbon Project taking place in Mozambique.

3.2.1. Mozambique

According to Hall (2014), this REDD+ project in the Sofala region of Mozambique, (originally established in 2002 by Environtrade), represents a bitter case of a REDD+ project gone bad. A study of the project found that farmers had signed contracts that committed them and their children to grow and tend trees for a period of 99 years, albeit the fact that payments would be made for only a seven-year period. Many of the farmers claimed to be oblivious of any such commitments after the seven-year period and assumed that new contracts would be signed thereafter. An examination of a farmer's contract by Via Campesina (the world's largest peasant movement) found that a farmer was to be paid $128 per annum over seven years, for tree planting (Hall 2014). It is clear at this point that the farmer is being paid a pittance for protecting trees for such a long period and this will not result, to any improvement in his standard of living. Some farmers mentioned that they had lost the motivation to tend the trees due to late payments and discounts. Many of the farmers attested to the fact that they might cut down the trees after the contracts ended for use, or market as timber or fuelwood and charcoal, as they regarded the wood as a future benefit of the project (Hall 2014). The project was also exposed to financial difficulties due to the decrease in carbon prices. Despite the negative impacts of the project on their livelihoods, some farmers noted that there had been some benefits in relation to fruit trees as food supplements and incomes, as well as the construction of health centres, derived as additional indirect benefit (Hall, 2014).

3.2.2. Kenya

Individuals in Binga, one of the four rural communities covering the Kariba REDD+ Project in Kenya have claimed that they have not received any payments and therefore accuse Carbon Green Africa (CGA) of taking advantage of them (Gogo 2014). According to Gogo (2014), CGA, provides inputs to farmers and trains them in alternative livelihoods such as beekeeping and conservation farming. One distraught councilor noted that only 20 farmers out of about 4000 people have benefited from the input supply. The REDD+ programme in this community is characterized by lack of consultation with the locales in that the latter is kept in the dark in matters relating to the amount of credits sold by the project managers and the amount that is due the community from those sales. The CGA chief however retorted that the community gets the least amount of money because it has the least portion of land under the project (Gogo, 2014).

Furthermore, the UNEP-funded Mau forest (REDD+) project in Kenya has been engulfed in serious allegations of human rights abuse. The use of the Mau forest for REDD+ projects was necessitated by the forceful removal of inhabitants from the Mau Forest Complex which led to the displacement of Ogiek original inhabitants, from their traditional abodes. The Mau forest covers 40,000 ha in the Narok District, Rift Valley Region of Kenya (Anaya, 2010). Some of the reasons why the people have failed to embrace this REDD+ programme are that they were not consulted, nor were their consent sought; as negotiations were made behind closed doors. Also they were not given any alternative livelihood options or compensation for their removal despite the fact that they have occupied those lands for centuries (Anaya, 2010). This project ignored an important cultural and social aspect of the indigenous people's relationship with the forest, in this case, their hunter-gatherer lifestyle. It is a well-known fact that the forest industry in most African countries is weighed down by weak governance, poor land tenure systems and corruption. Hall (2014), stresses that REDD+ projects are likely to exacerbate existing land disputes especially in cases where REDD+ project sites overlap with land rights of forest communities.

3.2.3. Indonesia

The Ulu Masen Project in Indonesia was primarily aimed at ensuring the sustainable use of forests and ultimately selling millions of tonnes of carbon on the international voluntary carbon market. Payments were projected to be reach $26 million over the first five years (Wertz-Kanounnikoff and Kongphan-Apirak, 2008). However, according to some locales who were questioned about the scheme, no benefits nor payments have been attained from the project; and the scheme has failed to sell a single carbon credit even after five years. Also, according to Chris Lang, there are still some people who have no idea what the project is all about (Pye, 2012).

Another major problem of REDD+ projects is the conflict between indigenous people and the government concerning land ownership and the failure to consult locales on activities relating to the project (Henshaw and Fyneface, 2014). There has been an increasing trend of decentralized forest governance in the tropics, where local communities and users of forest resources control an additional 200 million ha of forests as compared to the 1980s (Agrawal et al., 2008; Sandbrook et al., 2010).

3.2.4. Cameroon

The forestry law in Cameroon for instance, enables rural communities to access forest resources in and around their village (Minang, 2010).

Despite this feat however, there continues to be growing disquiets about the adverse outcomes of forest carbon projects like REDD+ schemes that weaken the resource rights of forest-dependent communities, and possible displacement from the fringed community. This is reiterated by Barbier and Tesfaw (2012), who noted that REDD+ initiatives could dispossess and marginalize forest communities especially in cases where tenure security is weak.

3.2.5. Vietnam

The nature of the land tenure system in Southeast Asia makes it difficult for the local populace to fully enjoy the benefits of REDD+ scheme. The ignorance of traditional land rights by government policies make schemes such as REDD+ since local landowners have only limited scope to enter such contracts.

According to Wertz-Kanounnikoff and Kongphan-Apirak (2008), the uses to which land is put to in Vietnam is largely determined by the government, and thus, even though households can enter into such contracts, freedom to make land-use decisions is limited. This highly restricted nature of land rights and land use means that the success of any REDD+ project in this region will be dependent on long term government cooperation.

3.2.6. Nigeria

According to Rita Osarogiagbon's presentation at the Cross River State forum in Nigeria, the operations of REDD+ projects can be likened to the concept of "a sheep in wolf's clothing" in that, rather than the project reducing emissions as the name depicts, it simply moves emissions around and at times triggers the release of more emissions through leakage. This is usually the case when the rights of forest-dependent communities are undermined (Osarogiagbon, 2011).

In a related instance, residents of Mbe Mountain Cluster Communities have classified REDD+ as another form of land grab, in that, the REDD+ project failed to take consideration of the people's exclusive dependence on the forests for their livelihood. Henshaw and Fyneface (2014), argues that aside $3000 made available to them, the community has not benefited from the scheme in any way. Furthermore, the people have been banned from all activities within the protected area with no alternative livelihoods provided. A similar situation in Akampkpa- the Ekuri Cluster and Akpabuyo has engendered a tremendous rise in illegal logging and other criminal activities. Henshaw and Fyneface (2014) further noted that big companies have benefited largely from REDD+ initiatives in terms of large scale land acquisitions at the expense of local communities.

The World Bank has clearly acknowledged that recognizing the rights of indigenous people and forest communities to land resources is fundamental to the success of any REDD+ intervention. However, most REDD+ projects seem to do the opposite as governments seek to forcibly take control of lands owned by local communities (Norton Rose, 2010; Dooley et al., 2011).

3.2.7. Tanzania

Many forest-based communities find it difficult to entrust their governments with such projects. This is emphasized by a socioeconomic survey conducted by Kajembe et al (2013) in the Kilosa District REDD+ Project in Tanzania which reported that a few respondents felt that compensation was not the appropriate approach to make them stop clearing and using forests because of the lack of trust amongst others (Kajembe et al. 2013).

3.2.8. Uganda

In Uganda, a project in the Bukaleba forestry reserve, aimed at offsetting GHG (greenhouse gas) emissions from a coal-fired power plant to be built in Norway is a clear indication of a case of *'Carbolonialism'* according to a report by journalist Harald Eraker. The once-off fee of US$410 coupled with a paltry $4.10 per hectare of plantation received by the Ugandan government, was extremely low lease in comparison the the volumes of carbon emission envisaged to emanate from the said coal plant. The project is also responsible for the eviction of about 8,000 people living on the land, dispossessing them of their livelihoods and possibly driving them to degrade lands elsewhere (Lohmann, 2006; Institute for Security Studies, 2011).

It is a truism that communities' engagement in REDD+ plans would help to control leakage since they will be incentivized to maintain healthy forests, thereby ensuring permanence and sustainability of the project (Larrazabal et al., 2012; Hess, 2014; Newton et al., 2015). This opinion is further supported by Chhatre et al (2012), who notes that effective participation at the local level may go a long way to engender or stimulate transparency and accountability in terms of corruption and weak enforcement of the law. Ribot (2011) argues that measures to ensure participation of forest-based communities in REDD+ implementation is not enough and that much attention must be turned towards ensuring local democratic decision-making.

3.2.9. Papua New Guinea

The situation in Papua New Guinea make for an interesting case that highlights the consequences of not being aware of one's tenure rights. Landowners who are unaware of their rights are easily exploited (with some signing over carbon rights) by certain carbon project developers - popularly known as 'carbon cowboys' by the media- (Larson et al., 2013). According to Babon et al (2012), one of the 'carbon cowboys' claimed to have made negotiations concerning 90 different carbon deals with various landowners, notwithstanding the absence of a national REDD+ strategy. This scandal brought attention to the dangers of REDD+ for communities and highlighted the need for REDD+ projects to protect the rights of customary landowners.

It is also essential that REDD+ developers make a note of the many ways in which forest communities exchange information and resources in order to ensure proper planning and design of REDD+ systems (ASEAN Swiss Partnership on Social Forestry and Climate Change, 2015). Local communities in Indonesia, Vietnam and Laos have been found to have significant knowledge of forest resources and management, and can measure carbon effectively (ASEAN Swiss Partnership on Social Forestry and Climate Change, 2015). Workshops were held in Kenya and Uganda in 2009 as part of stakeholder consultation process to draw up REDD+ Readiness proposals for the respective countries. A report of the workshop by Julian Sturgeon and Sena Kanyinke, showed that the participants had indeed acquired massive knowledge about the REDD+ process and therefore appreciated that the project had strategic potential not only for carbon sequestration but also for poverty alleviation (IPACC, 2011).

3.3 Gendered dimension of REDD+

According to Demetriades and Esplen (2008), UN Women Watch (2009), Brown (2011), Annecke and Koelle (2011), women in rural parts of developing countries are regarded as the most vulnerable to climate change. This is largely, consequential from the pervasively high rates of poverty among women, as compared to men; thereby exacerbating gender inequalities. According to UNDP (1995), women represent 70% of the poor across the world, and this trend has remained unchanged. Research has found that women comprise about 43% of the agricultural labour force both globally and in developing countries (Doss, 2014), and produce more than 50% of the world's food (FAO, 2011). UN-FAO (as cited in Nhantumbo and Chiwona-Karltun, 2012), estimates that women constitute nearly half the agricultural labour force but own only 15% of the productive land.

Thus, since women play a role in deforestation and degradation especially through agriculture, it is only expedient that mitigation strategies such as REDD+, mechanisms are gender mainstreamed and integrated into their agroforest resources management, aside their adaptation responses to the impacts of climate change.

Gender mainstreaming is explained by United Nations Economic and Social Council (UN ECOSOC) to mean the co-ordinated mode of political, social economic strategies of identifying men and women as integral part of the design, implementation, monitoring and evaluation of policies in all societal spheres, for equal benefits (Campese, 2011). According to Setyowati (2012), women have a right to access REDD+ benefits and partake in decision making processes on the basis of the vital role they play in forest management of which include tendering of trees on farms, gathering of wild plants for food and medicines, collection of honey and firewood for household consumption and even forest patrolling and monitoring for conservation and preservation of tree species. This is because, men and women use the forests and forest resources differently and may, thus, experience differently, the effects of climate change and REDD+ policies (Sills et al., 2014).

However, international agencies and governments largely underestimate women's contributions to forest management and ignore their needs and perspectives. But it must be noted that attention to REDD+ as a gender- responsive climate mitigation policy has increased considerably over the last decade as women are being placed at the center or heart of climate change policy discourse (Rietbergen-McCracken, 2011; Okali and Naess, 2013). This attention has been largely based on the insecure access to and limited property rights of women over forests and productive land as well as the exclusion of women in decision making processes on climate change; all of which have been recognized as major drawbacks to gender equality in the forestry sector and the implementation success of REDD+ initiatives (Ndobe, 2010; Mwangi et al., 2011, Setyowati, 2012; Okali and Naess, 2013). In the case of Nepal for instance, women's authority over management and use of forest benefits are not recognised reiterating the fact that discrimination against women is not just limited to culture and norms but extends to institutional arrangements (Basik et al., 2012).

Some of the menaces of REDD+ for women according to Gurung et al. (2011), include the displacement from forest lands, the inability of women to receive their fair share of the benefits as well as the frequent exclusion of women from consultations and capacity building workshops. Gurung (2011) further stresses that the lack of integration of gender perspectives into REDD+ implementation will have deleterious effects on the long-term sustainability of REDD+. Studies show that the strong connection of women to natural resources such as forests makes them the most efficient managers of the environment (Adeniji, 2011; Mwangi et al., 2011; Okali et al., 2013). Women in rural areas - as primary forestry users - tend to have unique local knowledge systems that can provide both environmental and social benefits from REDD+ (Setyowati, 2012). Environmental benefits include ecosystem services such as conservation of forest biodiversity, soil nutrient and moisture conservation, water regulation and many others, whist expected social and economic benefits include poverty reduction, improved livelihoods, gender equality and inclusive in decision making and governance (UN-REDD, 2013).

Whilst some countries in Africa have not done much to improve the incorporation of gender perspectives in its REDD+ policies, others like Mozambique, have made considerable attempts at integrating gender in the context of REDD+ implementation. The survey conducted by Micaia foundation, an NGO in Manica included gender-defined information on the causes of deforestation and degradation, in order to identify 'who does what?'; so as to establish policies that will help reduce emissions (Nhantumbo and Chiwona-Karltun, 2012). In similar manner, progress has been recorded in Nepal where community forestry policies were amended in 2009 to ensure that women accounted for 50% of community forestry executive committees.

In another vein, the country has been actively taking measures to address gender equality and exclusion in the forestry sector (FAO and RECOFTC, 2015). Accessibility to extension services, technological information and innovation, credits and markets have been hindered by low literacy rates among women in SSA, among other factors; all of which limits their ability to undertake activities that support REDD+ (Nhantumbo and Chiwona-Karltun, 2012). Also, the male and state dominated nature of tenure rights in SSA, makes it difficult for women to legally own lands and this has negatively affected them when it comes to sharing the benefits of REDD+ initiatives. This is because, in most cases, tenure rights determine who will be eligible for REDD+ benefits. Gender issues in land tenure must, therefore, be addressed in order to ensure sustainable REDD+ outcomes; otherwise women will continually be sidetracked.

In order for women to enjoy the benefits of REDD+, the benefit sharing mechanism used in REDD+ projects must be designed and suited to meet the specific needs of women in a bid to empower them economically and socially (Peskett, 2011). Gender analysis tools were designed to monitor the gender sensitivity of forestry related programmes such as the Community-Based Forest Management Strategy (CBFM) in the Philippines. These tools assess the specific needs of men and women, and the level of participation of stakeholders in all aspects of a project (design, implementation, monitoring and evaluation). The programme has received much accolade from the international community due to its gender-inclusive approach (FAO and RECOFTC, 2015). Poffenberger (2011) recognizes that the problem of making REDD+ payments accessible to women may be surmounted by linking REDD+ funds management to women-centered micro-finance supported activities.

Studies indicate that supporting women-centered groups in REDD+ project sites can help address gender equality, ultimately resulting in the effective functioning and long-term sustainability of forest management enterprises (Agrawal, 2001; CIFOR, 2012; Djoudi et al., 2012; Setyowati, 2012; WOCAN et al., 2013; Sills et al., 2014; Larson et al. ,2015). A case in point is the Kasigau Corridor REDD+ project in Kenya, according to Livingston (2015), which provided sustainable livelihoods for people to the extent that women have been recognized as effecting major changes in the communities whilst protecting forestlands.

According to the report by Livingston (2015), women's groups engage in alternative livelihood activities such as production of arts and crafts like basket weaving, stuffed animals, jewelry and many others; all of which are sold in the United States, via REDD+. Monies from these projects have been used to build water tanks for communities, bought for households, solar lighting and clean cook stoves as well as delivering education for their children. Additional opportunities include the inclusion of women into the Wildlife Works Ranger Corps to protect forests from slash and burn agriculture. Women have also joined the projects' eco-charcoal making team. This is a clear illustration of the consideration of women into the benefit sharing initiative of REDD+ which ultimately empowers them both economically and socially. A report by Wambugu et al (2015) on the other hand, indicated that there was low representation of women in decision making activities relating to the Kasigau REDD+ project.

Studies have also shown that women in forest communities are mostly ignored and restricted when it comes to making decisions concerning forest management (Gurung, 2011; Nhantumbo and Chiwona-Karltun, 2012). Several reasons account for this of which may include the fact that women spend most of their time taking care of their households and are thus disregarded in matters relating to forests whilst men are more mobile and have wide-ranging network (Annecke and Koelle, 2011). Consuelo Espinosa, a Senior Forests and Climate Change Officer of the international union for conservation of nature (IUCN), noted that women are usually marginalized on discussions pertaining to forest management, due to constraints to own forest lands (IUCN, 2010). Women, however, must have equal representation and influence in decision making as men in so as to obviate any adverse impacts of REDD+ on women's livelihoods.

Larson et al (2015), in a study of 20 REDD+ sites across six countries, pointed out that REDD+ projects only focused on the participation of women in numerical terms without considering whether the women feel included and if their participation influenced decisions. This was reiterated by Nhantumbo and Chiwona-Karltun (2012), who opined that women may not have the opportunity to freely express their views due to socio-cultural norms and this may discourage them from engaging in future meetings. In contrast to this, however, a study undertaken by Awono et al (2014), reported that women in REDD+ project sites were actually playing major roles in matters relating to the management of natural resources. According to the report, a considerable number of women in Mount Cameroon had statutory positions in local forest management committees.

REDD+ initiatives should be formulated to facilitate women's meaningful participation rather than simply inviting them to discussions. Campese (2011), highlights some of the strategies of which include organizing women only meetings to help them prepare for discussions; posting information in places where women congregate; framing information in ways that addresses their interests and rights; providing gender specific training in traditionally male activities as well as forming task-sharing groups in order to offer adequate participatory time for women to discussions without recourse to any overburden.

Countries across sub-Saharan Africa and Southeast Asia have taken into consideration the importance of gender mainstreaming into their respective REDD+ policies and implementation. The launch of the Gender and REDD+ roadmap in Ghana was viewed as a stepping stone towards the enhancement of the participation and contribution of women and other vulnerable groups to the forestry sector (IUCN, 2011b). Key stakeholders (such as gender officers from government ministries, representatives-women and men- from participating communities and women organizations in the forestry sector) were trained to understand the differentiated impacts of climate change on men and women, the different roles of men and women in forestry and what needs to be considered gender-wise for the success of REDD+ initiatives (IUCN 2011b). But many of them still remain gender-blind without much elaboration on what should be practiced (Smith, 2011). Mention can be made of National REDD+ proposals of countries like Kenya, Tanzania and Cambodia which does not mention specific strategies to integrate gender into proposed implementation strategies and fails to address the issue of women's participation in decision making respectively (Smith, 2011; WOCAN ,2011).

4. Observations and Conclusion

The arguments in the literature on the social economic benefits of REDD+ thus far, seem to point to a rather negative consequences of the potentials of the scheme in general. However, this paper counter-argues in sync with the views expressed from the positive experiences enumerated from the success stories emanating from the SSA and SEA sub-regions, that indeed REDD+ holds the potential to bring about sustainable community livelihood development. This would be particularly realized, when it is embraced by forest-fringed communities as an integral part of their subsistence livelihood engagements. This is because, the opportunities of an integrated human-ecosystem management from an agroforestry and other forestry programs perspectives hold potentials for climate change mitigation, using possible carbon emissions abatement as well as an adaptation strategy through the generation of alternative incomes from tree crops and carbon credits.

Despite the fact that women are the most vulnerable groups to climate change, measures must be taken to acknowledge the potential of women as effective agents or actors of climate mitigation and adaptation actions this would hold especially so, since they are the core users of natural resources. Thus the skills, roles and knowledge of women should be adequately tapped in considering holistic efforts to manage forest resources in a sustainable manner since the adage that *"when the last tree dies, the last man dies"* is not far from the reality.

Author Contributions

Divine Odame conceived of the study, helped in the design of the study, proofread and edited the final text. Stephanie Gbeddy reviewed literature from various sources and drafted the document. Both authors undertook discussions of the key issues raised in the paper.

Conflicts of Interest

The authors declare that they have no competing interests.

Acknowledgments

Acknowledgements are due to the Department of Geography and Rural Development (KNUST) for providing us with the opportunity to undertake this research project.

References

Adeniji, G. (2011). *Women as key players in climate adaptation*. Jotoafrika, issue 6

Agrawal, B. (2001). Participatory Exclusions, Community forestry, and Gender: An Analysis for South Asia and a Conceptual framework. *World development (29)*.10, p 1623-1630.

Agrawal, A., A. Chhatre., and Hardin, R. (2008). Changing Governance of the World's Forests. *International Forestry Resources and Institutions Program (IFRI) Working Paper #08I-4*

Anaya, J. (2010). *Kenya: Alleged eviction of the Ogiek Indigenous peoples from the Mau Forest Complex*. Support Project for the United Nations Special Rapporteur on Indigenous Peoples

Annecke, W. and Koelle, B. (2011). Including women in adaptation processes. *Indigo Development and Change, jotoafrika* issue 6

ASEAN Swiss Partnership on Social Forestry and Climate Change. (2015). Social forestry and climate change in Southeast Asia. CIFOR Research findings and goals. *Center for International Forestry Research (CIFOR).*

[ASFN] ASEAN Social Forestry Network. (2011). *ASEAN-Swiss Partnership on Social Forestry and Climate Change*. Program Document 2011–2013. Jakarta, Indonesia.

Asian Development Bank (ADB). (2010). National REDD+ Strategies in Asia and the Pacific: Progress and Challenges. Mandaluyong City, Philippines. p 1- 4.

Awono, A., Somorin, O. A., Atyi, R. E., Levang, P. (2014). Tenure and participation in local REDD+ projects: Insights from southern Cameroon. *Environmental Science & Policy. (35)*, p 76-84

Babon, A., Mcintyre D., and Sofe, R. (2012). REDD+ politics in the media: A case study from Papua New Guinea. *Center for International Forestry Research (CIFOR)*. Working Paper 97.

Barbier, E.B. and Tesfaw, A. T. (2012). Can REDD+ Save the forest? The Role of Payments and Tenure. *Forests, 3,* ISSN-1999-4907, P 882-885

Basik, N., Molnar, A. DiPaolo, J. and Qureshi, M. (2012). *Leadership for Forest Management: A Summary of the Asian Experience.* Rights and Resources Initiative. Jakarta, Indonesia.

Baxter, J. (2015). REDD+ to the rescue of Central Africa's forests? Not yet, study says. CGIAR Research Program on Forests, Trees and Agroforestry. *Center for International Forestry Research (CIFOR)*

Brockhaus, M., Luttrell, C. Wong, G. Pham, T. T. Dung, L. N. Tjajadi, J. S., Loft L. and Mvondo, S. A. (2013). Comparative Analysis of REDD+ benefit sharing mechanisms for Efficiency, Effectiveness and Equity. *Center for International Forestry Research (CIFOR)*

Brown, H. C. P. (2011). Gender, climate change and REDD+ in the Congo Basin forests of Central Africa. *International Forestry Review* Vol.13 (2), p 164-166

Campese, J. (2011). Gender and REDD+ in Tanzania: An overview of key issues. *Tanzania Natural Resources Forum*

Chhatre, A., Lakhanpal, S. Larson, A. M. Nelson, F. Ojha, H. and Rao, J. (2012). Social safeguards and co-benefits in REDD+: a review of the adjacent possible. *Current Opinion in Environmental Sustainability, Elsevier Limited*, p 657

CIFOR (2012). Forests, Trees and Agroforests. A Strategy for Gender-Responsive Research and Action. Bogor: CIFOR.

Demetriades, J. and Esplen, E. (2008). The gender dimensions of poverty and climate change adaptation. *IDS Bulletin* 39(4): 26-31.

Djoudi, H., Brockhaus, M., Brown, H. C. P., & Bandiaky-Badji, S. (2012). Forests: gender, climate change and women's representation. *CIFOR Infobrief*, (48).

Dooley, K., Griffith, T., Martone, F., and Ozinga, S. (2011). *Smoke and mirrors: A critical assessment of the Forest Carbon Partnership Facility*. FERN and Forest Peoples Programme

Dooley, E., and Chapman, S. (2014). *Climate-smart agriculture and REDD+ implementation in Kenya*, Cambridge Centre for Climate Change Mitigation Research (University of Cambridge)-REDD+ Law Project

Doss, C. (2014). If women hold up half the sky, how much of the world's food do they produce? In *Gender and Agriculture: Closing the Knowledge Gap*, edited by A.R. Quisumbing, R. Meinzen-Dick, T. L Raney, A. Croppenstedt, J.A. Behrman, and A. Peterman. Springer, Netherlands.

FAO. (2011). Women in Agriculture: Closing the gender gap for development. In: *The state of food and agriculture*. FAO, Rome, Italy.

FAO. (2009). *The special challenge for sub-Saharan Africa. High Level Expert Forum-How to Feed the World in 2050*. Rome-Italy

FAO. (2010). *Impact of the global forest industry on atmospheric greenhouse gases.* Food and Agriculture Organization of the United Nations, Rome, Italy. FAO Forestry Paper 159.

FAO and RECOFTC. (2015). *Mainstreaming Gender into forest policies in Asia and the Pacific*. Food and Agriculture Organization of the United Nations; The Center for People and Forests, p 3-6.

Forsell, N., Turkovska, O., Gusti, M. Obersteiner, M. Elzen, M. and Havlik, P. (2016). *Carbon Balance and Management.*

Funder, M., Fjalland, J., Ravnborg, H.M., and Egelyng. (2009). *Low Carbon Development and Poverty Alleviation: Options for development cooperation in energy, agriculture and forestry*. Danish Institute for International Studies (DIIS) Report: Climate Change, Copenhagen, Denmark.

Gockowski, J., Sonwa. D. (2011). *Cocoa intensification scenarios and their predicted impacts on CO2 emissions, biodiversity conservation and rural livelihoods in the Guinea rainforest of West Africa*. Environ Manage

Gogo, J., (2014), Zimbabwe: Outcry over Kariba REDD+ Project, No REDD in Africa Network

Gurung, J. (2011). Women's exclusion from forestry. In IUCN Arborvitae: Attending to gender. *The IUCN Forest Conservation Programme Newsletter*, issue 43, p 8-9

Gurung, J., Giri K., Setyowati A. and Lebow, E. (2011). *Getting REDD+ Right for women: An analysis of the barriers and opportunities for women's participation in the REDD+ sector in Asia.* USAID, Washington DC

Hall, R. (2014). *The great REDD gamble: Time to ditch risky REDD for community-based approaches that are effective, ethical and equitable*. Friends of the Earth International (FOEI), Amsterdam

Henshaw, K. and Fyneface, D. F. (2014). Seeing REDD: Communities, Forests and Carbon Trading in Nigeria. Social Development Integrated Centre (Social Action)

Hess, J. S., (2014). Is REDD+ the right approach to reducing deforestation in tropical forest countries? Retrieved: August, 23, 2016 from https://climate-exchange.org/2014/02/02/is-redd-the-right-approach-to-reducing-deforestation-in-tropical-forest-countries-3/ .

Institute for Security Studies. (2011). Carbon trading in Africa: A critical review. Institute for Security Studies (ISS) No. 184/ ISBN- 978-1-920422-62-2, p 21

Intergovernmental Panel on Climate Change (IPCC). (2007a). Summary for Policymakers. In: *Climate Change 2007: Synthesis Report.* Contribution of Working Groups I, II and III to the Fourth Assessment Report of the Intergovernmental Panel on Climate Change, [Core Writing Team, Pachauri, R.K. and Reisinger, A. (Eds.)], Geneva, Switzerland.

IPACC. (2011). *African Indigenous Peoples and REDD+ Human Rights, equity and forest carbon capture in climate mitigation*. Indigenous Peoples of Africa Coordinating Committtee (IPACC)-ISBN 978-0-9814477-4-2, p 29

IPCC. (2007b). *Climate Change 2007: Synthesis Report*. Contribution of Working Groups I, II and III to the Fourth Assessment Report of the Intergovernmental Panel on Climate Change Core Writing Team, Pachauri, R.K. and Reisinger, A. (Eds.) IPCC, Geneva, Switzerland

IPCC. (2014). *Climate Change 2014: Impacts, Adaptation, and Vulnerability.* Inter-governmental Panel on Climate Change.

IUCN. (2010). *Women in REDD critical for climate action.* International Union for Conservation of Nature

IUCN. (2011a). *IUCN Arborvitae-Forests: a legal challenge*. The IUCN Forest Conservation Programme Newsletter, issue 44

IUCN. (2011b). *Mainstreaming Gender Considerations in REDD-Plus Policy Implementation in Ghana.* International Union for Conservation of Nature

Kanninen, M., Brockhaus, M., Murdiyarso, D. and Nabuurs G. (2010). Harnessing Forests for Climate Change Mitigation through REDD+: challenges and opportunities. *Forests and society-responding to global drivers of change*. Vienna, Austria, IUFRO World Series Volume 2, chapter 3, p 44-45.

Kajembe, G. C., Silayo, D. S. A., Mwakalobo, A. B. S., and Mutabazi, K. (2013). *The Kilosa District REDD+ pilot project, Tanzania. A socioeconomic baseline survey*. International Institute for Environment and Development (IIED), London, p 33-34

Kill, J. (2013). Carbon Discredited: Why the EU should steer clear of forest carbon offsets. FERN and Les Amis de la Terre, ISBN: 978-1-906607-30-2

Kissinger, G., Herold, M., and De Sy, V. (2012). Drivers of Deforestation and Forest Degradation: A Synthesis Report for REDD+ Policymakers. *Lexeme Consulting*, Vancouver Canada

Kula, N., Haines, A., and Fryatt, R. (2013). Reducing Vulnerability to Climate Change in Sub-Saharan Africa: The Need for Better Evidence. *PLOS Medicine*. Volume 10, Issue 1,

Lang, C. (2013). *Launch of No REDD in Africa Network: "REDD could cause genocide"*. REDD Monitor

Lang, C. (2015). The REDD+ Mechanism will not be enough to curb deforestation in Central Africa: *Center for International Forestry Research (CIFOR), REDD Monitor*

Larrazabal, A., McCall, M.K., Mwampamba, T. H. and Skutsch, M. (2012). The role of community carbon monitoring for REDD+: A review of experiences. *Current Opinion in Environmental Sustainability,* p 707-716.

Larson, A. M., Brockhaus, M., Sunderlin, W. D., Duchelle, A., Babon, A., Dokken, T., Pham, T., Resosudarmo, I. A. P., Selaya, G., Awono, A., & Huynh, T. B. (2013). Land tenure and REDD+: The good, the bad and the ugly. *Global Environmental Change*, *23*(3), 678-689.

Larson, A. M., Dokken, T., Duchelle, A. E., Atmadja, S., Resosudarmo, I. A. P., Cronkleton, P., Cromberg, M., Sunderlin, W., Awono, A. & Selaya, G. (2015). The role of women in early REDD+ implementation: lessons for future engagement. *International Forestry Review*, *17*(1), 43-65.

Lawlor, K., Madeira, E. M., Blockhus, J., & Ganz, D. J. (2013). Community participation and benefits in REDD+: A review of initial outcomes and lessons. *Forests*, *4*(2), 296-318.

Lee, D., Seifert-Granzin, J., Neeff, T., Göhler, D., Liss, B. M and Busch, A. (2011). Maximizing the Co-benefits of REDD-Plus Actions. Discussion paper for a Regional Expert Workshop supported by the German International Climate Initiative, September 27-29 2011, Subic, Philippines.

Leventon, J., Kalaba, F. K., Dyer, J. C., Stringer, L. C. and Dougill, A. J. (2014). Delivering community benefits through REDD+: Lessons from Joint Forest Management in Zambia. *Forest Policy and Economics.* p 1-2.

Livingston, G. (2015). Opinion: How Women Are Effecting Change in Kenya's Kasigau Corridor. Ecosystem Marketplace-A Forest Trends Initiative

Lohmann, L. (2006). Carbon trading: a critical conversation on climate change, privatisation and power. Development Dialogue no 48. Uppsala: Dag Hammarskjöld Foundation, p 234-238

Madeira, E. C. M. (2008). Policies to Reduce Emissions from Deforestation and Degradation (REDD) in Developing Countries: An examination of the issues facing the incorporation of REDD into market-based climate policies. *Resources for the future (RFF).*

Madeira, E. M., Kelley, L., Blockhus, J., Ganz, D., Cortez, R., and Fishbein, G. (2012). Sharing the Benefits of REDD+-Lessons from the field. *The Nature Conservancy*

Metzel, R. (2015). *How to translate REDD+ benefit sharing practice into policy*. International Union for the Conservation of Nature (IUCN).

Minang, P. A. (2010), Forestry and REDD in Africa, *jotoafrika,* issue 4.

Minang, P. A., Duguma, L. A., Bernard, F., Mertz, O., and M. Noordwij. (2013). Prospects for agroforestry in REDD+ landscapes in Africa. *Elsevier Limited*. 1877-3435, p 3-4

Murthy, I. K., Gupta, M., Tomar, S., Munsi, M., Tiwari, R., Hegde, G. T., and Ravindranath, N. H. (2013). Carbon Sequestration Potential of Agroforestry Systems in India. *Earth Science and Climate Change*- ISSN: 2157-7617

Mutabazi, K. D., George, C. K., Dos Santos, A. S., & Felister, M. M. (2014). Livelihood implications of REDD+ and costs-benefits of agricultural intensification in REDD+ pilot area of Kilosa, Tanzania. *Journal of Ecosystem & Ecography*, *4*(2): 144.

Mwangi, E., Meinzen-Dick, R., & Sun, Y. (2011). Gender and sustainable forest management in East Africa and Latin America. *Ecology and Society*, *16*(1): 17

Nabanoga, G., Namaalwa, J., and Ssenyonjo, E. (2012). The Ongo Community Forest REDD+ pilot Project, Uganda. A socioeconomic baseline survey. IIED, London, p 21-23

Nakhooda, S., Caravani, A., Bird, N., and Schalatek, L. (2011). Climate finance in Sub-Saharan Africa. Climate finance policy brief. *Overseas Development Institute and Heinrich Böll Stiftung North America.*

Naoto, J. (2006). International trade and terrestrial open-access renewable resources in a small open economy. *Canadian Journal of Economics*, Vol. 39, No. 3

Ndobe, S. N. (2010). REDD and Climate Change: Implications for the African Woman. Centre for Environment and Development (CED). Presentation for the REFACOF planning meeting.

Newton, P., Schaap, B., Fournier, M., Cornwall, M., Rosenbach, D. W., DeBoar, J., Whittemore, J., Stock, R., Yoders, M., Brodnig, G., Agrawal, A. (2015). *Community forest management and REDD+. Forest Policy and Economics.*

Nhantumbo, I. and L. Chiwona-Karltun.(2012). His REDD+, Her REDD: how integrating gender can improve readiness. *International Institute for Environment and Development (IIED) Briefing*. United Kingdom. IIEC

Norton Rose. (2010). Forest carbon rights in REDD+ countries: a snapshot of Africa. Norton Rose LLP Edition No. NR9190, p 9

OECD. (2017). Building Food Security and Managing Risk in Southeast Asia. OECD Global Forum on Agriculture. OECD Publishing, Paris.

Okali, C. and Naess, L. O. (2013). Making Sense of Gender, Climate Change and Agriculture in sub-Saharan Africa: Creating Gender-Responsive Climate Adaptation Policy. *Future Agricultures Consortium,* Working Paper 057

Olivier, J. G. J. (PBL), Janssens-Maenhout, G. (IES-JRC), Muntean, M., (IES-JRC) and Peters, J. A. H. W. (PBL). (2013). Trends in global CO2 emissions: 2013 Report. *PBL Netherlands Environmental Assessment Agency, Institute for Environment and Sustainability (IES) of the European Commission's Joint Research Centre (JRC)*

Osarogiagbon, R. (2011). REDD and its implication on Community people. A presentation made at Cross River State stakeholders forum on Climate change, REDD and Forest Dependent Community Rights. Nigeria. Environmental Rights Action (ERA)/ Friends of the Earth Nigeria (FOEN).

Pan African Climate Justice Alliance (PACJA). (2012), Carbon trading in Africa. Climate Finance Africa Briefing Paper Series, 3. PACJA and Institute for Security Studies

Parrotta, J. A., C. Wildburger and S. Mansourian. (2012). Understanding Relationships between Biodiversity, Carbon, Forests and People: The Key to Achieving REDD+ Objectives. A Global Assessment Report prepared by the Global Forest Expert Panel on Biodiversity, Forest Management, and REDD+. *IUFRO World Series* Volume 31, p 16, 54.

Peskett, L. (2011). Benefit sharing in REDD+: exploring the implications for poor and vulnerable people. World Bank and REDD-net. Trust Fund for Environmentally and Socially Sustainable Development (TFESSD)

Poffenberger, M. (2011). Women's micro-finance and forest conservation in India. In IUCN Arborvitae: Attending to gender. The IUCN Forest Conservation Programme Newsletter, issue 43, p 11

Pokorny, B, Scholz, I., and de Jong, W. (2013). REDD+ for the poor or the poor for REDD+? About the limitations of environmental policies in the Amazon and the potential of achieving environmental goals through pro-poor policies. *Ecology and Society* 18(2): 3.

Population Action International. (2011). Why Population matters to Forests. *Population Action International, Washington DC, USA*

Population Reference Bureau (PRB). (2013). *2013 World Population Data Sheet.* PRB-USAID.

Pye, O. (2012). Carbon Markets and REDD in South-East Asia: An Interview with Chris Lang from REDD Monitor. *ASEAS - Austrian Journal of South-East Asian Studies*, 5(2), 352-358.

Ribot, J. (2011). *Seeing REDD for Local Democracy: A Call for Democracy Standards*

Rietbergen-McCracken, J. (2011). EDITOR. Gender in Cancun. In *IUCN Arborvitae: Attending to gender.* The IUCN Forest Conservation Programme Newsletter, issue 43

Romero, M. Z., Trærup, S., Wieben, E., Møller, L. R., and Koch, A. (2013). Economics of forest and forest carbon projects- Translating lessons learned into national REDD+ implementation. *UNEP Risø Centre*-ISBN: 978-87-92706-66-9

Sandbrook, C., Nelson, F., Adams, W. M., and Agrawal, A. (2010). Carbon, forests and the REDD paradox. *Fauna and Flora International*

Scriven, J. H. N. and Malhi, Y. (2012). Smallholder REDD+ strategies at the forest-farm frontier: a comparative analysis of options from the Peruvian Amazon. *Carbon Management* 3 (3), future science group, p 265

Setyowati, A. (2012). Ensuring that women benefit from REDD+. *Women Organizing for Change in Agriculture and Natural Resource Management (WOCAN), Unasylva 239*, Vol. 63

Shames, J. (2013). How can small-scale farmers benefit from carbon markets? CCAFS Policy Brief no. 8. *CGIAR Research Program on Climate Change, Agriculture and Food Security (CCAFS)*. Copenhagen, Denmark

Shanahan, M., Shubert, W., Scherer, C., Corcoran. T. (2013). Climate Change in Africa: A Guidebook for Journalists. *United Nations Educational, Scientific and Cultural Organization (UNESCO) Series on Journalism Education*, France

Sills, E. O., Atmadja, S. S., de Sassi, C., Duchelle, A. E., Kweka, D. L., Resosudarmo, I. A. P., & Sunderlin, W. D. (Eds.). (2014). *REDD+ on the ground: A case book of subnational initiatives across the globe*. Bogor, Indonesia: CIFOR. https://www.cifor.org/library/5202/redd-on-the-ground-a-case-book-of-subnational-initiatives-across-the-globe/

Smith, H. (2011). *REDD+ in East Africa: The gender dimension of the Readiness Plans in Uganda, Kenya and Tanzania*. University of California, San Diego

Spratt, S. (2016). *Forest taxation and REDD+*. Institute of Development Studies, University of Sussex. ICTD Annual Meeting, 10-13 February 2016. Addis Ababa.

Streck, C. (2010). An African Agricultural Carbon Facility, Ford Foundation

The Forests Dialogue (TFD). (2008). Beyond REDD: The Role of Forests in Climate Change. The Forests Dialogue, USA

UNDP. (1995). Human Development Report, 1995. New York, USA

UNFCCC. (2007a). *Climate Change: Impacts, Vulnerabilities and Adaptation in Developing Countries.* Climate Change Secretariat (UNFCCC), Bonn, Germany

UNFCCC. (2007b). *Investment and Financial Flows to Address Climate Change*. United Nations Framework Convention on Climate Change, Bonn, Germany.

UNFCCC, (2010). Decision 1/CP.16 The Cancun Agreements: Outcome of the Work of the Ad Hoc Working Group on Long-term Co-operative Action under the Convention. In: *UNFCCC (Ed.)*, (UNFCCC/CP/2010/7).

UN-REDD. (2013). *Guidance Note on Gender Sensitive REDD+.* United Nations REDD Programme

UN-REDD. (2016). *UN-REDD Programme*. Retrieved: July, 14, 2016 from: http://www.un-redd.org/

UN WomenWatch. (2009). *Women, Gender Equality and Climate Change*. Retrieved: September, 6, 2016 from http://www.un.org/womenwatch/feature/climate_change/.

USAID. (2017). Greenhouse Gas Emissions in Indonesia. United States Agency for International Development (USAID).

Virgilio, N. and Marshall, S. (2009). Forest Carbon Strategies in Climate Change Mitigation: Confronting Challenges Through On-the- Ground Experience. The Nature Conservancy. Arlington, Virginia.

Virgilio, N. R., Marshall, S., Zerbock, O., and Holmes, C. (2010). Reducing Emissions from Deforestation and Degradation (REDD): A Casebook of On-the-Ground Experience. *The Nature Conservancy, Conservation International and Wildlife Conservation Society*. Arlington, Virginia.

Wambugu, S. W. Chomba, S. W. and Atela, J. (2015). Institutional arrangements for climate-smart landscapes. In Minang, P. A., van Noordwijk, M., Freeman, O. E., Mbow, C., de Leeuw, J., & Catacutan, D. (Eds.) *Climate-Smart Landscapes: Multi-functionality in Practice*, 257-273. Nairobi, Kenya: World Agroforestry Centre (ICRAF).

Wertz-Kanounniko, S. and Kongphan-Apirak, M.(2008). Reducing forest emissions in Southeast Asia: A review of drivers of land-use change and how payments for environmental services (PES) schemes can affect them. *Center for International Forestry Research (CIFOR)*. Working Paper No. 41 p 5-6, 13.

WFSE-IUFRO. (2009). *Making Sub-Saharan African Forests Work for People and Nature: Policy approaches in a changing global environment.* Special Project on World Forests, Society and Environment (WFSE) of the International Union of Forest Research Organizations (IUFRO).

WOCAN. (2011). *Getting REDD+ Right for Women: An analysis of the barriers and opportunities for women's participation in the REDD+ sector in Asia.* Prepared by Women Organizing for Change in Agriculture and Natural Resource Management (WOCAN) and the United States Forest Service and produced by the United States Agency for International Development (USAID), p 19.

WOCAN, UN-REDD and USAID. (2013). *Women's Inclusion in REDD+ in Cambodia: Lessons from Good Practices in Forest, Agriculture and Other Natural Resources Management Sectors. Joint Regional Initiative for Women's Inclusion in REDD+.* Women Organizing for Change in Agriculture and Natural Resource Management (WOCAN), the United Nations Collaborative Programme on Reducing Emissions from Deforestation and Forest Degradation (UN-REDD) and the United States Agency for International Development (USAID).

World Bank Africa. (2017). *Forests in Sub-Saharan Africa: Challenges and Opportunities.* PROFOR Innovation and Action for Forests. World Bank Africa Region.

12

Transforming forest landscape conflicts: the promises and perils of global forest management initiatives such as REDD+

Seth Kane[1], Ahmad Dhiaulhaq[1], Lok Mani Sapkota[1,*], David Gritten[1]

1 RECOFTC – The Center for People and Forests, Kasetsart University, Bangkok, 10903, Thailand; e-mail: seth.kane100@gmail.com, ahmad.dhiaulhaq@gmail.com, lok.sapkota@recoftc.org, david.gritten@recoftc.org

* Corresponding author

Abstract: Implementation of Reducing Emissions from Deforestation and Forest Degradation (REDD+) is designed to relieve pressure on tropical forests, however, many are concerned that it is a threat to the rights of forest communities. These potential risks need serious attention as earlier studies have shown that the Asia-Pacific region is a forest conflict hotspot, with many economic, environmental and social implications at global (e.g. climate change) to local levels (e.g. poverty). Drawing on an analysis of nine case studies from four countries (Cambodia, Myanmar, Nepal and Vietnam) this paper examines why and how REDD+ can be a driver for forest conflict and how it also has the potential to simultaneously transform these conflicts. The analytical framework, "sources of impairment", applied in the study was developed to increase understanding and facilitate the resolution of forest landscape conflicts in a sustainable manner (i.e. transformation). The main findings are that REDD+ can be a source of conflict in the study sites, but also had transformative potential when good practices were followed. For example, in some sites, the REDD+ projects were sources of impairment for forest communities by restricting access to forest resources. However, the research also identified REDD+ projects that enabled the participation of traditionally marginalized groups and built local forest management capacities, leading to strengthened tenure for some forest communities. Similarly, in some countries REDD+ has served as a mechanism to pilot Free, Prior and Informed Consent (FPIC), which will likely have significant impacts in mitigating conflicts by addressing the sources at local to national levels. Based on these findings, there are many reasons to be optimistic that REDD+ can address the underlying causes of forest landscape conflicts, especially when linked with other governance initiatives such as Forest Law Enforcement, Governance and Trade – Voluntary Participation Agreements (FLEGT-VPA).

Keywords: forest landscape; conflict transformation; forest governance; REDD+; FLEGT VPA; FPIC

---

1. Introduction

REDD+ (reducing emissions from deforestation and forest degradation) is a performance-based mechanism that tries to compensate tropical countries for reducing deforestation and forest degradations in their territories. The initiative is generally portrayed in the literature and by international development organizations as having the potential for significant positive outcomes (e.g. Brown et al., 2008). Targeted benefits include mitigating climate change, supporting sustainable livelihoods, maintaining vital ecosystem services and preserving global biodiversity. However, REDD+ has also been depicted by some NGOs and activists (e.g. REDD Monitor, Indigenous Environmental Network), and within certain bodies of academic literature (e.g. Yasmi et al., 2012; Patel et al., 2013) as including significant risks as a driver of conflict.

These potential risks deserve serious attention as forest landscape conflicts are ubiquitous in tropical countries, even before REDD+ (e.g. Mola-Yudego and Gritten, 2010; Gritten et al., 2013). The negative environmental and social impacts of these types of conflicts are increasingly reported in the news media and analyzed in the academic literature (e.g. Gritten et al., 2012; Lester and Hutchins, 2012; Munden Project, 2012; Dhiaulhaq et al., 2014) along with the causes of these conflicts. One direct cause that is consistently highlighted involves an outside actor accessing a forest landscape with a project - whether it be establishment of protected areas by the government (e.g. Redpath et al., 2013) or establishment of a plantation by a company (e.g. Dhiaulhaq et al., 2014; 2017), that fails to respect the rights of local communities. REDD+ also fits with this model.

In the context of REDD+, efforts have been made to address some of these social impact risks (e.g. weakening the rights of forest communities) through the establishment of REDD+ "safeguards" (McDermott et al., 2012) which have the overall goal of ensuring that REDD+ initiatives do not cause environmental and social harms (UNFCCC, 2011; McDermott et al., 2012). The United Nations Framework Convention on Climate Change (UNFCCC) safeguards (UNFCCC Decision 1/CP.16), for example, calls upon the REDD+ implementers to ensure "full and effective participation of relevant stakeholders, in particular, indigenous peoples and local communities" and "respect for the knowledge and rights of indigenous peoples and members of local communities" in REDD+ implementation. Despite these efforts REDD+ is still viewed with concern given the complexity of ensuring adherence to the safeguards (Poudyal et al., 2016). Against this backdrop, one needs to consider that the focus of REDD+ in tropical countries with high levels of deforestation and degradation means that it is often implemented in conflict prone contexts that include weak tenure and rights of forest communities and poor governance. On the other hand academic literature often sidelines the potential for REDD+ to transform conflicts (e.g. Patel et al., 2013).

Drawing on an analysis of case studies in four countries (Cambodia, Myanmar, Nepal and Vietnam), the paper aims to examine why and how REDD+ can be a driver for forest conflict and how it also has the potential to simultaneously transform these conflicts. The paper expects to flag several critical conflict-related issues that need greater attention in the development and implementation of REDD+, which may help provide crucial information to prevent and transform conflicts. Additionally, the work will also put forward recommendations to ensure that REDD+, and other related international initiatives, such as the European Union's Forest Law Enforcement Governance and Trade (FLEGT) Voluntary Partnership Agreements (VPA) can deliver on its transformative potential particularly for the forest landscape conflict. "Conflict transformation" here is defined as a process for addressing conflict which promotes long-term cooperation and justice (see Dhiaulhaq et al., 2015; Kane et al., 2016). Conflict, if transformed effectively, can serve as an opportunity and catalyst for social change in which parties are empowered and structural inequalities are addressed (Reimann, 2004; Bush and Folger, 2005; Dhiaulhaq et al., 2015).

2. Analysing conflict mediation through a transformative mediation framework

In this paper, "conflict" is defined as a situation in which one party or more pursues goals and interests through behavior or actions that impairs another party (Glasl, 1999). This is the starting point for the sources of impairment analytical framework that was developed to facilitate the identification of possible causes of conflict related to forest landscape management (Table 1). The framework was created by Patel et al. (2013) and refined through participant input from trainings and workshops on forest conflict transformation organized by RECOFTC (the Center for People and Forests). Participants in these capacity development activities were mainly government and NGO staff from across the Asia-Pacific region. The framework's emphasis on REDD+ considers how this initiative can exacerbate existing, including latent, conflicts or create new ones.

3. Material and Methods

3.1. Data collection

The research was conducted in stages between April 2011 and July 2015. The research included 296 semi-structured interviews (SSI) and 47 focus group discussions ([FGD] with 445 people) with representatives from local communities, concerned government departments and agencies, NGOs, and REDD+ networks. Additionally, six national level expert workshops were held (one each in Cambodia and Vietnam, and two each in Myanmar and Nepal) with a total of 98 representatives from relevant government departments, implementing agencies, research institutes, and NGOs.

The same guiding questions were used in all of the country sites, though these were modified to fit the interests of the participants (e.g. community members, NGO staff, government officials). The questions followed the nine sources of impairment framework (Table 1) to capture REDD+ and non-REDD+ issues related to forest landscape management and conflict. The responses in one activity (e.g. SSI) were then further explored in another activity (e.g. FGD), with the expert meetings being used to identify key issues (if workshop held before fieldwork) or to discuss the main findings (if held after fieldwork).

The responses in each activity were aggregated with special attention to common perceptions within each respondent group and any differences between the groups allowing for comparative analysis. The research was not designed to determine the relative importance or intensity of the different sources of impairment according to any quantitative criteria.

Table 1. Sources of impairment framework (based on Patel et al. 2013)

Source	Examples of impairment	In REDD+ context
1. Access and use restriction	Regulations limiting local communities' access to or use of forests due to creation of protected forest areas and/or granting of concessions to companies	Policies or practices that limit local access and ability to e.g. harvest forest products can cause conflict. REDD+ may come with such restrictions that may alter the relationship that local communities have with forests.
2. Benefit distribution	Unclear or inequitable arrangements for distributing benefits from forest management	The introduction of resources through REDD+ must be factored into this already complex equation of benefit generation and distribution.
3. Competing demands	Contradictions in economic and development agendas, conservation and cultural importance of forest areas	Alternate land use options might generate more income, making REDD+ the less favorable option to communities thereby undermining government initiatives.
4. Conflict transformation capacity	Lack of capacity among key stakeholders for sustainably addressing conflict	The absence of effective grievance mechanisms or processes challenging top-down decision-making processes, like Free, Prior and Informed Consent (FPIC), could make REDD+ a conflict driver.
5. Leadership	Leadership is not representative, accountable, or transparent	The approach to and content of REDD+ implementation may strengthen prevalent power imbalances and practices or cause conflict by challenging them.
6. Legal and policy frameworks	Dominance of state law over local and/or customary traditions; multiple, ambiguous and overlapping regulations related to forest management; and poor enforcement	The commoditization of carbon through REDD+ will add complexity to existing regulatory frameworks for forest management.
7. Participation and information	Lack of understanding and access to information, limited opportunities for stakeholders to meaningfully participate in forest management	Even where REDD+ implementation is equipped with grievance mechanisms and processes to ensure that affected parties understand and agree with the implications, the effective use of such tools is complex and context specific.
8. Quality of resources	Actual and perceived decrease or increase in the condition of forest resources caused by an external actor	The pursuit of REDD+ benefits may lead to intentionally skewed assessments of forest quality and leakage.
9. Tenure security	Overlapping and contested boundaries, lack of recognition of customary rights and traditional uses of the land	REDD+ creates further complexity about carbon ownership and resulting benefit entitlements.

3.1. Study sites

The study countries were selected to reflect a range of levels of progress in the development of REDD+, from leaders such as Nepal and Cambodia where results- based payments have been made to pilot sites, to Myanmar which is still in the early stages of REDD+. The case study sites (Table 2, Figure 1) were selected based on presence or potential presence of REDD+ activities, occurrence of forest conflict in the landscape, site accessibility and costs of conducting research. All the sites are home to forest communities with a high dependence on forests for their well-being, with these forests being threatened by various issues

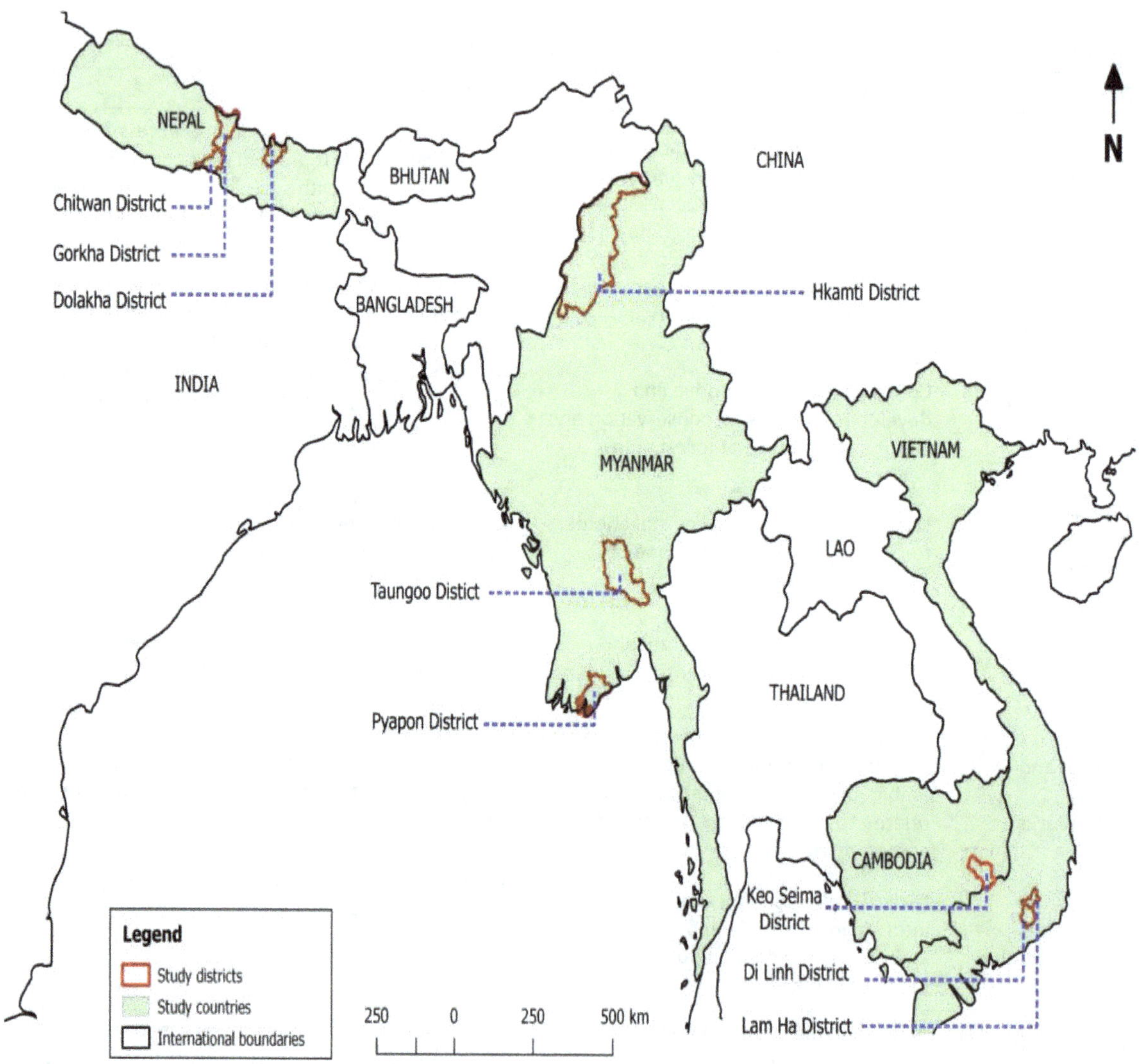

Figure 1. Location of the nine case studies

Table 2. Overview of the case study sites and countries

Case	Description
Cambodia: Keo Seima District	• The research was conducted in Seima Protection Forest (SPF) which is predominantly in Mondulkiri Province with a small area extending into Kratie Province. SPF is home to one of several REDD+ pilot projects in Cambodia. The SPF is a remote forested area of 292,690 hectares, divided into a Core Protection Forest Area and two Buffer Protection Forest Areas. SPF is home to 20 villages (roughly 5,000 people) of mainly Bunong ethnic communities who have been living in the area for many generations. The residents depend upon resources inside the SPF for their livelihoods, particularly for cash crop farming, collection of non-timber forest products (NTFPs) such as resin, and subsistence shifting agriculture. • Among the 20 villages, six villages (Pu Char, O Chrar, Pu Kong and Gati in Sre Preah Commune, Sre Khtumin in Sre Khtum Commune, Pu Haim in Sen Monorom Commune) were selected for the study sites. • Cambodia developed its REDD+ Readiness Plan (roadmap) in 2010, signed a UN-REDD National Programme and started the inception period in 2011. The National REDD+ Strategy for 2017 to 2025 was finalized in late 2017.
Myanmar: Hkamti District Pyapon District Taungoo District	• The research was conducted in three sites. The first site is in Yedashay township, Taungoo District in Bago region, in which the Forest Department conducted REDD+ pilot projects with the Korean Forest Service. The second site was in Hkamti Township, Hkamti District, Sagaing Region. The area was a potential REDD+ project site for the United Nations Development Programme (UNDP). The third site is a CF in Pyapon Township, Pyapon District, Ayeyarwady Region. While the last site is not a REDD+ site, it was selected on the basis of it being a community forest that is working to address the drivers of deforestation. • Myanmar joined the UN-REDD Programme in 2011. Its REDD+ Readiness Roadmap was published in 2013, implemented in 2015, and then migrated to a full National Programme for UN-REDD in 2016.
Nepal: Chitwan District Gorkha District Dolakha District	• The research was conducted in three REDD+ pilot sites: Kayarkhola (Chitwan District), Ludhikola (Gorkha District) and Charnawati (Dolakha District) watersheds. These watershed areas cover a total of 27,789 ha of land in which 13,970 ha (50%) is forest. 10,265 ha (73%) of the forest area is community forest (104 communities). The population (93,000) includes a wide range of caste categories such as Brahmin, Chhetri, Dalit and ethnic groups (Tamang, Chepang, Thami, Gurung, Magar etc.) [ANSAB 2010]. • The REDD+ projects were initiated by the International Centre for Integrated Mountain Development (ICIMOD), the Asia Network for Sustainable Agriculture and Bioresources (ANSAB), and the Federation of Community Forestry Users, Nepal (FECOFUN) in 2009. At the time of data collection, the three pilot projects were testing community forest-based governance and payment mechanisms for REDD+. • Nepal's Readiness Preparation Proposal (RPP) was approved by Forest Carbon Partnership Facility (FCPF) of the World Bank in 2010. Nepal has also been a member of the UN-REDD Programme since 2010. While the country's Readiness Package was approved in 2016, the draft REDD+ National Strategy is at final stage pending its approval after the federal governance process is clear[1].

[1] Promulgating a new constitution in September 2015, Nepal is in the process of adopting a federal structure from the previously centralized one. As of January 2018, the country has conducted elections for selecting local government and provincial and federal assembly members, but the federal processes are being worked out.

Vietnam: Di Linh District Lam Ha District	• The research was conducted in Di Linh and Lam Ha districts in Lam Dong province, both UN-REDD Programme pilot sites. The districts cover 260,000 ha with a population of 300,000. The districts contain over 23 ethnic minorities including K'Ho, Tay, Nung, Thai, Hoa and Kinh whose livelihoods come mainly from farming. Forest comprises 56% of the land area in Di Linh and 36% in Lam ha. Approximately 60,000 ha of forest areas are under the management of Commune Forest Management Boards[2]. Among the four communes studied, (two from each district) two communes were majority K'Ho ethnic group, one majority Kinh ethnic group, and the other no majority ethnic group. • Vietnam was selected to participate in the UN-REDD Programme based on the acceptance of its National Programme Document prepared by the United Nations Development Programme (UNDP), the Food and Agriculture Organisation (FAO), the United Nations Environment Programme (UNEP), and the Vietnamese government. Vietnam officially entered the inception and implementation phase of REDD+ in September 2009, moving into the second phase of the programme in 2013. Vietnam has a National REDD+ Action Program (NRAP)[3] as well as provincial action plans (at least 10 for provinces) and site-based implementation plans to facilitate REDD+ implementation (for at least 35 sites) (Huynh and Keenan, 2017).

[2] Data provided by People Committees in Di Linh and Lam Ha districts extracted from the 'Report on socio-economic development-2010'.

[3] The NRAP in Vietnam was first introduced in 2012 and subsequently revised in 2016 to make it more inclusive and add implementation guidance to 2016-2020 and a vision to 2030 (Huynh and Keenan, 2017).

4. Results

4.1. Access and use restriction

The findings from the nine study sites in the four countries indicate that access and use restrictions of forest resources prior to REDD+ were widespread (e.g. in conservation areas or areas designated for land concessions), and that REDD+ can also further restrict access and use. In some of the sites in all the countries, the restrictions have resulted in negative impacts on people's livelihoods through, for example, constraining traditional practices (e.g. shifting agriculture, NTFP collection) that lack legal recognition. This occurred, for example, in protected areas (e.g. Hugaung valley tiger reserve) in Myanmar's Hkamti Township and within economic land concessions in Cambodia's SPF.

In Nepal, following the implementation of the REDD+ pilot project, the Executive Committees of the Community Forest User Groups (CFUGs) placed restrictions on the extraction of forest products such as fuelwood and fodder, as well as livestock grazing. REDD+ piloting also encouraged CFUG leaders to increase enforcement and monitoring restrictions in general which have disproportionately affected the forest dependent poor, who have no alternate sources of energy, and vulnerable groups such as women, who are mostly tasked with managing firewood and fodder for livestock. For example, one of female community forest users in in one of the Nepal study sites stated: "I have been rearing goats for 10 years and grazing them in the forest. It is the main source of my livelihood which has been threatened by the CFUG decision to ban gazing in the forest. They (Executive Committee members) have told that if they find any one grazing in the CF, they will impose a fine of NRs 200 (~2 USD) per goat. Those without private land are facing challenges due to these new decisions."

In Vietnam's Lam Ha District, interview participants reported that after the implementation of the government's program of Payments for Forest Environmental Services (PFES)[4] and the UN-REDD activities, they encountered increased restrictions on accessing and using forest resources. While many local community members have benefited from the PFES program, as patrol members and/or through payment for forest management, some local community members expressed frustration with the restrictions, especially not being able to extract timber for house construction.

In Cambodia, after a series of FPIC processes, agreements on the "Cooperative Implementation of the REDD+ project" in SPF were signed by relevant government agencies and local communities in 2012. Based on these agreements, collection of NTFPs are allowed only for subsistence needs, with collection of timber for house building, for example, requiring government permission. Elderly people in SPF in particular expressed concerns that these restrictions would impact livelihoods and that REDD+ may progressively introduce even stricter rules in the future.

4.2. Benefit sharing

The findings in all four countries show that benefit sharing is a widespread and significant source of impairment that potentially leads to conflict. Community level REDD+ stakeholders asserted their desire to participate in decision making regarding benefit sharing. However, some stakeholders, such as government and local elites, often have more power to influence decision making processes. The concerns raised by local community members regarding the possibility of elite capture and corruption are rooted in experience.

The study found that the diversity of stakeholders and overlapping jurisdiction of government ministries and departments have slowed the development of REDD+ benefit sharing mechanisms, while intra-community socio-economic diversity further complicates benefit distribution. Including socio-economic criteria in determining REDD+ payments can enhance benefit sharing equity, but

[4] The PFES program aims to generate funds for forest conservation and improving livelihoods of forest owners and local people engaged in forest conservation. It was first piloted in Lam Dong and Son La province since 2008. Since 2011 PFES has been up-scaled to the national level. It has estimated to have collected about USD 142 million (Trung et al., 2015).

can also be a source of contention. Problems can also arise given a lack of valid and reliable data, which is sometimes the result of deliberate actions to increase their own benefits.

In Myanmar there are different expectations of how REDD+ money should be distributed. For example, in Hkamti District, some villagers proposed that payment should be distributed equally while others argued that households contributing to conservation activities should get more. Some villagers also expressed concerns that the eldest leaders often dominate the decision making due to patriarchal community social structures.

Nepal established the Forest Carbon Trust Fund (FCTF) in 2009 as a performance based financial mechanism for local communities protecting forests in REDD+ pilot sites. The trust fund was considered a key outcome of the pilot project with CFUG representatives expressing satisfaction with the REDD+ payments. However, they, along with NGO representatives, raised concerns regarding the distribution of money among and within CFUGs. This was linked to perceived manipulation by CFUG leaders (some CFUGs overestimated the number of poor households to increase payments). Additionally many upper caste households objected to a distribution system that favored certain minority ethnic groups or lower castes.

4.3. Competing demands

Competing stakeholder interests and management objectives regarding forest resources were found to be an important source of impairment. The increasing price of commodities such as coffee, rubber and luxury grade timber often lead to conflict with other forest management objectives (e.g. conservation). Moreover, in areas with especially valuable resources (e.g. minerals or cash crops) the number of stakeholders and overlapping claims are increasing. Likewise, changes in socio-economic conditions (e.g. migration, market access) and availability of technology (e.g. chainsaws, motorbikes, mobile phones) at the local level are creating complex and competitive dynamics regarding the relative benefits of REDD+ to illegal logging.

In Cambodia, the SPF and surrounding area are subject to high levels of competing interests. For example, in forest areas surrounding the SPF site, many economic land concessions (ELCs) have been granted to private companies; in addition, population growth, including through increased migration, is creating new land pressures, motivating particularly the youths to alternative livelihood sources (e.g. hunting and illegal logging of high value timbers), with some also finding employment with the ELCs.

In the Vietnam study sites, some of the respondents reported that to access more land for food cultivation, they have to enter the remaining forest area (production forest category). If there are no other options for rural livelihoods, the pressure to convert forest (e.g. crops cultivation) will increase. Many members of the Bao Thuan commune felt they have no alternatives to harvesting forest products and slash-and-burn farming for their living.

4.4. Conflict transformation capacity

The REDD+ readiness proposals and/or roadmaps in the countries covered have recognized the potential conflict from REDD+ and stipulate steps for redressing grievances. However, the proposed grievance mechanisms (in Cambodia, Nepal and Vietnam) are still in the early stages of development and there are few opportunities for conflict transformation capacity development.

In the site in Phya Pon District (Myanmar), conflict management is typically handled by the CF Management Committee (CFMC). If a conflict involves people from the same village, but not CFUG members, conflict is normally handled internally by the CFMC through traditional negotiation. However, if conflict arises between the community and outsiders CFMC members normally ask for help from local government authorities. Involvement of police and the court were not found in the study villages partly due to costs, including for travel. Moreover, no special conflict management strategy, such as professional mediation, was observed in the area.

In Nepal, the key institution within CFUGs for addressing conflict is the Executive Committee (EC). When grievances are reported, the EC meets the conflicting parties to seek resolution. If this is not possible, the EC may seek help from the local Federation of Community Forestry Users Nepal (FECOFUN), an NGO, or the District Forest Office (DFO). Normally, when the conflict is between CFUGs, representatives from the affected CFUGs meet. If they cannot effectively address the issue, they file the case with the DFO and follow the legal procedures. The DFO attempts to address conflicts through consensus building and only in the worst cases, takes legal measures. REDD+ pilot projects have seen the creation of two new institutions at the local level both with a mandate to address conflict issues related to REDD+.

4.5. Leadership

Access to and control over forest resources and associated financial and political capital makes local leadership positions increasingly attractive, with REDD+ raising the stakes. This competitive environment means local leaders are increasingly scrutinized; however, concerns of elite capture of REDD+ were present in all focal countries. In some communities, concern was expressed for how and whether REDD+ might expedite the decline in influence of traditional (i.e. indigenous) leaders. Nevertheless, REDD+ was perceived by some community members as an opportunity to strengthen governance and leader accountability (including through the use of FPIC).

In Cambodia, most of the community leaders are men. The reported reason by villagers was men were better educated and more confident to speak out in public as well as the societal tendency society to have more respect for male leaders. Similarly, the voluntary nature of work and need to take part in patrols, often at night, also discouraged women from leadership positions, further contributing to gender inequality.

In Myanmar, notwithstanding political reforms at the national level, decision making at the subnational level was often perceived as slow and bureaucratic, with frequent elite capture. Since the village leader is a paid position and income is expected to increase with REDD+, many respondents expected the village head position will become increasingly contested and political. Similarly, the involvement of household heads, men in most instances, without consideration of gender in collective decision making also poses challenges to equity in REDD+ in the country.

4.6. Legal and policy frameworks

Key challenges relating to legal and policy frameworks in REDD+ include ambiguous provisions in legal documents, overlapping jurisdiction of government agencies, and insufficient consultation with affected communities. Weak coordination during policy making across line ministries resulted in unclear provisions and increased the potential for conflict. Furthermore, insufficient consultation with affected communities during policy formulation sidelines customary practices. Often communities have little awareness of the laws that affect them, leading to confrontations with government authorities when they inadvertently violate laws.

The expert workshop in Cambodia highlighted that 'carbon rights' and 'benefit sharing' are particularly important issues regarding national legal and policy frameworks for REDD+. There is still no explicit definition or provision in the law regarding the carbon rights from REDD+ because in the 1993 Constitution (Article 58) 'state property' is defined to include almost all of the country's resources, including natural resources (and forests) with the ownership and use rights of the carbon going to the state (Yeang et al., 2014). This reflects the concern that the legal framework does not provide a strong basis for communities to claim carbon rights and associated benefits.

In Nepal, forest-related policy making processes, such as the REDD+ Strategy and the Forest Sector Strategy tend to prioritize conservation over local utilization (e.g. Bampton and Cammaert, 2007; Gritten et al., 2015). This has created tension between organizations advocating for community rights, the government and conservation organizations. In 2012 the government-proposed amending the 1993 Forest Act which was strongly opposed by CFUGs on the basis that it would curtail local rights. The proposed amendment was perceived by many as driven by the potential financial benefits of REDD+.

4.7. Participation and information

Concerns were raised in all countries, in the expert workshops and study sites, of the challenges regarding capacity, particularly of government staff, to conduct effective participatory processes. This may be compounded by the feeling of community members and NGOs that state actors are disproportionately powerful, which undermines the full and meaningful participation of communities and indigenous people. Nevertheless, the expert workshops in all countries also emphasized the opportunities that REDD+ is providing, including efforts to develop the capacity of stakeholders to facilitate participatory approaches.

One benefit of REDD+ has been the increased mainstreaming of FPIC, particularly in Cambodia and Vietnam. In Cambodia, the implementation of FPIC in SPF was the first instance of full FPIC in the country. The FPIC process engaged the residents of all 20 villages through a lengthy process of awareness-raising, participation, consultation and consent seeking for REDD+. The three-phase process was devised and implemented to ensure compliance with international standards (Anderson, 2011). However, the FPIC facilitation team faced problems related to the lack of strong legally-recognized local representative organizations and in some cases the FPIC process depended on non-elected village chiefs.

In Nepal, the REDD+ pilots have promoted inclusivity in REDD+ and developed leadership capacity of marginalized groups. Dalit households were motivated to participate in forest management activities after receiving REDD+ payments. A Dalit woman from Birenchowk CFUG, Ludhikhola Watershed, said in an interview that "after the REDD program, we are aware of our roles and responsibilities and can express our concerns and voices without any hesitation. Now I am a member of Executive Committee (EC). With this, Dalits have got more opportunities to attend meetings and workshops. Now in EC meetings, women also speak out which is a relatively new practice in our CFUG." However, not all the increased participation is self-motivated. Some representatives from the CFUGs felt that community leaders have pressured marginalized groups to participate in order to meet project requirements and there is some backlash by more affluent groups against positive discrimination.

4.8. Quality of resources

The performance-based payment mechanism of REDD+, with the amount of payment contingent on the extent and quality of forest resources, is a source of contention given data issues around the accuracy of different forest measurements. The respondents in all four countries said that private sector companies often manage high or the highest quality forest resources, for example, in the form of concessions and often convert forest areas to other uses (e.g. rubber plantations in Mondulkiri province in Cambodia). In Vietnam, while the government has attempted to improve local people's forest rights though Forestland Allocation (FLA) since the early 1990s, forest management is still dominated by state organizations whose interests often conflict with forest-dependent communities. Most of the high quality forest areas are controlled by Forest Management Boards, State Forest Enterprises and forest companies with local communities allocated poor forest areas (e.g. Sikor et al., 2013). If this remains unchanged, benefits from REDD+ will go primarily to non-community actors. Despite uneven government efforts to protect forests, forest loss and degradation continue due to logging, agricultural land expansion, and shifting cultivation.

In Cambodia, forest clearance in the REDD+ project site has been undertaken by smallholders, medium scale farmers and by speculators selling to large landholders, however their relative contribution to deforestation is unclear. Some forest is cleared for traditional subsistence crops (rice, maize, etc.), but most is cleared for cash crops such as cashew, soy and cassava. In Cambodia ELCs are the most important driver of deforestation (Delux, 2015). While this is not the case in the SPF itself, surrounding forest areas have been granted to ELCs (WCS, 2013) and the upland grasslands are rapidly being converted to tree crops and cassava (Evans et al., 2013). Whereas in February 2016 the Government confiscated 503,531 ha of forest land from 35 ELCs in ten provinces, as of June 2016, the Royal Government of Cambodia (RGC) had granted 1,552,700 ha of forest land (including inside protected areas) to 223 companies in 18 provinces.

4.9. Tenure security

Tenure security is strongly linked to policy and legal frameworks and competing demands for forest resources. Contested tenure and claims over forests and land have been identified in most of the case study sites, with communities still subject to weak tenure security. Local communities often face uncertain tenure security due to overlapping responsibilities of government offices and ambiguous policies. The research found that while there have been efforts to use REDD+ as an opportunity to clarify land tenure in the project sites and elsewhere, numerous challenges remain including undefined boundaries, conditional or short-term tenure and lack of recognition for customary institutions.

In Cambodia, one of the highlights of the REDD+ project in SPF are efforts to support indigenous community land rights by obtaining Indigenous Communal Land Titles. Andoung Kraloeng village, for example, received its land title in March 2012, which was only the third village in the country to reach this stage. However, the efficacy of such a title remains to be seen as land and forest seizures by large land concessions and mines remains prevalent in the country (Evans et al. 2013). Communal ownership also has a different and less secure set of rights compared to private ownership. Unclear guidelines and lack of transparency in granting ELCs and policies over forestland use were causes of conflict in many of the studied communities.

In all the study countries there has been notable progress in development of community forestry, in terms of area (e.g. CF in Vietnam covers over 4 million ha [RECOFTC, 2017]), but also its legal foundations (e.g. revision to Community Forestry Instruction (2016) in Myanmar moving from emphasis on subsistence needs of community members to commercialization of CF). However, policies and laws do not take into account characteristics of their ownership relations, traditional norms and rules, and farming systems, making tenure claims with regard to REDD+ complex and tenuous.

5. Discussion

This paper set out to examine areas of REDD+ implementation that can potentially be a driver for forest landscape conflict, but also help to identify how REDD+ may address existing conflict issues. Numerous sources of impairment that may exacerbate existing, and create new conflicts regarding the development and implementation of REDD+ were found in the case studies in the four countries. These sources of impairment often build on existing issues under the broad umbrella of poor governance (including weak tenure and rights of rural communities, top-down decision making, and lack of concordance in laws and policies) that afflict forest landscapes in the region (Barr and Sayer, 2012). This is reflected in the fact that the sites have suffered from existing conflicts, in various forms, prior to REDD+. The work also found that, in some instances, the implementation of REDD+ has also introduced new sources of impairment which in turn may increase the likelihood of future conflict. It is also notable that REDD+ has introduced some initiatives (e.g. FPIC) that help to address conflict drivers.

Disaggregating the interview and focus group discussion results according to gender and economic status, shows that REDD+ implementation affects forest stakeholders differently. Women and the poorest community members, for example, suffered the most when there are forest access and use restrictions due to conservation and REDD+ given their lack of alternative livelihoods. In this regard, equity and gender considerations in REDD+ are particularly important, especially in countries that are highly dependent on forest resources (e.g. Khadka et al. 2014, Larson et al., 2018). Furthermore, in many areas access, use and control of forest resources are influenced by gendered power relations (Khadka et al., 2014; Larson et al., 2018). Pham et al. (2016) provide an insightful example from Vietnam where there are significant barriers to the participation of women in decision making on a national level. While progress has been made in the study countries (including being signatories to the Convention on the Elimination of All Forms of Discrimination Against Women (CEDAW)), often there is little tangible change, for example women are still generally absent from the highest levels of decision making (RECOFTC, 2015).

In terms of conflict intensity, the conflicts in the case study sites were at a low level, indicated by contentious debates and fears of future tensions. However, no coordinated community actions such as protests, lawsuits or physical confrontations were present in the study sites according to the interview participants or local media reporting. This does not mean, however, that the conflict issues are insignificant or escalation will not occur. It is often the case that the conflict parties, specifically local communities, realize the risks if the conflicts were to escalate (Yasmi et al., 2012), and therefore avoid direct confrontation. Nevertheless, left smoldering and/or improperly addressed, low intensity conflicts can intensify with impacts ranging from project failure to violence (Engel and Korf, 2005; Yasmi et al., 2006). In this context therefore identifying sources of impairment preemptively helps create a deeper understanding, with the goal of identifying and preventing potential conflicts and transforming conflicts that exist or will likely emerge. Using participatory tools and processes designed to facilitate mutual understanding would be of value (Gritten et al., 2009).

This study also found that certain REDD+ initiatives had significant conflict mitigation and transformation impacts and potential. Some new developments such as increased participation of traditionally marginalized groups, including through the use of FPIC in Cambodia, Nepal and Vietnam (but also looking forward in Myanmar where political changes and policy shifts are creating unprecedented opportunities – reflected in strengthening of CF in the country) may help address existing or potential conflicts. Similarly, the emphasis on traditionally disadvantaged social groups (e.g. Dalits) in the REDD+ payment criteria in Nepal has helped to address some social injustices and other deep structural challenges. Additionally, the development of grievance redress mechanisms (although still in early stages of development) in the four countries studied could positively contribute to reducing conflicts in and around forestlands in REDD+ sites.

The recognition and appreciation of communities of the REDD+ projects are attributable, in part, to the efforts and good faith in implementing FPIC. The FPIC processes in the case study sites were some of the first ever applied in the country regarding forest management. These are big steps towards effective participation in REDD+, there is, however, still room for improvement (e.g. AIPP and IWGIA, 2012), with one of the biggest challenges being perceptions and capacities of FPIC among government and company staff. This is coupled with the fact that FPIC's effectiveness is strongly linked to the strength of tenure and legal rights (Mahanty and McDermott, 2013).

Ensuring the effective implementation of social and environmental safeguards into REDD+ strategies and implementation is essential for the effectiveness of REDD+, particularly to ensure that REDD+ projects 'do no harm' (Peskett and Todd, 2013). For example, if unclear benefit sharing mechanisms, elite capture and lack of accountability are not addressed within REDD+, the influx of funds could create perverse incentives and deepen economic and social inequity while having adverse impacts on forest climate change mitigation (Doherty and Schroeder, 2011). REDD+ design is actively seeking to address these issues. There are concerns, however, that safeguards do not go far enough and are too open to interpretation by governments and REDD+ project proponents while NGOs serve as the only watchdogs (WRM, 2015). Others have pointed out that the 'do no harm' principle erroneously implies that forest communities already have secure rights (Marion Suiseeya, 2015; 2016), weakening the transformative potential of REDD+.

The different initiatives that may result from the introduction of REDD+ should also consider the drivers of conflict, and how the sources of impairment may vary according to their prominence. This has implications not only for the impacts of the conflict, but also the focus of the interventions to address them (Figure 2).

Figure 2. Mapping the sources of impairment, their causality and conflict transformation implications

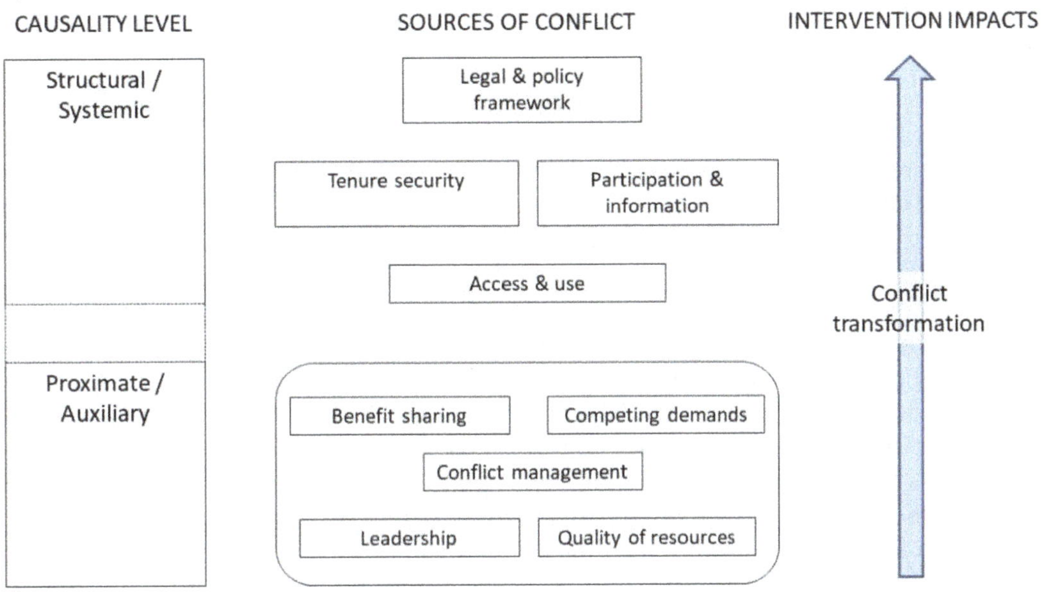

Figure 2 is an illustration based on the research findings of how the sources of impairment can be structured to reflect structural (systemic) and proximate (auxiliary) levels of causality. As such it enables the identification of focus areas when aiming for conflict prevention and transformation in implementing REDD+ through focusing on structural causes. Addressing the legal and policy framework, tenure security and participation in REDD+ requires significant investment, including time and political will from the decision makers and project implementers. The emphasis on these sources of impairment is also recognized in the literature on the issue (e.g. Ribot and Larson, 2012; Dhiaulhaq et al., 2015). The proximate / auxiliary conflict causes are listed in no particular order as they were found to have mixed levels of importance in the cases studied.

The findings have several implications for preventing and transforming conflict in the study sites and beyond, including the necessity of multilevel (local, subnational, and national) efforts. In particular, to prevent and transform conflict in REDD+, requires going beyond party level conflicts (e.g. restoring relationships and reaching agreements) to include the larger conflict context (i.e. structural issues). This effort will require long term engagement, trust building and cooperation between the government, communities and NGOs at local, subnational and national levels starting at the earliest stage of REDD+ development, as well as synergizing with other forest governance initiatives such as FLEGT-VPA. For the sustainable and socially equitable implementation of these and similar initiatives, progress on fundamental issues of land reform will be required.

The EU's FLEGT initiative, particularly the VPAs, is comparable to REDD+ in many ways including the fact that it is an international forest governance regime, it mandates non-governmental stakeholder involvement, seeks to use market mechanisms to incentivize improved forest governance, and requires consistent and credible monitoring (Tegegne et al., 2017). Most significantly, similar to REDD+, FLEGT has the potential to both benefit and harm local communities (including smallholders), and thus cause and mitigate conflict, depending on its effective implementation. This study on the sources of impairment in REDD+ provides some key transferable insights into guiding FLEGT initiatives to ensure it is implemented in a manner which mitigates and transforms conflicts. This includes the primacy of the three structural sources of impairment, namely legal and policy frameworks, tenure security, and participation and information in determining the success and transformative potential of FLEGT. As both FLEGT and REDD+ create incentives for governmental and non-governmental stakeholders to address these three issues sustainably, synergies should be sought wherever possible to ensure these initiatives are mutually reinforcing, especially where they can mainstream norms of effective forest governance.

6. Conclusion

The findings suggest that to anticipate and transform conflict in the context of REDD+, multifaceted approaches which address, or at least consider, all nine sources of impairment are required. This is supported by a large number of respondents in the four countries being able to identify concrete examples of all nine sources of impairment in the selected communities. If one were to prioritize, the current research study, and the literature greatly emphasize the importance of clarifying tenure and rights. Many countries in the Asia-Pacific region have programs for strengthening community forestry. However, they often fail to recognize the importance of ensuring that the local communities can tangibly benefit from their tenure. If their tenure and rights are weak, it means their starting position for REDD+ is also weak (also undermining FPIC). One way forward would be the mainstreaming of the UN Voluntary Guidelines on the Governance of Tenure (VGGT) in forest landscape related policies. The developing FLEGT-VPA process in many countries in the region also provides opportunities for this. Additionally, the mainstreaming of FPIC in projects is also highly recommended. This would require legal and regulatory support, but also requires a capacity development program for national and sub-national government staff, as well as staff of companies and NGOs. Furthermore, forest governance monitoring systems need to be strengthened, harmonized, and made accessible to local communities.

REDD+ and FLEGT VPA have significant potential to strengthen forest landscape governance across the Asia-Pacific region, with transformative outcomes for rural communities. Renewed efforts by international organizations such as EFI FLEGT and REDD Facilities, academics, civil society and donors to maximize the synergies from these initiatives, as well as linking with other global (e.g. The UN Sustainable Development Goals) and national initiatives (e.g. the peace process in Myanmar) is vital to institutionalize and eventually scale up lessons learned, including to address the structural drivers of forest landscape conflicts in the region.

Acknowledgements

Support from the Swedish International Development Cooperation Agency (Sida), the Norwegian Agency for Development Cooperation (Norad) and the Swiss Agency for Development and Cooperation (SDC), as well as through the "Conflict and cooperation over REDD+ in Mexico, Nepal and Vietnam" (CoCooR) project, funded by the Netherlands Organisation for Scientific Research (NWO) and the United Kingdom Department for International Development (DFID) is gratefully acknowledged.

The following organizations and their staff have contributed in the development of this study: Cambodian Center for Mediation (Cambodia), Forest Action (Nepal), and Da Lat University (Vietnam), as well as colleagues past and present from RECOFTC. Finally, we are extremely grateful to the invaluable advice from Poshendra Satyal, Pham Thu Thuy and Donal Yeang, on an earlier version of this paper, as well as to the journal's anonymous reviewers.

Conflicts of interest

The authors declare no conflict of interest.

References

Anderson, P. (2011). *Free, prior, and informed consent in REDD+: principles and approaches for policy and project development*. RECOFTC and GIZ, Bangkok.

ANSAB. (2010). *Report on Forest Carbon Stock of Community Forests in three watershed- Ludhikhola, Kayerkhola and Charnawati. REDD+ pilot Project*: Norad, ICIMOD, ANSAB, FECOFUN, Kathmandu, Nepal.

Asia Indigenous Peoples Pact (AIPP) & International Work Group for Indigenous Affairs (IWGIA). (2012). *Briefing paper on REDD+, Rights and Indigenous Peoples: Lessons from REDD+ Initiatives in Asia*. Chiang Mai, Thailand

Bampton, J. & Cammaert, B. (2007). Can timber rents better contribute to poverty alleviation through community forestry in the Terai region of Nepal? In Oberndorf, R., Durst, P., Mahanty, S., Burslem, K. & Suzuki, R. (Eds). A Cut for the Poor. *Proceedings of the international conference on managing forests for poverty reduction: Capturing opportunities in forest harvesting and wood processing for the benefit of the poor*. pp. 85-99 FAO and RECOFTC Report No. 19. FAO and RECOFTC– The Center for People and Forests. Bangkok, Thailand.

Barr, C. M., & Sayer, J. A. (2012). The political economy of reforestation and forest restoration in Asia–Pacific: Critical issues for REDD+. *Biological Conservation, 154*, 9-19.

Brown, D., Seymour, F. & Peskett, L. (2008). How do we achieve REDD co-benefits and avoid doing harm in A. Angelsen (ed.) Moving ahead with REDD: issues, options and implications. Bogor, Indonesia, *Center for International Forestry Research (CIFOR)*, pp.107-118.

Bush, R., & Folger, J. (2005). *The Promise of Mediation: The Transformative Approach to Conflict*. John Wiley & Sons, San Francisco.

Delux, C. (2015). *Drivers of Deforestation in the Greater Mekong Subregion Cambodia Country Report*. Lowering Emissions in Asia's Forests (USAID LEAF).

Dhiaulhaq, A., Gritten, D., De Bruyn, T., Yasmi, Y., Zazali, A., & Silalahi, M. (2014). Transforming conflict in plantations through mediation: Lessons and experiences from Sumatera, Indonesia. *Forest Policy and Economics, 41*:22-30.

Dhiaulhaq, A., De Bruyn, T., & Gritten, D. (2015). The use and effectiveness of mediation in forest and land conflict transformation in Southeast Asia: Case studies from Cambodia, Indonesia and Thailand. *Environmental Science & Policy, 45*, 132-145.

Dhiaulhaq, A., McCarthy, J.F. & Yasmi, Y. (2017). Resolving industrial plantation conflicts in Indonesia: Can mediation deliver?. Forest Policy and Economics.

Doherty, E. & Schroeder, H. (2011). Forest tenure and multi-level governance in avoiding deforestation under REDD+. *Global Environmental Politics, 11*(4):66-88.

Engel A. & Korf B. (2005). *Negotiation and mediation techniques for natural resource management. Training manual.* Food and Agriculture Organization of the United Nations, Rome, Italy.

Evans T., O'Kelly, H., Soriyun, M., Meng Hor, N., Phaktra, P., Pheakdey, S. & Pollard, E. (2013). Seima Protection Forest. In T.C. Sunderland, J. Sayer, & M.H. Hoang eds. *Evidence-based conservation: lessons from the Lower Mekong*. Oxon, Routledge.

Glasl, F. (1999). *Confronting Conflict: A First-aid Kit for Handling Conflict.* Hawthorn Press, Stroud, UK.

Gritten, D., Saastamoinen, O. & Sajama, S. (2009). Ethical analysis: A structured approach to facilitate the resolution of forest conflicts. *Forest Policy and Economics 11*(8): 555–560.

Gritten, D., Mola-Yudego, B., & Delgado-Matas, C. (2012). Media coverage of forest conflicts: A reflection of the conflicts' intensity and impact?. *Scandinavian Journal of Forest Research,*

27(2), 143-153.

Gritten, D., Mola-Yudego, B., Delgado-Matas, C., & Kortelainen, J. (2013). A quantitative review of the representation of forest conflicts across the world: resource periphery and emerging patterns. *Forest Policy and Economics, 33*, 11-20.

Gritten, D., Greijmans, M., Lewis, S.R., Sokchea, T., Atkinson, J., Quang, T.N., & Paudel, N.S. (2015). An uneven playing field: regulatory barriers to communities making a living from the timber from their forests–examples from Cambodia, Nepal and Vietnam. *Forests, 6(*10):3433-3451.

Huynh, T.B. & Keenan, R.J. (2017). Revitalizing REDD+ Policy Processes in Vietnam: The Roles of State and Non-State Actors. *Forests, 8*(3):53.

Kane, S., Gritten, D., Sapkota, L.M., Bui, L.T., & Dhiaulhaq, A. (2016). Getting the positives out of forest landscape conflicts. *Unasylva, 67*(247/248),45.

Khadka, M., Karki, S., Karky, B.S., Kotru, R. & Darjee, K.B. (2014). Gender equality challenges to the REDD+ initiative in Nepal. *Mountain Research and Development, 34*(3):197-207.

Larson, A., Solis, D., Duchelle, A., Atmadja, S., Resosudarmo, I., Dokken, T & Komalasari, M. (2018). Gender lessons for climate initiatives: A comparative study of REDD+ impacts on subjective wellbeing. *World Development 108*:86-102.

Lester, L., & Hutchins, B. (2012). The power of the unseen: environmental conflict, the media and invisibility. *Media, Culture & Society*, 34(7):847-863.

Mahanty, S., & McDermott, C.L. (2013). How does 'Free, Prior and Informed Consent'(FPIC) impact social equity? Lessons from mining and forestry and their implications for REDD+. Land *Use Policy, 35*, 406-416.

Marion Suiseeya, K.R. (2015). A retreat from justice in global forest governance: REDD+ and the "do no harm" principle. *Presented at the 3rd Annual UCSB Environmental Politics Conference*, Santa Barbara, California, June 4, 2015.

Marion Suiseeya, K.R. (2016). Transforming Justice in REDD+ through a Politics of Difference Approach. *Forests, 7*(12):300.

McDermott, C.L., Coad, L., Helfgott, A., & Schroeder, H. (2012). Operationalizing social safeguards in REDD+: actors, interests and ideas. *Environmental Science & Policy, 21*, 63-72.

Mola-Yudego, B., & Gritten, D. (2010). Determining forest conflict hotspots according to academic and environmental groups. *Forest Policy and Economics, 12*(8):575-580.

Munden Project. (2012). *The financial risks of insecure land tenure: An investment view*. Prepared for the Rights and Resources Initiative by The Munden Project.

Patel, T., Dhiaulhaq, A., Gritten, D., Yasmi, Y., De Bruyn, T., Paudel, N.S., & Suzuki, R. (2013). Predicting future conflict under REDD+ implementation. *Forests, 4*(2):343-363.

Peskett, L. & Todd, K. (2013). Putting REDD+ safeguards and safeguard information systems into practice. *UN-REDD Policy Brief No.3.* The UN-REDD Programme.

Pham, T.T., Mai, Y.H., Moeliono, M., & Brockhaus, M. (2016). Women's participation in REDD+ national decision-making in Vietnam. *International Forestry Review, 18*(3):334-344.

Poudyal, M., Ramamonjisoa, B. S., Hockley, N., Rakotonarivo, O. S., Gibbons, J. M., Mandimbiniaina, R., ... & Jones, J. P. (2016). Can REDD+ social safeguards reach the 'right' people? Lessons from Madagascar. *Global Environmental Change, 37*:31-42.

RECOFTC. (2015). *Mainstreaming gender into forest policies in Asia and the Pacific*. RECOFTC – The Center for People and Forests, Bangkok, Thailand.

RECOFTC. (2017). *Social forestry and climate change in the ASEAN region: Situational analysis 2016.* RECOFTC – The Center for People and Forests, Bangkok, Thailand.

Redpath, S.M., Young, J., Evely, A., Adams, W.M., Sutherland, W.J., Whitehouse, A., & Gutierrez, R. (2013). Understanding and managing conservation conflicts. *Trends in ecology & evolution, 28*(2):100-109.

Reimann, C. (2004). Assessing the state-of the-art in conflict transformation: reflecting from a theoretical perspective. In A. Austin, M. Fischer & N. Redpers, eds. *Transforming ethno-*

political conflict: the Berghof handbook. Berlin, Germany, Verlag fur Sozialwissenschaften.

Ribot, J., & Larson, A.M. (2012). Reducing REDD risks: affirmative policy on an uneven playing field. *International Journal of the Commons, 6(*2):233-254.

Sikor et, T., Gritten, D., Atkinson, J., Huy, B., Dahal, G.R., Duangsathaporn, K., Hurahura, F., Phanvilay, K., Maryudi, A., Pulhin, J., Ramirez, M.A., Win, S., Toh, S., Vaz, J., Sokecha, T., Marona., S., & Yaqiao, Z. (2013). *Community forestry in Asia and the Pacific: Pathway to inclusive development.* RECOFTC-The Center for People and Forests, Bangkok, Thailand.

Tegegne, Y.T., Ramcilovic-Suominen, S., Fobissie, K., Visseren-Hamakers, I.J., Lindner, M., & Kanninen, M. (2017). Synergies among social safeguards in FLEGT and REDD+ in Cameroon. Forest Policy and Economics, 75:1-11.

Trung L.Q., Phuong V.T., Yang A.L. & Hai V.D. (2015). The distribution of powers and responsibilities affecting forests, land use, and REDD+ across levels and sectors in Vietnam: A legal study. *Occasional Paper 137*. Center for International Forestry Research (CIFOR), Bogor, Indonesia.

UNFCCC. (2011). The Cancun Agreements: Outcome of the work of the Ad Hoc Working Group on Long-term Cooperative Action under the Convention. *Report of the Conference of the Parties on its sixteenth session held in Cancun 2010*. FCCC/CP/2010/7/Add.1.

WCS. (2013). *Reduced Emissions from Deforestation and Degradation in Seima Protection Forest, Cambodia. Wildlife Conservation Society* (available at: https://s3.amazonaws.com/CCBA/Projects/Reduced_Emissions_from_Deforestation_and_Degradation_in_Seima_Protection_Forest/Seima_Protection_Forest_08132013.pdf).

WRM. (2015). *REDD: A collection of conflict, contradictions and lies. World Rainforest Movement*, Montevideo, Uruguay.

Yasmi, Y., Schanz, H. & Salim, A. (2006). Manifestation of conflict escalation in natural resource management. *Environmental Science & Policy, 9*(6):538-546.

Yasmi, Y., Kelley, L., Murdiyarso, D., & Patel, T. (2012). The struggle over Asia's forests: An overview of forest conflict and potential implications for REDD+. *International Forestry Review, 14*(1):99-109.

Yeang, D., Sherchan, K., Heffernan, J., Chapman, S.M., Dooley, B. & Engbring, G. (2014). *Carbon rights and benefit sharing in Cambodia.* Policy Brief. REDD+ Asia-Pacific Community Carbon Pools Programme, Cambodia.

13

Productivity and Cost Analysis of Forest Harvesting Operation in Matang Mangrove Forest, Perak, Malaysia

Albert Empawi Tindit[1], Seca Gandaseca[2,] *, Laurna Nyangon[3] and Ahmad Mustapha Mohamad Pazi[4]

[1,2,3,4] Department of Forest Production, Faculty of Forestry, Universiti Putra Malaysia, 43400 Serdang, Selangor Darul Ehsan Malaysia.

[1] aempawi@gmail.com

[2] seca@upm

[3] laurnanyangon@gmail.com

[4] muss.pazi@gmail.com

* Correspondence author: seca@upm.edu.my

Abstract: Matang Mangrove Forest is under systematic management since 1902 and still considered as the best managed mangrove forest in the world. This research was conducted to measure the time and productivity of forest harvesting operation and also to analyze the cost and revenue of mangrove forest harvesting operation at Matang mangrove forest. This project had been carried out in cooperation with Seri Sepetang Enterprise, one of the harvesting licenses in Kuala Sepetang, Perak. Data collections were taken in every station starting from standing tree until to the Kiln-Drying jetty. The data then calculated by using the formulas of productivity and cost analysis. As the result, the productivity for felling, bucking and debarking, the manual skidding using wheel-barrow and the water transportation are 1.84 tan/hour, 3.82 tan/hour and 4.64 tan/hour respectively. The cost for each operation of 9 tan log volume for felling, bucking and debarking, the manual skidding using wheel-barrow and the water transportation are RM 56.88, RM 10.80 and RM 36.72 respectively. As the revenue, the company paid RM 260 per 9 tan of log for the in-forest operation (felling, manual skidding and loading to the ship) and pay RM 80 per 9 tan for the water transportation, and they gained the net profit of RM 192.32 and RM 43.28 respectively. The average of forest harvesting operation is twice operation in a day (equal with 2 x 9-ton volume of log production a day), so they will gain a double profit. In conclusion, the forest harvesting operation is sustainably managed for supplying the raw material of charcoal industries in Matang mangrove forest. Since, they work manually and spend much energy in this forest harvesting operation, so for further study it recommends to conduct the ergonomics evaluation during forest harvesting operation at Matang Mangrove Forest.

Keywords: Time Study; Productivity; Unit Cost; Cost Analysis; Mangrove; Forest Harvesting

---

1. Introduction

Matang mangrove forest known as one of the most sustainable management system around the world especially in forestry resources and have an important role in supplying constant yield of forest product (Jusoff, 2009). Borges et al. (2003) reported that mangrove is among the most productive coastal intertidal ecosystem in the world, confined to the tropics and subtropics, which dominate approximately 75% of the world's coastline between 25o N and 25o S. Indonesia, Thailand, Malaysia, and India are such country in Asia where the mangrove population distributed (Alongi, 2008). Matang Mangrove Forest is under systematic management since 1902 and still considered as the best managed mangrove forest in the world. The present study on silvimetrics assessed the ongoing forest management, which includes a first thinning after 15 years, a second thinning after 20 years and clear-felling of 30-year old forest blocks, for its efficiency and productivity in comparison to natural mangroves (Goessens et al., 2014).

Forest harvesting is among the contributor to the socioeconomic activities in Matang mangrove forest beside aquaculture, agriculture and ecotourism. The main product of forest harvesting in mangrove forests are fuel wood, charcoal, and poles. Mangrove forest is the most productive ecosystem for services to the environment and people as the demand is increasing nowadays (WWF, 2013). Forest harvesting refers to the process of cutting down trees and the process of delivering them to a mill or log pond. Most forest operation will aim for high productivity and gain high profit. What make the forest harvesting is challenging when it comes to the term of efficient in productivity and cost? Each forest operation has different types of system they used depends on the area of the operation done. Besides that, the productivity incurred is less compare to other type forest production like dipterocarp forest, hill forest and others. Due to the challenge in forest harvesting activities, work study including time studies and productivity studies can improve the work performance to increase the efficiency and effectiveness.

Most of the works or operations are manually done by human and the use of machine is very limited due to the conditions in the mangrove forest. The tides also give effect to the operation hour. If all of these restraints are not well managing into an efficient forest operation system, both productivity and efficient cost of the operation are cannot be achieved. This research was focused on the time study, productivity and total cost of the mangrove forest harvesting operation for a certain number of logs.

2. Materials and Methods

2.1. Study Site

This research project was conducted in one of the forest harvesting area at Matang Mangrove Forest, Perak. Seri Sepetang Enterprise is one of the charcoal's producers at Matang Mangrove Forest, where this research project had been carried out; in one of their license compartment. The Matang Mangrove Forest Reserve located on the northwest coast of Peninsular Malaysia as shown in Figure 1.

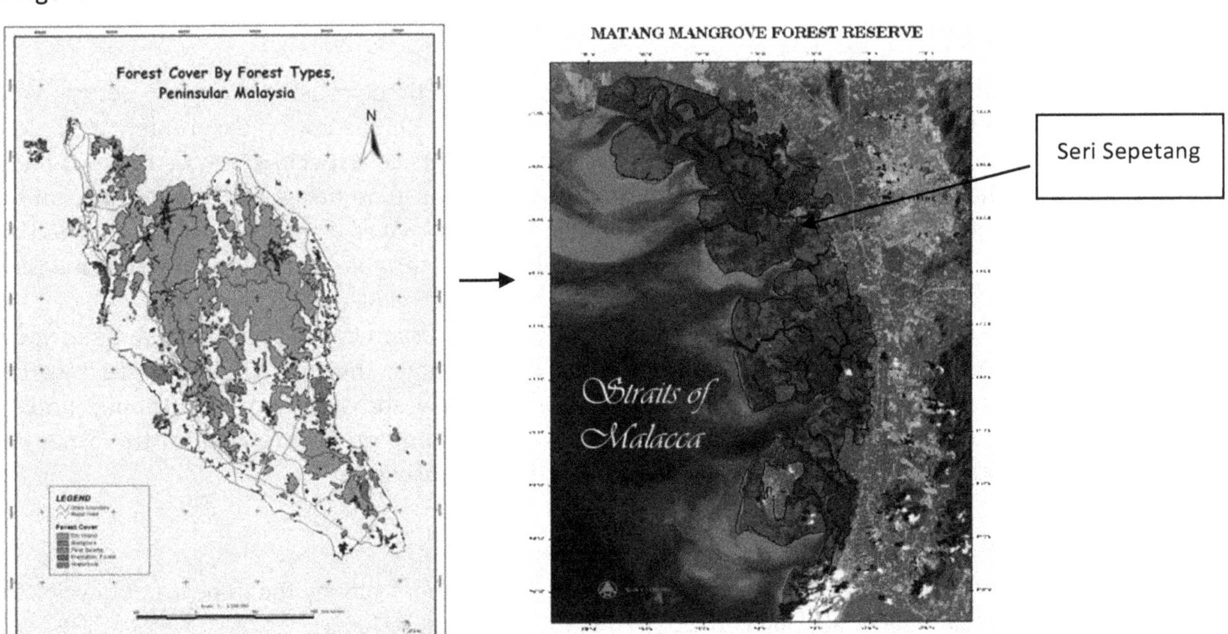

Figure 1: The Study Site at Matang Mangrove Forest Reserve

2.2. Work Element and Data Collection of Forest Harvesting Operation

Figure 2 shows the forest harvesting operation at Matang Mangrove forest. This operation consists of 3 main operations, such as: i) felling, bucking and debarking; ii) Loading, manual skidding using wheel-barrow and unloding at the log pond at the river-side; and iii) Loading on boat, water transportation, and unloading log at the Kiln-drying jetty.

Figure 2: Work Element of Forest Harvesting Operation

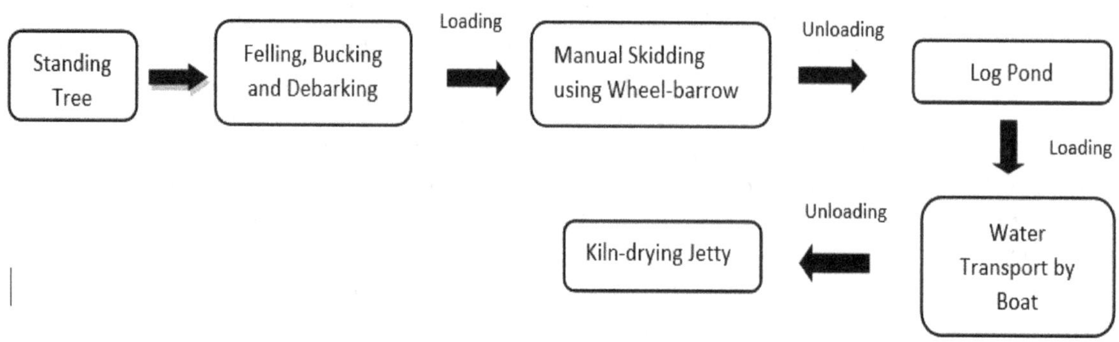

Usually on bucking activity, one felled tree is divided into 11 to 12 small logs, with length of each log is 1.6 m and has the average weight of 0.07ton/log. Then about 5-6 logs are loaded on the wheel-barrow per trip. Data collections were taken in every station starting from standing tree, felling, bucking and debarking, manual skidding, loading on the boat, water transport and unloading at the Kiln-Drying jetty.

2.3. Time Study and Productivity of Machine/Equipment

Time consumption and productivity during forest harvesting operation at Matang Mangrove forest was recorded using stop watch, digital camera and distance tape. This baseline data was used for determining the time consumption of each element of work by following certain formula.

2.3.1 Time Study

Time study is the systematic observation of the time consumption to define work performance during harvesting activities (Currie & Faraday, 1989). Work measurement technique for recording the times and rates of working for the elements of a specified job carried out under specified conditions and for analyzing the data to obtain the time necessary for carrying out the job at a defined level of performance. This baseline time data was used for determining the time consumption of each element of work by following certain formula, including time for chainsaw, skidding, loading and unloading, and hauling (water transport) (FAO, 1992).

For time study, the time taken for each activity was recorded based on the actual time, speed or effort of the worker during their work daily. For example, the time of felling, bucking and debarking were recorded once the feller start the chainsaw and the time was stopped until all these three elements of operation were done. It was the same procedure also for the other two main operations where 3 replications were done for each processed.

- Felling, Bucking and Debarking Time

Felling production is estimated by dividing the volume per cycle by the minutes per cycle. The time per Tree for felling, bucking and debarking, follow the equation 1:

$$T = a1 + b1 + c1$$

where; a is the felling time per cycle, b is the bucking time per cycle and c is the debarking time per cycle

- *Skidding Time*

Skidding production is estimated by dividing the volume per load by the time per around trip. The round-trip time, T, is the sum of the time for travel unloaded, hooking, travel loaded, and unhooking, follow the equation 2:

$$T = a\,N + b1\,x1 + b2\,x2$$

where; a is the combined time for hooking and unhooking per log, b1 is the minutes per meter for unloaded travel, b2 is the minutes per meter for loaded travel, x1 is the distance from the landing to load pickup point and x2 is the distance from the load pickup point to the landing.

- *Loading and Unloading Time*

Loading production is estimated by dividing the volume per cycle by the minutes per cycle. The time per log for loading single logs, follow the equation 3:

$$T = a1 \text{ and } T = b1$$

where; a is the loading time per cycle and b is the loading time per cycle.

2.3.2. Productivity:

The productivity (P) was calculated according to equation 4 (Norizah, 2012):

$$P = \text{Volume} / \text{Time}$$

where; V = volume per log (m^3)

2.4. Cost *Analysis of Logging Operation at Matang Mangrove Forest*

2.4.1. Machine Rate

The machine rates for a chainsaw, a manual skidding using wheel barrow and water transport by boat are shown on Table 1 (FAO, 1992).

Table 1: Cost Component and Equations for determining the Machine Rate.

No.	Cost Component	Equation 5 -12
1	Depreciation	= (delivery cost) x 0.9 / life in hours
2	Interest	= (delivery cost) x 0.6 x 10% / average hours per year
3	Insurance	= (delivery cost) x 0.6 x 3% / average hours per year
4	Taxes	= (delivery cost) x 0.6 x 2% / average hours per year
5	Labor	= Labor cost per year x (1 + f) / average hours per year where: f is social costs of labor as decimal
6	Fuel	= 0.2 x GHP X LF X CL ; Where: GHP is gross engine horsepower, CL is cost per liter for fuel, and LF is load factor
7	Oil and Grease	= 0.1 x fuel cost
8	Servicing and Repair	= 1.0 x depreciation
9	Others	Certain contingency amount needed

Although the machine rates share the same general format, there is flexibility to represent costs that are specific to the machine type, particularly in the calculation of the operating costs as shown on Table 1.

2.4.2. Cost Analysis

For analysis of the operation cost, first is to determine the unit cost (RM/ton), then if for the cost operation is depending on how many log volume produced on certain time or day or period (FAO, 1992)

- *Unit Cost*

The unit cost of machine is the total expenditure by a machine to produce one unit of production, so it is equal with machine rate divide by productivity, as follow the equation 13:

Unit Cost = Machine Rate / Productivity, where: Unit Cost is in RM/ton.

- *Cost Operation*

The cost operation is the expenses which are related to the operation to produce certain units of production, and it is calculated follow the equation 14:

Cost Operation (RM) = Unit Cost (RM/ton) x Volume of Log Production (ton)

3. Results and Discussion

3.1. Time and Productivity

Table 2 shows that the time and productivity of Matang mangrove forest harvesting operation after the study of time and productivity were conducting based on the worker's daily work. Repetition of each operation was done as much as 3 times to get the average time of the operations. For felling, bucking and debarking, the systems were operated by chainsaw where after the allowable tree was cut down, it was continued with bucking. Bucking was the process where the tree was cut according to the predetermined length by Forestry Department, 1.6 meter. A man-made pole which length 1.6 m was used to mark the measure by an axe first and then it can cut into 1.6 m length log. At the same time, debarking process was done where the bark was removed using man-made hammer, which made from the mangrove root. All these processes were done continuously by 2 or 3 man.

In Matang Mangrove forest operation, the buttress of the tree was left or not removed for many purposes. The purposes were to maintain the stability of the soil in the operation area after harvested and also as the shield to the new seedling for re-plantation. The buttress cannot be destroyed directly or been removed because it was the habitat of many life in mangrove forest.

Table 2: Time Study and Productivity of Forest Operation at Matang Mangrove Forest

Operations	Time (Minutes)	Productivity (Ton/hour)
Felling, Bucking, Debarking	28.67	1.84
Manual Skidding (Wheel Barrow)	6.60	3.82
Water Transportation	116.4	4.64
TOTAL	151.67	10.22

Then, manual skidding was done by using man-made wheel barrow. A wheel barrow can load around 5 to 6 log that had cut into 1.6 m length. The wheel barrow then push on the track which specially made to skid it from the felling area to the log pond. The idea of using wheel barrow in mangrove forest harvesting was due to the condition of soil is soft which cause the potential development of flotation vehicles suited to the operations without road constructions (Silversides and Sundberg, 2013). For water transportation, the log transport by boat was the best way to access the mangrove operation area. A boat can load 9 tan of log per trip which was equal to 130 – 150 logs. The boat was powered by Yanmar engine that can speed 9 km/hour.

3.2. Cost Analysis of Forest Operation at Matang Mangrove Forest

3.2.1. Machine Rate

The result of machine rates for a chainsaw, manual skidding using wheel-barrow and the water transportation by boat are shown on Table 3. The economic life time is the period over which the equipment can operate at an acceptable operating cost and productivity (FAO, 1992). In this study, the economic life time for chainsaw, wheel-barrow and boat are 5,000 hours, 5,000 hours and 15000 hours, respectively.

Table 3: Cost Component and Machine/Equipment Rate (RM/hour)

No.	Cost Component	Rate of Machine/Equipment (RM/hour)		
		Chainsaw	Wheel Barrow	Boat
1	Depreciation	0.45 ± 0.08	0.015 ± 0.01	0.9 ± 0.21
2	Interest	0.15 ± 0.05	-	1.8 ± 0.42
3	Insurance	0.045 ± 0.016	-	0.54 ± 0.13
4	Taxes	0.03 ± 0.01	-	0.36 ± 0.08
5	Labor	3.58 ± 0.72	3.58 ± 0.27	3.58 ± 0.26
6	Fuel	4.48 ± 1.39	-	8.06 ± 0.32
7	Oil and Grease	0.448 ± 0.139	-	0.806 ± 0.032
8	Servicing and Repair	0.45 ± 0.08	1	0.9 ± 0.21
9	Other	2	-	2
	Total	11.63 ± 2.17	4.6 ± 0.47	18.95 ± 0.13

Since, the manual skidding using wheel-barrow, so the equation for interest, insurance, taxes, fuel, oil and grease are not taken into calculation, and the servicing and repair in average is about RM 1 per hour.

3.2.2. Cost Analysis

Table 4 shows that the cost analysis forest operation in the mangrove forest per trip. A boat can transport of 9 ton of log from the log pond to the jetty of the Kiln-Drying. As shown on the Table 2, the forest operation in the forest was handled by 2-3 of workers. The cost operation from felling until manual skidding is RM 77.86, while they were paid RM 260 per 9 ton for these forest operations. They gained RM192.32 for each trip or 9 ton of log started from felling until loaded the log into the boat. So, each worker gained approximately RM64.11/person (if done by 3 workers) and RM96.16/person (if only done 2 workers). By the way, a boatman gained RM43.28 per trip for transported the log from log pond to the Kiln-Drying jetty. The boatman job was just transported the log with his own boat to the jetty, meanwhile the unloading and loading of log were handled by other persons.

Table 4: Cost Analysis of Forest Operation (per 9 tan of log volume) at Matang Mangrove Forest

Operations	Unit Cost (RM/ton)	Log Volume (ton)	Cost RM/9 ton	Wages per 9 ton	Net Profit per 9 ton
Felling, Bucking, Debarking	6.32		56.88	260	192.32
Manual Skidding	1.20	9	10.80		
Water Transportation	4.08		36.72	80	43.28
TOTAL			104.40	340	235.60

Note: One wage per 9 ton production of forest operations (felling, bucking, debarking & skidding).

Based on the Table 4, this forest harvesting operation was well managed in term of the net profit of the workers gained because they saved more profit compare to the cost operation. The time study and productivity of forest operation at Matang mangrove forest were showed that they managed the forest operation efficiently where as we know that mangrove forest operation was challenging in term of hard to access by machinery. So, they had to handle all of the works there manually and this proved that they were very experienced and strong workers. Many studies were reported that the skidding distance is the most significant variable for the productivity (Kluender et al. 1997; Wang et al. 2004). It shows that the workers here were efficiently managed their work even without the help of machine to do the hard work like the skidding. This also proved that they managed to settle down their work in a short time by only depend fully on their own strengths and competences, based on the time study result. They also proved their works were managed well when they had to race with the tide phenomenon in the mangrove forest to settle down their work in time to avoid losses.

According to Naghdi (2005), there were some variables that can influenced the time consumption of skidding for examples, not well managed felling time (felling just a few days before skidding), skid trail with poor design and also workers were keeping in pressure. This matter was not just only affect the skidding process, but it was also affect the other activities productivity and time consumption which lead to loss. Nurminen et. al (2006) were also said that the productivity and time consumption of harvesting were depend on the conditions of the standing tree, the skills of operator and the working techniques. Moreover, this mangrove forest operation was also managed to avoid unnecessary loss for each of their works, even they managed to gain more profit compare to the cost operation they had to bear. It was proved after the analysis of cost and revenue had shown a great net profit for the forest operation. In conclusion, Seri Sepetang Enterprise had produced a good productivity and managed the cost operation well.

Since, they work manually and spend much energy in this forest harvesting operation, so for further study it recommends to conduct the ergonomics evaluation during forest harvesting operation at Matang Mangrove Forest. Besides that, this study was conducted to provide a baseline for the time study and productivity in the mangrove forest operation since that nowadays there were a few areas only that still actively operate in mangrove area in Malaysia. In the future, maybe this project can be the benchmarking for other mangrove forest operation for comparison on the efficiency of the system they use, maybe with new technology and future operation system.

5. Conclusions

In Matang mangrove forest, the forest harvesting operation is sustainably managed for supplying the raw material of charcoal industries. In this study, the productivity for felling, bucking and debarking, the manual skidding using wheel-barrow and the water transportation are 1.84 ton/hour, 3.82 ton/hour and 4.64 ton/hour respectively. All these operations were done continuously by 2 or 3 workers. The cost operations per 9-ton log volume for felling, bucking and debarking, the manual skidding using wheel-barrow and the water transportation are RM56.88, RM10.80 and RM36.72 respectively. As the revenue, the company paid RM260 per 9 ton of log for the in-forest operation (felling, manual skidding and loading to the ship) and paid RM80 per 9 ton for the water transportation, and they gained the net profit of RM192.32 and RM43.28 respectively. The average of forest harvesting operation is twice operation in a day (equal with 2 x 9-ton volume of log production a day), so they will gain a double profit. Since, they work manually and spend much energy in this forest harvesting operation, so for further study it recommends to conduct the ergonomics evaluation during forest harvesting operation at Matang Mangrove Forest.

Author Contributions: Albert Empawi Tindit and Seca Gandaseca conceived and designed the experiments; Albert Empawi Tindit, Seca Gandaseca, Laurna Nyangon and Ahmad Mustapha Mohamad Pazi performed the experiments, analyzed data and contributed reagents/material/analysis tools; Albert Empawi Tindit and Seca Gandaseca wrote the paper.

Acknowledgments

The appreciation goes to Forestry Department of Kuala Sepetang for providing the study site and data collection in Matang Mangrove Forest. I am so grateful to Faculty of Forestry for providing laboratory facilities and staff during this study.

References

Alongi, D. M. (2008). Mangrove forests: resilience, protection from tsunamis, and responses to global climate change. *Estuarine, Coastal and Shelf Science*, *76*(1), 1-13.

Borges, A. V., Djenidi, S., Lacroix, G., Théate, J., Delille, B., & Frankignoulle, M. (2003). Atmospheric CO2 flux from mangrove surrounding waters. *Geophysical Research Letters*, *30*(11).

Currie, R.M. & Faraday, J.E., 1989. Work Study. Pitman Publishing, London, UK, pp. 47-48.

FAO, (1992). *Cost Control in Forest Harvesting and Road Construction.* ISBN 92-5-103161-4. Publication Division, FAO, Rome, Italy.

Goessens, A., et al., (2014). Is Matang Mangrove Forest Sustainably Rejuvenating after More than a Century of Conservation and Harvesting Management?. *Journal of Plos One*. http://dx.doi.org/10.1371/journal.pone.0105069.

Kluender, R., Lortz, D., McCoy, W., Stokes, B., & Klepac, J. (1997). Productivity of rubber-tired skidders in southern pine forests. *Forest Products Journal*, *47*(11/12), 53.

Naghdi, R., RAFATNIA, N., Sobhani, H., Jalali, G., & Hosseini, M. (2005). A survey of the efficiency of Timberjack C450 wheeled skidder in Shafaroud forests in Guilan province.

Nurminen, T., Korpunen, H., & Uusitalo, J. (2006). Time consumption analysis of the mechanized cut-to-length harvesting system. *Silva Fennica*, *40*(2), 335.

Norizah, K., Hasmadi, M. I., Kamaruzaman, J. (2012). Operational efficiency of rimbaka timber harvester in hilly tropical forest. *Journal of Tropical Forest Science. Vol* 24(3), 368-378.

Silversides, C. R., & Sundberg, B. (2013). *Operational Efficiency in Forestry: Vol. 2: Practice* (Vol. 32). Springer Science & Business Media.

Wang, J., Long, C., & McNeel, J. (2004). Production and cost analysis of a feller-buncher and grapple skidder in central Appalachian hardwood forests. *Forest products journal*, *54*(12), 159.

WWF (World Wide Fund). (2013). *Mangrove Forest*. Retrieved from http://www.wwf.org.my/. in September 8, 2015.

14

Mainstreaming community-based forest management in West Sumatra: Social forestry arguments, support, and implementation

Ferdinal Asmin [1,*], Dudung Darusman [2], Iin Ichwandi [2] and Didik Suharjito

[1] West Sumatra Forestry Office, Indonesia; ferdinalasmin76@yahoo.com
[2] Bogor Agricultural University, Indonesia; dudungdarusman@gmail.com, iichwandi@yahoo.com, dsuharjito@gmail.com
* Correspondence author: ferdinalasmin76@yahoo.com; Tel.: +62-812-903-50978

Abstract: Although social forestry in Indonesia is envisioned as a policy for recognizing local practices to forest management, research is still limited. This research describes conditions of social forestry policy in West Sumatra Province as a form of mainstreaming community-based forest management. This paper provides the context of social forestry arguments, its support, and subsequent implementation. The research approach is qualitative, using a case study method. Data collection was conducted through unstructured interviews, field observations, and document studies. The analysis used categorization and coding, historical analysis, document analysis, and descriptive policy analysis. The findings revealed that the arguments for social forestry schemes were based on the persistence of state forest conflicts, forest degradation and deforestation threats, as well as human resource limitations of forestry officers. The Provincial government then initiated stakeholder support, mainly from non-governmental organizations. Social forestry implementation at the site in West Sumatra thus focused on providing development assistance programs after granting management rights to local people, as well as initiating similar schemes in other villages. Our discussions considered challenges that should be addressed, including the approach to granting management rights to secure a management area, the process of developing participatory institutions, synchronizing provincial government policies to overcome forest degradation and deforestation, and initiating activities for strengthening community solidarity and agency.

Keywords: management rights; institutions; social forestry; development assistance programs

---

1. Introduction

Several studies show that community-based forest management (CBFM) has succeeded in some developing countries, such as India, Thailand, Nepal, Vietnam, Ghana, Chinese Taipei, Tanzania, Guatemala, Sudan, Peru, Nicaragua, and Mexico (Tole, 2010). Wiersum (2004) also concluded that CBFM achieved a significant level of support in the early 21th century as a strategy for tropical countries, and thereafter spread to other developed countries in Europe, North America, and Australia. Furthermore, researchers believe that an effective CBFM strategy can streamline the government's work administratively and ensure equitable benefit distribution for local and indigenous people (Nayak & Berkes, 2008; Suharjito, 2009).

CBFM is a forest management system on any lands within local community territory and state forests in form of collaborative management (Wiersum, 2004). The concern with CBFM development has encouraged various research on CBFM activities. Researchers have described a number of CBFM activities, such as those focused on institutions (Pagdee, Kim, & Daugherty, 2006), conservation (Vodouhê et al., 2010; Sawitri and Subiandono, 2011; Pietrzyk-kaszy et al., 2012), knowledge (Ito et al., 2005), decision-making (Hujala, Tikkanen, Hanninen, & Virkkula, 2009), and goal attainment (Masozera, Alavalapati, Jacobson, & Shrestha, 2006). Gilmour (2016) also revealed that CBFM evolved concurrently with other development events related to the forestry sector. Recent issues include climate change policy, illegal logging eradication, and payment for environmental services, which are generating widespread attention among the national and regional governments in Indonesia.

The West Sumatra Provincial Government (WSPG) through the West Sumatra Forestry Agency (WSFA) has encouraged social forestry (SF) schemes as a form of CBFM mainstreaming. The WSPG has allocated 500,000 hectares of the state forest for SF schemes as a CBFM development target in West Sumatra since 2012. The target could fulfill 20% of the national target. There is no other provincial government in Indonesia that has set such an ambitious target. As a result, WSPG has become a pilot province of SF development in Indonesia.

In the last five years, lessons have been learned from the WSPG's policy in SF development regarding their arguments, support, and implementation frameworks. However, there has been no scientific research to reveal the lessons learned from SF development in West Sumatra. Such research is important to ensure the achievement of SF objectives at local, regional, and national levels. The lessons learned from the WSPG also provide a wealth of research opportunities due to the positive track record of local forest management schemes instituted by the Minangkabau people, the dominant ethnic group in West Sumatra (Michon et al., 1986; Martial et al., 2012; Hamzah et al., 2015; Asmin et al., 2016; Asmin et al., 2017a).

WSPG's policies are therefore seeking to mainstream CBFM as a broader development strategy. According to Dunn's (1981) theory, the arguments in favor of a policy are related to the rationale for that policy. Policy reform over time might be influenced by various events that relate to forestry development at regional and national levels. Furthermore, current policies form the rationale for shaping policies elsewhere in Indonesia and a basis for developing a broader policy approach (Dunn, 1993; Sutton, 1999). Meanwhile, support for SF development include governmental activities (such as program, actions, and schemes) and regulations. Regarding the policy definition from Hill and Hupe (2002), governmental activities are also considered as policy. Simultaneously, the implementation of the activities is part of the policy process (Dunn, 1981; Hoppe, 1993). In relation to SF development, implementation could involve all efforts to realize programs, actions, and schemes as well as regulations.

According to Ministerial Regulation of Environment and Forestry No. P.83/MenLHK/Setjen/Kum.1/10/2016, SF is a forest management system by a local community outside or inside a state forest. There are four main forms of SF schemes, i.e. community plantation forest (*hutan tanaman rakyat*, HTR), community forest (*hutan kemasyarakatan*, HKm), village forest (*hutan desa*, HD), and customary forest (*hutan adat*, HA). The SF schemes are allocated for three formal categories of state forest in the Indonesian system, i.e. production, protection, and conservation forests. The HTR scheme is only applicable in production forests, while HKm and HD schemes are provided for protection and production forests. Meanwhile, the HA scheme occurs in communal forests, including in all three formal categories of state forest areas.

Conceptually, SF initiatives have evolved from an initial focus on the biological constraints of trees and forests, to the socio-economic constraints, to the local institutional constraints, and most recently to the government policy constraints (Dove, 1995; Wiersum, 1999; Oliver, 2014). Wiersum (2004) has also stated that all SF initiatives are forms of CBFM mainstreaming. Based on these concepts, this research aims to describe SF initiatives in West Sumatra by elaborating the provincial government's arguments, support, and implementation. We start with the history of CBFM policy and the characteristics of SF development in West Sumatra, because the essence of CBFM history in the province contain the governance background to initiate all forms of SF schemes. We then explain WSPG support for SF initiatives by identifying all forms of programs, actions, schemes, and regulations. The paper further explains government policy efforts to implement SF initiatives. Government activities indicate the processes for initiating an SF scheme. Related to the arguments, support, and implementation, we also summarize the CBFM policy in other countries. We then consider our results to discuss four challenges that should be overcome to ensure the objectives of SF initiatives.

2. Materials and Methods

Our research uses a qualitative approach, which refers to meanings, concepts, characteristics, and description (Berg, 2001). The research process is characterized as inductive, interpretive, and constructivist (Bryman, 2004). Therefore, this research requires dialectical interaction between researcher and research object (Guba & Lincoln, 1994), allowing empathy, intentionality, and interpretation of subject statements and actions which are known as hermeneutics (Suharjito, 2014). Our method is a case study of governmental activities and SF schemes related to CBFM policy in West Sumatra. This research was conducted from September 2015 to August 2016.

Data were collected using various techniques, including unstructured interviews with 13 key informants from governmental officers and non-governmental organization (NGO) activists, field observations, and document studies. The determination of informants is based on their roles within governmental and NGO structures, their experiences related to CBFM or SF activities, and their understandings of local history and culture for ensuring reliability (Davis & Wagner, 2003). For governmental officers, we interviewed the Head of WSFA with 30 years of experience in West Sumatra, the Head of Social Forestry Task Force with 26 years of experience, and some key informants from WSFA officers with minimal experience 15 years, both in provincial and district offices. For NGO activists, we interviewed a project manager with 10 years of experience in West Sumatra and some community facilitators who facilitated communities SF development that had minimal 5 years of experience.

Field observations aimed to observe the implementation of SF schemes, such as the village forest in Simancuang, as well as formal and informal meetings related to SF programs and activities. Meanwhile, document review consisted of governmental regulations, reports, statistical information, photos, recordings, and maps. Our analysis used categorization and coding, historical analysis, document analysis, and descriptive policy analysis, and was further strengthened by our own knowledge about the system and local language.

This research focuses on the WSFA's arguments, their support, and implementation that led the vast proliferation of SF schemes in West Sumatra. The argument related to WSFA's reasons emerged from a historical review of CBFM development in West Sumatra, which is further evident from from the selection of particular SF schemes. To develop the SF schemes, WSFA created and implemented various programs, actions, and regulations that that were also assisted by NGOs and other governmental institutions. Meanwhile, WSFA's implementation mechanisms were studied by reviewing the existence of West Sumatran CBFM task forces and development assistance programs. We compared the research findings with the research reports from other developing countries and discussed it in the context of: (1) the granting of management rights, (2) institutions, (3) challenges of forest degradation and deforestation, and (4) initiation of development assistance programs.

3. Results and Discussion

3.1. Government arguments to encourage SF development

3.1.1. The history of CBFM in West Sumatra

Government and corporate roles have dominated forest resource management in West Sumatra since the 1970s. Based on the concession permits of the forestry sector (Brown, 2001), the period from 1980 to 1998 was one of timber extraction in West Sumatra. Almost 40% of the state forest of West Sumatra was extracted by timber corporations (Brown, 2001; KemenLHK, 2015).

Meanwhile, local community roles tended only to be recognized in the non-state forest areas through community empowerment programs, which were funded by the government. According to our informants, there were various activities of community empowerment, such as reforestation and afforestation of pine in the 1970s, followed by community forests in the early 1990s, micro credit for agro-silvo-pastoral activities, and sericulture in the 1990s.

The forest cover of West Sumatra declined steadily since 1990, especially in state forest areas. The degradation and deforestation rates of state forests are high, although the forest cover percentage is still higher than other provinces in Sumatra (KemenLHK, 2015; FWI/ DFW, 2001). Based on land use change data from the WSFA, approximately 4 % of primary and secondary forest areas have changed into shrubs in the period from 2000 to 2013. Degradation and deforestation problems have become a priority issue in WSFA's strategic planning since 2000. Regarding WSFA's financial documents between 2006 and 2015, the financial allocation for land and forest rehabilitation (LFR) and forest protection and security (FPS) activities were higher than other forestry administration activities.

LFR activities were more intense after the Indonesian Government initiated the National Movement of Land and Forest Rehabilitation in 2003. The Indonesian Government also initiated other movements to strengthen LFR activities with familiar slogans like "One Person One Tree" and "One Billion Indonesian Trees" between 2010 and 2014. These movements provided free seedlings and established many nurseries managed by local communities. However, LFR activities were only able to handle circa 20% of 500,000 hectares of degraded lands in West Sumatra (Dinas Kehutanan Provinsi Sumatera Barat, 2011). There are various challenges to rehabilitate degraded lands, especially related to LFR outcomes and impacts on forest ecosystem improvement, local participation, and local community welfare (Nawir et al., 2008; Dinas Kehutanan Provinsi Sumatera Barat, 2011).

The WSFA response to degradation and deforestation issues included increasing law enforcement against illegal logging activities as a part of FPS activities. Law enforcement was strengthened with the Presidential Instruction on Illegal Logging Eradication in 2004. WSFA conducted repressive and preventive actions that involved other law enforcement agencies, including the police, army, and prosecutors. Furthermore, since 2005, WSFA initiated a village-based forest protection program that encouraged the participation of village governments by establishing a task force of forest protection at the village level. To provide legal support for FPS activities, WSFA also formulated the Provincial Regulation on Community Participation in Forest Protection in 2015.

The implementation of LFR and FPS activities deals with two main challenges. Based on our interviews, the first challenge is the lack of WSFA's technical officers to facilitate all LFR and FPS activities (Dinas Kehutanan Provinsi Sumatera Barat, 2012a). The number of technical officers, such as extension officers and forestry police officers, is inadequate, given the extent of state forest and the number of villages in West Sumatra. According to WSFA's statistics in 2015, the number of extension officers and forestry police officers was only 75 and 150 persons, respectively. If we compare the number of extension officers with the total villages near the state forest areas, the extension officers could only cover 10% of these villages. Meanwhile, if we compare the number of forestry police officers with the extent of state forests, one forestry police officer must protect ten thousand hectares of state forest.

Another challenge is communal claims (*ulayat*) on state forests by local communities. Our informants revealed that, in formal and informal meetings, local communities frequently contested government considerations in the context of state forest status and function. Local people have criticized the unilateral actions of the government in the designation of their communal lands as state forest. This complaint is often made by other communities and communal claims that represent tenurial problems throughout Indonesia (Nurrochmat, Darusman, & Ruchjadi, 2014). Our informants also revealed that communal claims have complicated law enforcement, one of the jobs of the government.

To cope with these challenges, WSFA began to encourage more intensively mainstreaming CBFM activities when the community plantation forest (HTR) scheme was introduced in 2007. HTR was based on Forestry Ministerial Regulation No. P.23/Menhut-II/2007. Regarding the WSFA reports between 2009 and 2012, WSFA frequently organized the explanation of policies, extension, and training on HTR in Sijunjung District and South Pesisir District. Both districts have acquired 5,345 hectares of state forest for HTR allocation through Forestry Minister Decree No. SK.402/Menhut-II/2009 dated July 6, 2009, and SK.356/Menhut-II/2009 dated June 18, 2009. However, based on government documents and our interviews, the licensing process took three years, while only one third of the HTR allocations were distributed due to a slowdown in the process, i.e. limited extension capabilities, weak mentoring, and complicated administration processes.

When village forest (HD) and community forest (HKm) schemes were introduced through Forestry Ministerial Regulation No. P.49/Menhut-II/2008 and No. P.37/Menhut-II/2007, respectively, WSFA also began facilitating the HD and HKm schemes since 2011. There are two pioneer villages for the HD scheme, i.e. Simanau Village in Solok District and Simancuang Alam Pauh Duo Village in South Solok District. Based on West Sumatra Governor Decree No. 522-43-2012 and 522-44-2012 dated January 19, 2012, the first HD area in West Sumatra included 1,783 hectares of protected forest. According to our informants, the NGO KKI Warsi facilitated the licensing for both villages, partnering with them since 2009. The licensing process was relatively fast: the verification process took only six months after the local community submitted its proposal, while the permits issuance took three months.

Thereafter, WSFA began intensively engaging with two NGOs (KKI Warsi and Qbar) to expand social forestry licensing preparations. Based on meeting documents from 2010-2016, both NGOs organized meetings and proposal documents. KKI Warsi intensively facilitated some local communities for HD schemes in Solok District and South Solok District, meanwhile, Qbar worked intensively to facilitate local communities for HKm schemes in Pasaman District and West Pasaman District. During the workshop on CBFM (on mainstreaming greenhouse gas reduction) on May 31, 2012, the West Sumatra Governor instructed WSFA to expand the implementation targets of HTR, HD, and HKm schemes to 500,000 hectares within five years (2012-2017).

To realize these ambitious targets, WSFA compiled a working plan of SF development to delineate state forests for SF schemes (Dinas Kehutanan Provinsi Sumatera Barat, 2012b) and established a task force under WSFA Head's Decree No. 522.4/1602/RHL-2012. The task force consists of stakeholders from government institutions and NGOs. The duties of the task force are to coordinate all actions to expand CBFM and facilitate local communities to propose HTR, HKm, and HD schemes. The existence of the task force also gained the interest of all districts in West Sumatra through routine intergovernmental meetings and appeals for the support of NGOs, such as KKI Warsi, Qbar, Kemitraan/UNDP, ICS, YCM, Walhi, SSS-Pundi, and FFI. However, based on recent WSFA reports, the task force was only able to facilitate 25% of the total implementation targets (see Table 1).

Table 1. The progress of social forestry schemes in West Sumatra

Schemes	Proposal from local communities (ha)	Working area from government (ha)	Management license (ha)
HTR	7,745	7,745	2,246
HKm	12,118	6,296	2,595
HD	106,874	50,086	38,598
Total	126,737	64,127	43,440

Source: The WSFA report in 2013 and result of social forestry coordination meeting on February 23-24, 2017

3.1.2. The characteristics of social forestry in West Sumatra

State forest allocations for SF schemes in West Sumatra are in protection and production forests, according to recent regulations of the Environmental and Forestry Minister No. P.83/MenLHK/Setjen/Kum.1/10/2016. The four priority schemes mentioned above (HTR, HKm, HD, and HA) differ in their features (see Table 2).

According to our informants, WSFA provided intensive support for the HTR, HKm, and HD schemes because the regulation for these schemes are more operational than the regulation for HA. Based on WSFA meeting documents, governmental support for the HA scheme began in 2013 after the Indonesian Constitutional Court accepted the judicial review of the *adat* forest definition in 2012. However, the complex procedure, which involves executive and legislative decision-making processes to recognize an *adat* community, is a major obstacle to HA implementation. The government recognizes an *adat* community when the *adat* community has complied with provincial regulations. More efforts at collaborative governance to recognize *adat* communities, as documented by Fisher et al. (2017), is needed.

Table 2. The features of social forestry schemes

Features	HTR	HKm	HD	HA
Forest status and function	Production forest	Production and protected forest	Production and protected forest	Production, protected, and conservation forest
Organization	Farmers group/ cooperatives/ individual	Farmers group	Village institution	Adat, or customary institution
Utilization	According to forest status and function, mainly for timber production	According to forest status and function	According to forest status and function	According to forest status and function
License duration	35 years	35 years	35 years	No explanation
Form of rights	Management rights	Management rights	Management rights	Communal rights
Administration procedure	Only executive (government)	Only executive	Only executive	Executive and legislative (Provincial representatives)

*community plantation forest (HTR), community forest (HKm), village forest (HD), and customary forest (HA)
Source: Ministerial Regulation No. P.83/MenLHK/Setjen/Kum.1/10/2016 on Social Forestry

Based on the WSFA report, the allocation of management rights for HTR, HKm, and HD are more likely in protection forest than in production forest. Only 14% of the management rights areas are in production forest. HD and HKm have become the priority schemes of WSFA as shown in Table 3. According to the governmental regulations, HD involves management rights to state forest within a village territory, meanwhile, HKm and HTR are management rights of state forest for given farmer groups, on which their livelihoods depend. In accordance with the size of the management areas (see Table 3), HD tends to obtain intensive assistance. This is likely because of the government's past experiences encouraging the first village forests in Simancuang and Simanau Villages.

Table 3 The range of social forestry area by the schemes in West Sumatra

Range	Scheme		
	HD	HKm	HTR
> 2.000 ha	9	2	-
1.001-2000 ha	11	3	-
501-1000 ha	7	5	2
100-500 ha	17	18	3
<100 ha	2	-	-

Source: WSFA report 2013 and meeting report on social forestry progress on February 23-24, 2017 in Padang

A new local institution was also established to manage the state forest in the context of the HD scheme, i.e. village forest management institution, called *lembaga pengelola hutan nagari* (LPHN). The terminology of LPHN was then applied in establishing local management institutions for each village. For example, based on field observations and document reviews owned by the Simancuang community, LPHN members consist of local and *adat* leaders as well as other local representatives. LPHN was established by a village head decree and has responsibilities and obligations to manage the village forest.

3.2. The support for social forestry schemes

Our observations from formal and informal meetings, which were held by WSFA, indicated that WSFA officers implemented SF policy in accordance with governmental regulations. Their choices of support and the actions they took depended on their understanding of the regulations. If we review the history of CBFM in West Sumatra, government initiatives became more committed after the SF schemes were supported with clear regulations from the central government. WSFA officers consistently emphasize the legal aspects within the program explanation, extension, and facilitation.

According to our informants, WSFA support for HD initiatives in Simanau and Simancuang Villages began to improve in 2011. Their support was in accordance with the operational regulations from the Forestry Ministry through Ministerial Regulation No. P.49/Menhut-II/2008 about Village Forests. Government support for HTR and HKm initiatives were also strengthened after the Forestry Ministry published Ministerial Regulation No. P.23/Menhut-II/2007 about Community Plantation Forests and P.37/Menhut-II/2007 about Community Forests. Meanwhile, government support for HA initiatives has not received adequate support. In fact, WSFA officers have been involved in a series of discussions with the NGOs regarding HA initiatives and local wisdom regarding forest resource management, as in Koto Malintang Village as reported by KKI Warsi in 2004. Koto Malintang also obtained an award as a forestry village at the regional and national levels in 2006. Because there were no clear operational regulations to support local wisdom in HA, follow-up was conducted from a series of discussions, but delivering the recognition terms was still unclear.

The attitude and behavior of WSFA officers in implementing SF policy were generated from their understanding of regulations about social forestry as governed by the last ministerial regulation No. P.83/MenLHK/Setjen/ Kum.1/10/2016. Based on the 2013 WSFA report, the SF schemes include allocating the management rights in state forest to the local communities. The management rights are intended to resolve forestry conflicts. According to our informants, conflicts involved governmental officers and local communities in cases where the government unilaterally determined state forest area boundaries. Consequently, government officers perceived that all illegal activities in the state forest are forestry crimes and violations. Meanwhile, local communities claimed that the state forest is their communal land in accordance with Minangkabau *adat* law, as applied in Koto Malintang and Simancuang villages (Asmin et al., 2016; Asmin et al., 2017a). In every governmental coordination meeting, implementing the SF schemes is also intended to legalize the socio-economic activities of local communities, previously considered illegal.

Government support for SF development was also in accordance with the limited human resources of WSFA, especially for facilitating the LFR and FPS activities. Furthermore, SF development has become a strategic policy of REDD+ (*reducing emissions from forest degradation and deforestation)* implementation plans in West Sumatra (Hermansah et al., 2013), which focuses on LFR and FPS activities. In the governmental coordination meetings, WSFA expected that SF schemes would increase the LFR and FPS activities to reduce forest degradation and deforestation. The government believed that SF schemes could achieve this.

SF schemes also aim to clarify forest resource management at the village level. Our informants perceived that local management in the protection and production forests has not been allowed to function because state forests were still categorized as common property. WSFA expected that forest resource management would improve, delivering greater benefit by applying SF schemes. Although there is no comprehensive evaluation of their performance, our informants believed that the scheme would give legal certainty, create development assistance program opportunities, build community capacity, and resolve forestry conflicts. Development assistance programs included technical and financial assistance to local communities for infrastructures, business ventures (such as developing forest products, agricultural products, and ecotourism), as well as forest rehabilitation and protection.

The active roles of NGOs have also strengthened the support of WSFA for CBFM. WSFA designed various programs to develop SF schemes, such as program explanation, training, mentoring, and extension. WSFA cooperates with various NGOs. For example, KKI Warsi was one of the NGOs that has encouraged the extension and mentoring of SF schemes in West Sumatra since 2009 (Diana & Sukmareni, 2015). WSFA has also received institutional support from another NGO, Qbar. The existence of both NGOs is important because they have the technical and financial capacity to support the WSFA program. Lacking technical officers and financial support has also encouraged WSFA to cooperate with NGOs.

3.3 The Implementation of Social Forestry in West Sumatra

Regarding the WSFA's document on the 2012-17 working plan of SF development, formerly WSFA emphasized only two forms of SF schemes, HD and HKm. This is likely because those were the only SF schemes included before the government published Ministerial Decree P.83/MenLHK/Setjen/Kum.1/10/2016 on Social Forestry. In the previous regulations, HTR was not yet specifically mentioned as one of the SF schemes. Consequently, there is no HTR phrase in the document. However, the informant statements in the formal and informal meetings indicated that WSFA has also been considering HTR as part of the SF working plan in West Sumatra.

According to the above mentioned document, SF development became an important instrument to promote CBFM in accordance with the local wisdom of the West Sumatran people (Dinas Kehutanan Provinsi Sumatera Barat, 2012b). The objective of SF development is to create wise and prudent forest management based on local wisdom at the village level towards sustainable forests and the prosperity of communities. To realize this objective, WSFA formulated 13 strategies, two of which are unique, i.e. (1) to establish a service center known as the CBFM task force and (2) to consolidate the roles of other sectoral institutions, such as agriculture, fishery, plantations, food security, tourism, infrastructure, water resources, energy and mining resources, corporation (corporate social responsibility), etc.

Based on our interviews and field observations, the WSFA encouraged SF development by facilitating local community proposals, handling working area quotations and management rights permits, explaining the SF schemes, and allocating development assistance programs for HTR, HKm, and HD sites. Our informants stated that it is important to allocate development assistance programs after the local communities obtained the management rights from the government. The WSFA has been able to build cooperative agreements with other governmental institutions in West Sumatra. There are many forms of development assistance programs as a result of the agreements, such as a seed nursery, sericulture, mushroom cultivation, livestock, check dam, retaining dams, ponds, freshwater fish, and forest protection equipment.

WSFA also supported the NGOs to allocate the development assistance programs. For example, FFI is one of the NGOs which has supported HD schemes in South Solok District. FFI has designed multi-year activities as follows: forest protection patrol in the first year, (2) socio-economic activities in the second, (3) institutional roles in the third, and (4) LFR activities in the fourth. WSFA, then, found that the development assistance programs were able to change community behavior from previously destructive attitudes to conservation attitudes. Our key informants explained that the success of CBFM in West Sumatra was measured from the development assistance program initiatives after the local community obtained the management rights.

The NGOs were also encouraged to facilitate the local communities preparing their work plans. There are three work plan documents, i.e. annual work plan, five-year work plan, and 35-year work plan. The work plans contain activities to manage forest resources. In the work plans, WSFA expected that the local communities should develop business activities in accordance with, and not limited to the forestry sector, but also across other development sectors, such as tourism and plantations outside the state forest. Financial support from governmental and private institutions was expected in accordance with the community work plans.

The will of WSFA to facilitate the local communities with many business activities has been supported by various NGOs and related institutions. For example, after Simancuang people obtained HD management rights, they received various development assistance programs from WSFA and NGOs. The development assistance programs were coordinated by LPHN, the special local institution established after the local community obtained management rights. Nevertheless, there is no comprehensive evaluation of the benefits of the development assistance programs. Certainly, as the new management rights are always increasing, the WSFA will require various other development assistance programs, as well as financial resources. Therefore, the WSFA in various forums has always emphasized the importance of development assistance programs for CBFM sites that have been granted management rights. They are trying to leverage additional funding from elsewhere for these activities.

3.4. CBFM Policy in Other Countries

Some developing countries have recognized CBFM as their national forestry strategy. CBFM is an increasingly important form of forest management in developing countries (Agrawal, 2007; Baynes et al., 2015; Gilmour, 2016). We summarized CBFM development in some countries as shown in Table 4. We categorized them based on their arguments, support, and implementation.

Based on the table above, CBFM development is mainly related to forest degradation and deforestation, poverty alleviation, and local community access to state forests. Recent research has supported the view that CBFM can ensure lower and less variable rates of deforestation (Porter-Bolland et al., 2012; Pinyopusarerk et al., 2014; Baynes et al., 2015). However, some researchers were still questioning CBFM performance to ensure conservation efforts and achieve sustainable forest management (Meilby et al., 2014; Rasolofoson et al., 2015; Pokharel et al., 2015). Related to poverty alleviation, the effects of CBFM on poverty are also ambiguous because there are numerous reports of benefits from CBFM leading to elite capture (Mahanty et al., 2006; Gilmour, 2016).

Meanwhile, CBFM policy in developing countries does provide management rights over state forests to local communities, but the policy should provide clear rights and responsibilities allowing local community autonomy in the exercise of power (Moeliono et al., 2017).

Table 4. Arguments, support, and implementation of CBFM in some countries

Countries	Arguments	Support	Implementation
Vietnam	to increase forest cover and alleviate poverty (Sunderlin & Ba, 2005; To & Tran, 2014; Moeliono et al., 2017)	part of forest land allocation program in the form of: (1) village forest management and (2) forest management by groups of households and individuals (To & Tran, 2014; Moeliono et al., 2017)	contract-based allocation of forest land to households and individuals (Tan & Sikor, 2011; To & Tran, 2014; Moeliono et al., 2017)
Philippines	to abate forest degradation, ensure equitable access, and manage limited resources (Pulhin et al., 2007; Rebugio et al., 2010)	part of national strategy for sustainable forest management with three systems, i.e. central government initiated program, local government initiated program, and traditional forest management (Pulhin et al., 2007; Suharjito, 2009; Rebugio et al., 2010; Hlaing et al., 2013)	Contract-based program for central government initiative, co-management agreement for local government initiative, and self-initiated for traditional forest management (Rebugio et al., 2010; Hlaing et al., 2013)
Nepal	to address local livelihoods and abate forest degradation (Gurung et al., 2011; Uprety et al., 2012; Pandey & Paudyall, 2015)	part of national programs in the form of community forestry (CF), leasehold forest (LHF), collaborative forest management (CFM), buffer zone community forestry (BZCF), protected forest (PF) and religious forest (RF) (Pathak, Yi, & Bohara, 2017)	operational co-operation between government and forest user groups for CF and BZCF (Wakiyama, 2004), lease to pro-poor households for LHF (Pathak et al., 2017), collaboration with local people, local government and Department of forests for CFM (Bampton et al., 2007; Pathak et al., 2017), council formation for PF and forests around temples or other sacred religious places for RF (Pathak et al., 2017)

Bangladesh	to conserve protected area, abate deforestation, and improve socio-economic condition of local people (Chowdhury et al., 2009; (Jashimuddin & Inoue, 2012a; Rashid et al., 2013; Nath et al., 2016)	part of national programs in the form of community forest (CF), co-management of Protected Areas (PA), and Village Common Forest (VCF) (Jashimuddin & Inoue, 2012a; Nath et al., 2016)	land tenure certificate and lease for CF (Nath et al., 2016), co-management council and committee for PA (Chowdhury et al., 2009; Chowdhury & Koike, 2010; Nath et al., 2016), and local management for VCF (Jashimuddin & Inoue, 2012b; Nath et al., 2016)
Mexico	to abate forest degradation and deforestation, secure access to forest resources, and improve rural livelihoods (Klooster & Masera, 2000; Bray et al., 2006; Ellis et al., 2015)	Part of agrarian reform program which created *ejidos* (agrarian nuclei with communal ownership of land) for the landless and supported the restitution of their original communal lands to indigenous groups in the form of *comunidades indigenas* (Klooster & Masera, 2000; Antinori & Bray, 2005; Ellis & Porter-Bolland, 2008; Flores et al., 2016)	as much as 80% of Mexico's forests are in the hands of communities with collective land grants (Bray et al., 2003; Antinori & Bray, 2005; Ellis & Porter-Bolland, 2008; Cronkleton et al., 2011; Ellis et al., 2015)
Tanzania	to promote socio-economic development and protect forest reserves (Meshack et al., 2006; Treue et al., 2014; Scheba & Mustalahti, 2015)	Part of national forestry policy in the form of Village Land Forest Reserve (VLFR), Community Forest Reserve (CFR), and Private Forest (PF), as well as, Joint Forest Management (JFM) (Scheba & Mustalahti, 2015) or known as Participatory Forest Management (PFM) (Treue et al., 2014)	JFM areas are managed by co-management, as owners, between National or Local Government Forest Reserves and village governments, meanwhile, VLFR areas are managed by the entire community, CFR areas are managed by a particular designated group in the community, and PV areas are managed by individual households (Scheba & Mustalahti, 2015)

Madagascar	efforts to curb deforestation and conserve threatened and endangered species (Raik & Decker, 2007; Raik, 2007; Rasolofoson et al., 2015)	part of national forestry policy through decentralization of forest management known as GELOSE (*Gestion Locale Sécurisée*) (Raik & Decker, 2007; Raik, 2007; Fritz - Vietta et al., 2009; Cullman, 2015)	by contractual forest management known as GCF (*Gestion Contractualisée des Forêts*) (Raik & Decker, 2007; Raik, 2007; Fritz - Vietta et al., 2009; Cullman, 2015)
Indonesia	to abate deforestation, promote agrarian reform, alleviate poverty, and recognize community rights (Moeliono et al., 2017; Santika et al., 2017)	known as social forestry strategy in form of HTR, HKm, HD, HA, and partnership schemes (Moeliono et al., 2017; Myers et al., 2017)	By granting management rights for HTR, HKm, and HD schemes during 35 years and extendable, full ownership for HA scheme, and rights to collaborate in activities for partnership scheme (Moeliono et al., 2017; Myers et al., 2017)

Government support for CBFM across these examples indicate that mainstreaming CBFM policy is usually conducted alongside other agrarian reform agendas and state forest management strategies. There are various schemes in CBFM implementation. Those schemes focus on local community participation in state forest management. However, documented evidence of the success of CBFM implementation is mixed. Baynes et al. (2015) and Gilmour (2016) show that CBFM success in developing countries is variable. They suggest that policies need to address factors that influence the success of CBFM implementation, i.e. (1) security of property rights, (2) government support for enabling regulatory frameworks, (3) strong governance and effective local-level institutions, (4) equality of socio-economic benefits for marginalized individuals and groups, (5) viable technology and adequate market access, and (6) supportive bureaucratic culture.

3.5. Rationale for CBFM Development in West Sumatra

This research discussed the WSPG's steps in developing SF schemes. Our discussions were based on the above results. We discussed four focus areas, including: (1) the granting of management rights, (2) institutions, (3) challenges of forest degradation and deforestation, (4) initiation of the development assistance programs. These four focus areas provided a strong rationale for CBFM development in West Sumatra.

The management rights granted in protection and production forests provide a guarantee of management areas for local communities in and around state forests. However, the initiation of SF schemes does not necessarily cover the entire community management area. In Simancuang, the community only proposed certain protected forest areas, while other protected forest areas were not included (see Asmin et al., 2016). Even in Koto Malintang, the community did not agree to any schemes in accordance with government regulations because they believed that their *adat* system could maintain their natural resources (see Asmin et al., 2017a). This is in accordance with the Minangkabau's *nagari* philosophy (Kahn, 1980), i.e, the *nagari* usually has communal forests (*ulayat*), rice fields, dryland fields, houses, ponds, agroforestry lands (*parak*), mosques, and a meeting room (*balai adat*).

The spatial planning of the *nagari,* which is conducted by the community, demonstrates comprehensive and holistic considerations for achieving sustainable development. The Minangkabau people are generally familiar with the spatial concept of forest in three terms, namely forbidden forest (*rimbo larangan*), cultivated forest (*parak*), and reserve forest (*rimbo cadangan*). A similar concept has also been applied by other indigenous peoples in Indonesia such as Baduy, Rumahkay, Dayak, and Ammatoa (Ichwandi & Shinohara, 2007; Ohorella, Suharjito, & Ichwandi, 2011; Samsoedin, Wijaya, & Sukiman, 2010; Husain & Kinasih, 2010). Thus, the SF schemes should be a pioneer for the recognition of community spatial planning in accordance with their *adat* system.

SF schemes that have been developed also require strong institutional arrangements. Nevertheless, the government has tended to establish a new institution, such as LPHN. *Adat* institutions however, are also able to guarantee sustainable forest management as evidenced by various studies on Baduy, Rumahkay, Dayak and Ammatoa communities. Policy makers at any governmental level should consider the local knowledge and the *adat* system to ensure a participatory institution (Asmin et al., 2017a; Asmin et al., 2017b). This was also expressed by Uphoff (1992), who stated that participatory institutions can serve in private and bureaucratic affairs.

The WSFA's support for the implementation of SF schemes was also included as part of forest lands (especially state forest) that had been converted into agricultural lands. Some studies also supported such arguments as Mulyanto & Jaya (2004) and Weatherley-Singh & Gupta (2015), each of which show that these schemes only covered a single perspective through land use change analysis. Even, Hosonuma et al. (2012) concluded that 33% of the deforestation rate was due to utilizing forested lands for subsistence agricultural practices. The low quality of human resources was also considered as triggering deforestation rates (Salahodjaev, 2016). This may have led to our perception that forest-dependent farmers are most likely the cause of forest degradation and deforestation. However, Asmin et al. (2016) have provided an explanation from a sociological perspective to understand the causes of forest degradation in the case of the Simancuang community of South Solok District, West Sumatra. In the context of the state claim over the forest areas, Mutolib et al. (2017) also indicated that forest degradation and deforestation were encouraged by forest seizures by local communities against the state claim over the forest area. Indeed, forest degradation and deforestation could be due to the allocation of local lands to plantations and timber companies.

The will of the WSFA to help communities with various development assistance programs provides hope for community development in the context of forest resource management. The WSFA's role in embracing various stakeholders presents a unique willingness of formal institutions to collaborate with other stakeholders. Muttaqin et al. (2017), conducting research in Central Kalimantan's Buntoi Village Forest, suggested the need for development assistance programs to gain support from local government. However, disingenuous government approaches to realize their programs will lead to the failure of SF schemes.

SF policy should not only show government willingness to collaborate with local communities, and furthermore indicates increasing appreciation of local knowledge and *adat* systems. In the context of community development, SF schemes can thus be regarded as mainstreaming CBFM, and can continue to be a tool for strengthening community solidarity and agency (Bhattacharyya, 2004). Strengthening solidarity happens through the recognition of local community identities and norms, while strengthening their agency takes place through the optimum local participation to manage forest resources.

The implementation of community development should emphasize participation and empowerment as foundational elements. Participation and empowerment require devolution of control and accountability from the government to individuals, groups, and communities (Narayan, 1995). Participation and empowerment is encouraged by four key aspects (Parks et al., 2013), namely: (1) devolution of authority and resources to key stakeholders, (2) involvement in decision-making, (3) two-way flow of information between resource providers and recipients, and (4) community contributions to encourage local ownership. The outcome of community development is an empowered community with the following characteristics (Patel, 2011): (1) having access to information, (2) engaging in forums on issues and decision-making, (3) considering options and actions, and (4) having the capacity and resources to institutionalize local community interests and roles towards public institutions.

4. Conclusions

SF policy in West Sumatra consists of a form of CBFM mainstreaming taking place through the implementation of specific schemes on community plantation forests (HTR), community forests (HKm), village forests (HD), and customary forests (HA). The governmental arguments for the policy are based on interests in addressing state forest conflicts, reducing threats to forest degradation and deforestation, and conditions of lacking the human resources (forestry officers) in overall forest management. WSPG, through WSFA, has encouraged community involvement in forest management in the form of these SF schemes, especially for ensuring the LFR and FPS activities. WSFA has supported various SF schemes by coordinating the roles of the parties, whether governmental agencies, the private sector, or NGOs. Nevertheless, there are challenges to developing the SF schemes that should be addressed. These challenges include: (1) granting the management rights that ensure the recognition of community-based spatial planning, (2) developing a participatory institution, (3) synchronizing the government policies to control forest degradation and deforestation, and (4) initiating development assistance programs that strengthen local community solidarity and agency.

Government officials should consider adaptive ways of initiating SF schemes that appreciate local community norms and values through the recognition of local knowledge and *adat* systems. In the context of community development, the success of SF schemes should not only be measured by the realization of development assistance programs after the granting of management rights, but also the enhancement of local community solidarity based on its cultural strength and the recognition of community capacity within the established social system.

Acknowledgments

We would like to thank the Indonesian Endowment Fund for Education (LPDP) of the Ministry of Finance, for supporting this research. We would also like to thank Koto Malintang and Simancuang people, especially *adat* leaders, village leaders, and local informants for helping us to provide the data and information, as well as the non-governmental organization of KKI Warsi and West Sumatra Forestry Agency.

References

Agrawal, A. (2007). Forests, governance, and sustainability: common property theory and its contributions. *International Journal of the Commons*, *1*(1), 111–136. doi:https://doi.org/10.18352/ijc.10

Antinori, C., & Bray, D. B. (2005). Community forest enterprises as entrepreneurial Firms: Economic and institutional perspectives from Mexico. *World Development*, *33*, 1529–1543. doi: https://doi.org/10.1016/j.worlddev.2004.10.011

Asmin, F., Darusman, D., Ichwandi, I., & Suharjito, D. (2016). Local ecological knowledge on forest clearing: A case study of parak and rimbo practices in Simancuang Community, Indonesia. *Komunitas: International Journal of Indonesian Society and Culture*, *8*(2), 208–220. doi:https://doi.org/10.15294/komunitas.v8i2.5856

Asmin, F., Darusman, D., Ichwandi, I., & Suharjito, D. (2017a). Elaborating the attributes of local ecological knowledge : A case study of parak and rimbo practices in Koto Malintang Village.

Advanced Science Letters, *23*(4), 2812–2817. doi: https://doi.org/10.1166/asl.2017.7682
Asmin, F., Darusman, D., Ichwandi, I., & Suharjito, D. (2017b). Social capital of parak and rimbo management in West Sumatra. *Jurnal Manajemen Hutan Tropika*, *23*(3), 140–149.
Bampton, J. F. R., Ebregt, A., & Banjade, M. R. (2007). Collaborative forest management in Nepal's Terai : Policy , practice and contestation. *Journal of Forest and Livelihood*, *6*(2), 30–43.
Baynes, J., Herbohn, J., Smith, C., Fisher, R., & Bray, D. (2015). Key factors which influence the success of community forestry in developing countries. *Global Environmental Change*, *35*(November), 226–238. doi:https://doi.org/10.1016/j.gloenvcha.2015.09.011
Berg, B. L. (2001). *Qualitative Research Methods for the Social Sciences*. Boston: A Pearson Education Company. doi:https://doi.org/10.2307/1317652
Bhattacharyya, J. (2004). Theorizing community development. *Journal of the Community Development Society*, *34*(2), 6–34. https://doi.org/10.1080/15575330409490114
Bray, D. B., Antinori, C., & Torres-Rojo, J. M. (2006). The Mexican model of community forest management: The role of agrarian policy, forest policy and entrepreneurial organization. *Forest Policy and Economics*, *8*(4), 470–484. doi:https://doi.org/10.1016/j.forpol.2005.08.002
Bray, D. B., Merino-Pérez, L., Negreros-Castillo, P., Segura-Warnholtz, G., Torres-Rojo, J. M., & Vester, H. F. M. (2003). Mexico's community-managed forests as a global model for sustainable landscapes. *Conservation Biology*, *17*(3), 672–677. doi: https://doi.org/10.1046/j.1523-1739.2003.01639.x
Bryman, A. (2004). *Social Research Methods*. New York: Oxford University Press Inc.
Chowdhury, M., Koike, M., & Muhammed, N. (2009). Embracing collaborative protected area management for conservation: An analysis of the development of the forest policy of Bangladesh. *International Forestry Review*, *11*(3), 359–374. doi: https://doi.org/10.1505/ifor.11.3.359
Chowdhury, M. S. H., & Koike, M. (2010). An overview on the protected area system for forest conservation in Bangladesh. *Journal of Forestry Research*, *21*(1), 111–118. doi: https://doi.org/10.1007/s11676-010-0019-x
Cronkleton, P., Bray, D. B., & Medina, G. (2011). Community forest management and the emergence of multi-scale governance institutions: Lessons for REDD+ development from Mexico, Brazil and Bolivia. *Forests*, *2*, 451–473. doi:https://doi.org/10.3390/f2020451
Cullman, G. (2015). Community forest management as virtualism in Northeastern Madagascar. *Human Ecology*, *43*(1), 29–41. doi: https://doi.org/10.1007/s10745-015-9725-5
Davis, A., & Wagner, J. R. (2003). Who knows? On the importance of identifiying "expert" when researching local ecological knowledge. *Human Ecology*, *31*(3), 463–489. doi: https://doi.org/10.1023/A:1025075923297
Diana, E., & Sukmareni. (2015). *Hutan Nagari: Mengembangkan PHBM di Sumatera Barat*. Jambi: KKI Warsi.
Dinas Kehutanan Provinsi Sumatera Barat. (2011). *Laporan Akhir Tim Pembina Rehabilitasi Hutan dan Lahan Tingkat Provinsi Sumatera Barat Tahun 2011*. Padang.
Dinas Kehutanan Provinsi Sumatera Barat. (2012a). *Rencana Kehutanan Tingkat Provinsi (RKTP) Provinsi Sumatera Barat 2012-2032*. Padang: Dinas Kehutanan Provinsi Sumatera Barat.
Dinas Kehutanan Provinsi Sumatera Barat. (2012b). *Rencana Kerja Pengembangan Perhutanan Sosial di Provinsi Sumatera Barat*. Padang: Dinas Kehutanan Provinsi Sumatera Barat.
Dove, M. R. (1986). Peasant versus government perception and use of the environment: A case-study of Banjarese ecology and river basin development in South Kalimantan. *Journal of Southeast Asian Studies*, *17*(1), 113–136.
Dove, M. R. (1992). Foresters' beliefs about farmers: A priority for social science research in social forestry. *Agroforestry Systems*, *17*, 13–41. doi: https://doi.org/https://doi.org/10.1007/BF00122925
Dove, M. R. (1995). The theory of social forestry intervention: The state of the art in Asia.

Agroforestry Systems, *30*(3), 315–340. doi: https://doi.org/10.1007/BF00705217

Dunn, W. N. (1981). *Public Policy Analysis: An Introduction*. New Jersey: Prentice-Hall, Inc.

Dunn, W. N. (1993). Policy reforms as arguments. In F. Fischer & J. Forester (Eds.), *The argumentative turn in policy analysis and planning* (pp. 254–290). London: UCL Press Limited. doi: https://doi.org/10.1017/CBO9781107415324.004

Ellis, E. A., Kainer, K. A., Sierra-Huelsz, J. A., Negreros-Castillo, P., Rodriguez-Ward, D., & DiGiano, M. (2015). Endurance and adaptation of community forest management in Quintana Roo, Mexico. *Forests*, *6*, 4295–4327. doi: https://doi.org/10.3390/f6114295

Ellis, E. A., & Porter-Bolland, L. (2008). Is community-based forest management more effective than protected areas?. A comparison of land use/land cover change in two neighboring study areas of the Central Yucatan Peninsula, Mexico. *Forest Ecology and Management*, *256*(11), 1971–1983. doi:https://doi.org/10.1016/j.foreco.2008.07.036

Fisher, M. R., Workman, T., Mulyana, A., Moeliono, M., Yuliani, E. L., Colfer, C. J. P., & Adam, U. E. F. B. (2017). Striving for PAR excellence in land use planning: Multi-stakeholder collaboration on customary forest recognition in Bulukumba, South Sulawesi. *Land Use Policy*, (July 2016), 0–1. doi:https://doi.org/10.1016/j.landusepol.2017.09.057

Flores, I. D. H., Skutsch, M., & Mustalahti, I. (2016). Impacts of Finnish cooperation in the Mexican policy making process: From the community forest management to the liberalization of forest services. *Forest Policy and Economics*, *73*, 229–238. doi: https://doi.org/10.1016/j.forpol.2016.09.011

Fritz - Vietta, N., Röttger, C., & Stoll - Kleemann, S. (2009). Community - based management in two biosphere reserves in Madagascar – distinctions and similarities: What can be learned from different approaches ? *Madagascar Conservation & Development*, *4*(2), 86–97. https://doi.org/10.4314/mcd.v4i2.48648

FWI/DFW. (2001). *Keadaan Hutan Indonesia*. Bogor dan Washington DC: Forest Watch Indonesia dan Global Forest Watch.

Gilmour, D. (2016). *Forty Years of Community-Based Forestry*. Roma: FAO.

Guba, E. G., & Lincoln, Y. S. (1994). Competing paradigms in qualitative research. In N. Denzin & Y. Lincoln (Eds.), *Handbook of qualitative research* (pp. 105–117). Thousand Oaks, CA: Sage.

Gurung, A., Karki, R., & Bista, R. (2011). Community-based forest management in Nepal: Opportunities and challenges. *Resources and Environment*, *1*(1), 26–31.

Hamzah, Suharjito, D., & Istomo. (2015). Efektifitas kelembagaan lokal dalam pengelolaan sumber daya hutan pada masyarakat Nagari Simanau, Kabupaten Solok. *Risalah Kebijakan Pertanian Dan Lingkungan*, *2*(2), 117–128. doi: https://doi.org/10.20957/jkebijakan.v2i2.10979

Hermansah, Ranti, G., Aisyah, S., Jusmalinda, Asmin, F., Mutiara, V., ... Boyce, A. (2013). *Strategi dan Rencana Aksi Provinsi untuk Implementasi REDD+*. Padang: Dinas Kehutanan Provinsi Sumatera Barat dan Satuan Tugas REDD+ Indonesia.

Hill, M., & Hupe, P. (2002). *Implementation Public Policy: Governance in Theory and in Practice*. London: Sage Publication, Ltd. doi:https://doi.org/10.1017/CBO9781107415324.004

Hlaing, E. E. S., Inoue, M., & Pulhin, J. M. (2013). A property-rights approach to understanding regulations and practices in community-based forest management: Comparison of three systems in the Philippines. *Small-Scale Forestry*, *12*(4), 579–596. doi: https://doi.org/10.1007/s11842-012-9232-0

Hoppe, R. (1993). Political judgment and the policy cycle: The case of ethnicity policy arguments in the Netherlands. In F. Fischer & J. Forester (Eds.), *The argumentative turn in policy analysis and planning* (pp. 77–100). London: UCL Press Limited.

Hosonuma, N., Herold, M., De Sy, V., De Fries, R. S., Brockhaus, M., Verchot, L., ... Romijn, E. (2012). An assessment of deforestation and forest degradation drivers in developing countries. *Environmental Research Letters*, *7*, 1–12. doi:https://doi.org/10.1088/1748-9326/7/4/044009

Hujala, T., Tikkanen, J., Hanninen, H., & Virkkula, O. (2009). Family forest owners ' perception of decision support. *Scandinavian Journal of Forest Research*, *24*, 448–460. doi:https://doi.org/10.1080/02827580903140679

Husain, S., & Kinasih, S. (2010). Pasang sebagai modal sosial pengelolaan hutan pada masyarakat adat Ammatowa. *Masyarakat, Kebudayaan, Dan Politik*, *23*(3), 229–235.

Ichwandi, I., & Shinohara, T. (2007). Indigenous practices for use of and managing tropical natural resources : A case study on Baduy community in Banten, Indonesia. *Tropics*, *16*(2), 87–102. doi:https://doi.org/http://doi.org/10.3759/tropics.16.87

Ito, K., Oura, Y., Takeya, H., Hattori, S., Kitagawa, K., Paudel, D., & Paudel, G. (2005). The influence of NGO involvement on local people's perception of forest management: A case study of community forestry in Nepal. *Journal of Forest Research*, *10*(6), 453–463. doi:https://doi.org/10.1007/s10310-005-0172-9

Jashimuddin, M., & Inoue, M. (2012a). Community forestry for sustainable forest management : Experiences from Bangladesh and policy recommendations. *FORMATH*, *11*, 133–166.

Jashimuddin, M., & Inoue, M. (2012b). Management of village common forests in the Chittagong Hill Tracts of Bangladesh: Historical background and current issues in terms of sustainability. *Open Journal of Forestry*, *2*(3), 121–137. doi:https://doi.org/10.4236/ojf.2012.23016

Kahn, J. S. (1980). *Minangkabau Social Formations: Indonesia Peasants and the World Economy*. Cambridge: Cambridge University Press.

KemenLHK. (2015). *Statistik Kementerian Lingkungan Hidup dan Kehutanan Tahun 2014*. Jakarta: Kementerian Lingkungan Hidup dan Kehutanan. doi:https://doi.org/10.1017/CBO9781107415324.004

Klooster, D., & Masera, O. (2000). Community forest management in Mexico: Carbon mitigation and biodiversity conservation through rural development. *Global Environmental Change*, *10*(4), 259–272. doi:https://doi.org/10.1016/S0959-3780(00)00033-9

Mahanty, S., Gronow, J., Nurse, M., & Malla, Y. (2006). Reducing poverty through community based forest management in Asia. *Journal of Forest and Livelihood*, *5*(1), 78–89. doi:https://doi.org/10.3126/jfl.v5i1.1983

Martial, T., Helmi, Effendi, N., & Martius, E. (2012). Land and tree tenure rights on agroforestry (parak) system at communal land in West Sumatra, Indonesia. *Journal of Agricultural Extension and Rural Development*, *4*(19), 486–494. doi:https://doi.org/10.5897/JAERD11.120

Masozera, M. K., Alavalapati, J. R. R., Jacobson, S. K., & Shrestha, R. K. (2006). Assessing the suitability of community-based management for the Nyungwe Forest Reserve, Rwanda. *Forest Policy and Economics*, *8*(2), 206–216. doi:https://doi.org/10.1016/j.forpol.2004.08.001

Meilby, H., Smith-Hall, C., Byg, A., Larsen, H., Nielsen, Ø., Puri, L., & Rayamajhi, S. (2014). Are forest incomes sustainable? Firewood and timber extraction and productivity in community managed forests in Nepal. *World Development*, *64*, S113–S124.

Meshack, C. K., Ahdikari, B., Doggart, N., & Lovett, J. C. (2006). Transaction costs of community-based forest management: Empirical evidence from Tanzania. *African Journal of Ecology*, *44*(4), 468–477. doi:https://doi.org/DOI 10.1111/j.1365-2028.2006.00659.x

Michon, G., Mary, F., & Bompard, J. (1986). Multistoried agroforestry garden system in West Sumatra, Indonesia. *Agroforestry Systems*, *4*(4), 315–338. doi:https://doi.org/10.1007/BF00048106

Moeliono, M., Thuy, P. T., Bong, I. W., Wong, G. Y., & Brockhaus, M. (2017). Social forestry - why and for whom? A comparison of policies in Vietnam and Indonesia. *Forest and Society*, *1*(2), 1–20. doi:https://doi.org/10.24259/fs.v1i2.2484

Mulyanto, L., & Jaya, I. N. S. (2004). Analisis spasial degradasi hutan dan deforestasi: Studi Kasus di PT. Duta Maju Timber, Sumatera Barat. *Jurnal Manajemen Hutan Tropika*, *10*(1), 29–42.

Mutolib, A., Yonariza, Mahdi, & Ismono, H. (2017). Forest ownership conflict between a local community and the state: A case study in Dharmasraya, Indonesia. *Journal of Tropical Forest*

Science, *29*(2), 163–171.

Muttaqin, M. Z., Samsoedin, I., Subarudi, Nurtjahjawilasa, & Hamdani, F. A. U. (2017). Pemanfaatan jasa lingkungan di hutan desa Buntoi, Kecamatan Kahayan Hilir, Kabupaten Pulang Pisau, Provinsi Kalimantan Tengah. *Jurnal Analisis Kebijakan Kehutanan*, *14*(1), 1–16.

Myers, R., Intarini, D., Sirait, M. T., & Maryudi, A. (2017). Claiming the forest: Inclusions and exclusions under Indonesia's "new" forest policies on customary forests. *Land Use Policy*, *66*(July), 205–213. doi:https://doi.org/10.1016/j.landusepol.2017.04.039

Narayan, D. (1995). *Designing community based development*. Washington. Retrieved from https://openknowledge.worldbank.org/handle/10986/11662

Nath, T. K., Jashimuddin, M., & Inoue, M. (2016). *Community-Based Forest Management (CBFM) in Bangladesh*. *World Forests* (Vol. 22). Swiss: Springer International Publishing. doi:https://doi.org/10.1080/10549811.2016.1231615

Nawir, A., Murniati, Rumboko, L., Hiyama, C., & Gumartini, T. (2008). Potret rehabilitasi di Indonesia. In A. Nawir, Murniati, & L. Rumboko (Eds.), *Rehabilitasi Hutan di Indonesia: Akan Kemanakah Arahnya setelah Lebih Dari Tiga Dasawarsa?* (pp. 119–186). Bogor: Center for International Forestry Research (CIFOR).

Nayak, P. K., & Berkes, F. (2008). Politics of co-optation: Community forest management versus joint forest management in Orissa, India. *Environmental Management*, *41*(5), 707–718. doi:https://doi.org/10.1007/s00267-008-9088-4

Nurrochmat, D., Darusman, D., & Ruchjadi, D. (2014). Rekonstruksi sistem tenurial kehutanan. *Risalah Kebijakan Pertanian Dan Lingkungan*, *1*(1), 24–29.

Ohorella, S., Suharjito, D., & Ichwandi, I. (2011). Efektivitas kelembagaan lokal dalam pengelolaan sumber daya hutan pada masyarakat Rumahkay di Seram Bagian Barat, Maluku. *Jurnal Manajemen Hutan Tropika*, *XVII*(2), 49–55.

Oliver, C. D. (2014). Functional restoration of social-forestry systems across spatial and temporal scales. *Journal of Sustainable Forestry*, *33*(sup1), S123–S148. doi:https://doi.org/10.1080/10549811.2014.884003

Pagdee, A., Kim, Y., & Daugherty, P. J. (2006). What makes community forest management successful: A meta-study from community forests throughout the world. *Society & Natural Resources*, *19*(1), 33–52. doi:https://doi.org/10.1080/08941920500323260

Pandey, G. S., & Paudyall, B. R. (2015). *Protecting forests, improving livelihoods – Community forestry in Nepal*. *Regional Development Dialogue*. Nepal. doi:https://doi.org/10.3126/hjs.v1i1.190

Parks, T., Schuler, N., Chua, J., Magno, C., Johnson, K., Domado, H., & McCarthy, P. (2013). *Community-Based Development in Conflict-Affected Areas of the Philippines*. Pasig City: The Asia Foundation.

Patel, N. (2011). *Democracy: Decentralization and grassroot empowerment initiatives in Malawi*. Oslo.

Pathak, B. R., Yi, X., & Bohara, R. (2017). Community based forestry in Nepal: Status, issues and lessons learned. *International Journal of Sciences*, *6*(3), 119–129. doi:https://doi.org/10.18483/ijSci.1232

Pietrzyk-kaszy, A., Cent, J., Grodzi, M., & Szyma, M. (2012). Factors influencing perception of protected areas — The case of Natura 2000 in Polish Carpathian communities, *20*, 284–292. doi:https://doi.org/10.1016/j.jnc.2012.05.005

Pinyopusarerk, K., Tran, T. T. H., & Tran, V. D. (2014). Making community forest management work in northern Vietnam by pioneering participatory action. *Land Use Policy*, *38*, 257–263. https://doi.org/10.1016/j.landusepol.2013.11.019

Pokharel, R. K., Neupane, P. R., Tiwari, K. R., & Köhl, M. (2015). Assessing the sustainability in community based forestry: A case from Nepal. *Forest Policy and Economics*, *58*(June 1992), 75–84. doi:https://doi.org/10.1016/j.forpol.2014.11.006

Porter-Bolland, L., Ellis, E. A., Guariguata, M. R., Ruiz-Mallén, I., Negrete-Yankelevich, S., & Reyes-García, V. (2012). Community managed forests and forest protected areas: An assessment of their conservation effectiveness across the tropics. *Forest Ecology and Management*, *268*, 6–17. doi:https://doi.org/10.1016/j.foreco.2011.05.034

Pulhin, J. M., Inoue, M., & Enters, T. (2007). Three decades of community-based forest management in the Philippines: Emerging lessons for sustainable and equitable forest management. *International Forestry Review*, *9*(4), 865–883. doi:https://doi.org/10.1505/ifor.9.4.865

Raik, D. (2007). Forest management in Madagascar: An historical overview. *Madagascar Conservation & Development*, *2*(1), 5–10. doi:https://doi.org/10.4314/mcd.v2i1.44123

Raik, D. B., & Decker, D. J. (2007). A multisector framework for assessing community-based forest management: Lessons from Madagascar. *Ecology and Society*, *12*(1), 14. https://doi.org/14

Rashid, M. A. Z. M., Craig, D., Mukul, S. A., & Khan, N. A. (2013). A journey towards shared governance: Status and prospects for collaborative management in the protected areas of Bangladesh. *Journal of Forestry Research*, *24*(3), 599–605. doi:https://doi.org/10.1007/s11676-013-0391-4

Rasolofoson, R. A., Ferraro, P. J., Jenkins, C. N., & Jones, J. P. G. (2015). Effectiveness of community forest management at reducing deforestation in Madagascar. *Biological Conservation*, *184*, 271–277. doi: https://doi.org/10.1016/j.biocon.2015.01.027

Rebugio, L. L., Carandang, A. P., Dizon, J. T., Pulhin, J. M., Camacho, L. D., Lee, D. K., & Peralta, E. O. (2010). Promoting sustainable forest management through community forestry in the Philippines. In G. Mery, P. Katila, G. Galloway, R. I. Alfaro, M. Kanninen, M. Lobovikov, & J. Varjo (Eds.), *Forests and Society* (pp. 355–368). Vienna and Vantaa: International Union of Forest Research Organizations.

Salahodjaev, R. (2016). Intelligence and deforestation: International data. *Forest Policy and Economics*, *63*, 20–27. doi:https://doi.org/10.1016/j.forpol.2015.12.003

Samsoedin, I., Wijaya, A., & Sukiman, H. (2010). Konsep tata ruang dan pengelolaan lahan pada masyarakat Dayak Kenyah di Kalimantan Timur. *Jurnal Analisis Kebijakan Kehutanan*, *7*(2), 145–168.

Santika, T., Meijaard, E., Budiharta, S., Law, E. A., Kusworo, A., Hutabarat, J. A., ... Wilson, K. A. (2017). Community forest management in Indonesia: Avoided deforestation in the context of anthropogenic and climate complexities. *Global Environmental Change*, *46*(December 2016), 60–71. doi:https://doi.org/10.1016/j.gloenvcha.2017.08.002

Sawitri, R., & Subiandono, E. (2011). Karakteristik dan persepsi masyarakat daerah penyangga Taman Nasional Gunung Halimun-Salak. *Jurnal Penelitian Hutan Dan Konservasi Alam*, *8*(3), 273–285.

Scheba, A., & Mustalahti, I. (2015). Rethinking "expert" knowledge in community forest management in Tanzania. doi: *Forest Policy and Economics*, *60*, 7–18. https://doi.org/10.1016/j.forpol.2014.12.007

Suharjito, D. (2009). Devolusi pengelolaan hutan di Indonesia: Perbandingan Indonesia dan Philipina. *JMHT*, *XV*(3), 123–130.

Suharjito, D. (2014). *Pengantar Metode Penelitian*. Bogor: PT Penerbit IPB Press.

Sunderlin, W. D., & Ba, H. T. (2005). *Poverty Alleviation and Forests in Vietnam*. Bogor: Center for International Forestry Research (CIFOR). doi:https://doi.org/10.17528/cifor/001660

Sutton, R. (1999). *The policy process: An overview* (No. Working Paper 118). London. Retrieved from http://www.eldis.org/vfile/upload/1/document/0708/DOC7279.pdf%5Cnpapers://33432778-ef7b-4391-ae67-220f9c46620d/Paper/p2665

Tan, N. Q., & Sikor, T. (2011). Forest land allocation: An overview of policy framework and outcomes. In T. Sikor & N. Q. Tan (Eds.), *Realizing Forest Rights in Vietnam: Addressing Issues in Community Forest Management* (pp. 5–13). Hanoi: RECOFTC-The Center for People and Forests. Retrieved from http://www.recoftc.org/recoftc/download/4581/810

To, X. P., & Tran, H. N. (2014). *Forest Land Allocation in the Context of Forestry Sector Restructuring: Opportunities for Forestry Development and Upland Livelihood Improvement*. Hue City: Tropenbos International Viet Nam. Retrieved from http://www.forest-trends.org/documents/files/doc_4826.pdf

Tole, L. (2010). Reforms from the ground up: A review of community-based forest management in tropical developing countries. *Environmental Management*, *45*, 1312–1331. doi:https://doi.org/10.1007/s00267-010-9489-z

Treue, T., Ngaga, Y. M., Meilby, H., Lund, J. F., Kajembe, G., Iddi, S., ... Burgess, N. D. (2014). Does participatory forest management promote sustainable forest utilisation in Tanzania? *International Forestry Review*, *16*(1), 23–38. https://doi.org/10.1505/146554814811031279

Uphoff, N. (1992). *Local institutions and participation for sustainable development* (Vol. 31). London. doi:https://doi.org/10.1002/sd.201

Uprety, D. R., Gurung, A., Bista, R., Karki, R., & Bhandari, K. (2012). Community forestry in Nepal: A scenario of exclusiveness and its implications. *Frontiers in Science*, *2*(3), 41–46. doi:https://doi.org/10.5923/j.fs.20120203.05

Vodouhê, F. G., Coulibaly, O., Adégbidi, A., & Sinsin, B. (2010). Community perception of biodiversity conservation within protected areas in Benin. *Forest Policy and Economics*, *12*(7), 505–512. doi:https://doi.org/10.1016/j.forpol.2010.06.008

Wakiyama, T. (2004). *Community Forestry in Nepal: A Comparison of Management Systems between Indigenous Forestry and Modern Community Forestry. Policy Trend Report*. Retrieved from http://www.iges.or.jp/en/fc/pdf/report11/full_text.pdf

Weatherley-Singh, J., & Gupta, A. (2015). Drivers of deforestation and REDD+ benefit-sharing: A meta-analysis of the (missing) link. *Environmental Science & Policy*, *54*, 97–105. doi:https://doi.org/10.1016/j.envsci.2015.06.017

Wiersum, K. F. (1999). *Social forestry: Changing perspectives in forestry science or practice?* Wageningen Agricultural University. Retrieved from http://search.ebscohost.com/login.aspx?direct=true&db=lah&AN=19990608844&site=ehost-live

Wiersum, K. F. (2004). Social and community forestry. In J. Burley, J. Evans, & Youngquist (Eds.), *Encyclopedia of Forest Sciences* (pp. 1136–1143). Oxford: Elsevier Ltd.

15

Examining forest economies: A case study of silk value chain analysis in Wajo District

Andi Gunawan Pratama[1,], Supratman Supratman[1], Makkarennu Makkarennu[1*]

1 Forestry Faculty of Hasanuddin University, Jl. Perintis Kemerdekaan Km. 10, Makassar, Indonesia
E-mail: nmakkarennu@gmail.com

* Corresponding author

Abstract: Value chain analysis is a strategic analytical tool for understanding activities that represent the value of a product (or service), meanwhile providing opportunities for identifying competitive advantage. In this value chain analysis, the research aims to 1) map the actors of silk commodity business ventures, 2) determine the activities of these business actors, and 3) calculate and analyze financial ratios. This study was conducted in the Wajo District as a site historically well known for silk production, and which is undergoing efforts at revival. The method used in this study is observation, interview, and modeling. Data analysis was conducted as a classical value chain analysis, emphasizing the calculations of financial ratios. The results show that the main actors and activities in commodities trading are farmers, local traders, weavers, and silk clothing stores. The largest marketing margin is approximately IDR. 300,000, which is found in silk clothing stores. On the other hand, the smallest marketing margins amounted to about IDR. 100,000, which were identified among local traders. Finally, the largest profit margins were identified among farmers valued at IDR. 196,000, while the smallest profit margins were among local traders amounting to IDR. 68,000. Overall, this case study highlights that forest enterprises, in this case of the silk industry in Wajo, can improve household incomes for forest farmers, while maintaining their overall bargaining position against the market.

Keywords: silk; value chain; marketing margin; profit margin; Sulawesi; Indonesia; Social Forestry; livelihoods

---

1. Introduction

Natural silk takes place through a series of agro-industry activities, beginning with the mulberry plant, which consists of the feed for silkworm, requiring nursing and maintenance of silkworms (Bombyx mori, L), followed by the processing of cocoons into yarn (spinning yarn), the weaving of cloth, and finally to the marketing of silk fabrics. This business falls into the category of the home industry market, and can be a lucrative product for rural livelihoods. The industry is categorized as relatively easy to be taxed, applies a fairly simple technology, and while it is labor intensive, results in relatively rapid generating incomes of high economic value. Natural farming activities can also take place in parallel, presenting possibilities to rehabilitate land and improve soil conservation. Furthermore, the silk farming industry can help to increase the load capacity and productivity of land, especially on lands that have not been optimally utilized (Sihar, 2008).

One of the well-known silk-producing regions in Indonesia is Wajo District, South Sulawesi. The region is also site to social forestry programs in South Sulawesi. In its implementation, social forestry is based on community-based forest management by paying attention to the following principles: benefits and sustainability, self-reliance, togetherness and partnership, intersectoral integration, in approaches that are gradual, sustainable, and adaptive (Hakim, et al., 2010). The Wajo District is famous as a silk region because of the longstanding and skillful abilities of people in the region in trading silk. Silk is one of the leading commodities because managing production has been carried out from generation to generation, and are either managed as a side activity, as a home industry venture, or even as a medium-scale industry (Sihar, 2008). In the 1960s and 1970s the silk weaving industry in the Wajo district experienced its peak, but was followed by a period of precipitous decline due to the difficulty of obtaining high quality raw materials. This ultimately affected results on the production side of weaving and also affected market demand. Renewed weaving industry activities began to develop in the late 1990s when imported raw materials were allowed to re-enter and the market began to experience regeneration (Sihar, 2008).

Production from silk farmers has fluctuated in recent years (Harbi, et al., 2015). This not only has had an impact on improving the economy of these communities, but also interacts with the prevailing prices on the market. Natural silk production, for example, experienced a rapid increase in 2013, jumping to a level of 3,000 kg; but in 2014 experience a drastic decline falling below 500 kg.

The decline in silk production was caused by changes in crop prices, from which the mulberry croplands for silk food were being converted to corn production. This was due to a brief spike in corn prices presenting more promising possibilities in the short term in comparison to silk. In addition, the supply of silkworm seeds to farmers had also been of a fairly low quality, and as a result affected the quality of the yarn produced (Nuraeni, 2017). Farmers have so far only acted by accepting prices (price takers) from traders who control the market (as price makers), indicating the weak bargaining position of rural farmers.

Furthermore, constraints on silk are not only in terms of production, but also in terms of marketing. As with other farmers, they tend to be dependent on more rapid rates of return because of urgent needs. This broad trend is what collectors or other actors use as bargaining power to reduce the price of yarn in the market. At present, the price of silk yarn is quite low at the farmer level, whereas when compared to yarn at the market price, the trader receives a higher price. Farmers have so far only acted as price takers while traders who control the market become price makers indicating the weak bargaining position of farmers.

Based on this background of the silk industry and conditions observed in Wajo, this research seeks to take a closer look at the value chain to help understand how to reshape more efficient market processes by identifying the actors who are participating across the value chain. In particular the analysis seeks to identify those who benefit from production activities at stages from upstream to downstream. Thus, the purpose of this research is 1) to map the actors of the silk commodity business, 2) to determine the activities of business actors, 3) to analyze the calculation of financial ratios (marketing margins and profits) of business actors.

2. Method

2.1 Data analysis

Value chain analysis examines activities that produce additional value across the various stages of production (Widarsono, 2009). One of the basic assumptions in value chain analysis is the development of markets that aim to have a positive impact on a broader society by providing income or improved employment opportunity through their participation in developing the market. This means encouraging synergy between farmers and/or producers with post-harvest processing entrepreneurs so that they are unwittingly able to increase the income of the people in the governance of the value chain. With the integration of the silk sector with the community home industry sector, a value chain might form that will provide additional income and absorb labor with the presence of actors in the value chain. However, before beginning to conduct the value chain analysis it is necessary to decide what sub-sectors, products, or commodities should be prioritized for analysis (ACIAR, 2012).

In analyzing the value chain, especially in silk commodities in the Wajo District, the analytical method used is a descriptive method. The steps in using this method are (i) descriptively identifying data and information obtained from the results of in-depth interviews with stakeholders or related parties (farmers, local traders, weavers, and local government) and (ii) mapping the value chain to identify precisely the actors who are participating across the value chain. These research goals answer the overall research question: (1) who are the actors of the silk commodity business, (ii) what are the activities of business actors, and (iii) how do the marketing margins and profits of business actors get distributed?

Mapping the value chain is a major step in helping to answer questions about the linkages in value chain governance. Value chain mapping objectives include obtaining a fundamental review of the value chain as a framework for the overall value chain analysis carried out, identifying barriers and solutions that may be taken at various levels in the value chain, identifying the position of each actor in the value chain, demonstrating interdependence between actors and processes in the value chain, and creating awareness for actors to identify more than the scope of their involvement in the value chain. The steps include the following:

(i) *Identifying actors who play a role*: Identifying actor relationships that influence governance structures in the value chain is essential. The value chain does not only calculate the marketing margins of each actor but also determines the magnitude of the distribution of value added received by each actor along the Silk value chain in the Wajo District, which aims to find out where the biggest beneficiaries are located across the value chain. Each value chain has its own processes and activities, and knowing which activities each actor is involved with, helps to identify potential efforts to increase the value of a product. This was done by interviewing farmers, local traders, weavers and silk clothing stores.

(ii) *Calculating financial ratios (marketing and financial margins):* By interviewing the actors who are participating in the value chain, financial ratios could be determined. In addition, observation is also needed to better understand the activity of the farmer. The marketing margin can be determined by the mathematical formula as follows (Popoko, 2013):

$$\boldsymbol{Mji = Pri - Pfi \text{ atau } Mji = bi + ki}$$

Information:

Mji	= Marketing margin at the first institution level (Rp)
Pri	= Prices at a certain level (Rp)
Pfi	= Prices at the next level (Rp)
bi	= Marketing costs at the first level (Rp)
Ki	= Marketing advantage at the first institution level (Rp)

The value chain does not only calculate the marketing margins of each actor but also determines the magnitude of the distribution of value added received by each actor along the Silk value chain in the Wajo District, which aims to find out where the biggest beneficiaries are located. The profit margin formula is as follows:

$$\boldsymbol{\pi = Pa - Pb - C}$$

Information:

π	= Benefits received by every actor (actor)
Pa	= Selling price in each actor (Rp.)
Pb	= Price of buying products in each Actor (Rp)
C	= Marketing costs for each actor (Rp)

3. Results and Discussion

3.1 Map of silk business actors

The silk business actors have their respective roles in a series of activities in producing the products. Silk value chain actors are presented in Figure 1. The various roles include:

(i) *Farmers*. Farmers are a group of people who use land daily to support their needs. The number of farmer groups in Sabbangparu sub-district include 7 groups. Farmers as silk entrepreneurs act as coarse yarn producers. Farmer activities include maintenance and spinning the yarn into coarse yarn.

(ii) *Local Traders*: Local traders are buyers of coarse silk thread from farmer groups located in the Salojampu market (sub-district market). Local traders act as the producers of fine yarn silk and as fine yarn sellers. Local traders get raw materials from coarse yarn from farmers, then process coarse yarn into fine yarn through the process of bleaching and cooking. After becoming fine thread, local traders then sell it at the Salojampu market.

(iii) *Weavers / craftsmen*: Weavers / craftsmen are people who are engaged in making products from silk thread. Weavers develop products from fine yarn raw materials. There are various products including cloth, clothing, and others. Products are also developed by craftsmen in accordance with the order.

(iv) *Shops for silk clothing*: A silk clothing store is a place to buy and sell silk products. The silk shop offers a variety of woven products from craftsmen.

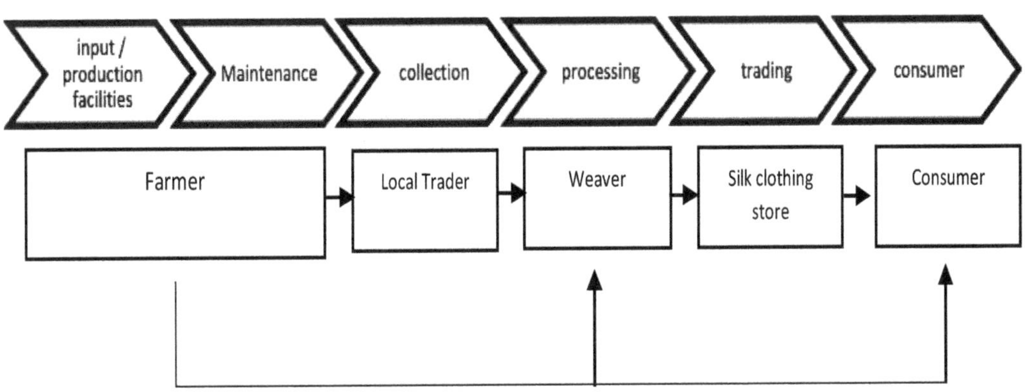

Figure 1. Map of Silk Commodity Business Actors

3.2 Activities of Silk Business Actors

3.2.1 Main Activities

The main activity is a process of physical product creation, marketing and transfer to buyers and after-sales service. The main activity of business actors in silk commodities begin from the farmers with their roles as providers or producers of coarse yarn from silkworms. At the next level of activity, are local traders where the main activity is buying silk thread from farmers and selling it to weavers / craftsmen. Furthermore, the activities of weavers / craftsmen manage the process of turning threads into finished goods such as sarongs and others. Then at the final level of the main activity of silk commodity business actors are the clothing stores with their activities as weavers. The activities of silk entrepreneurs are presented in Table 1.

Table 1. Main Activities of Business Actors

Inbound Logistics	Operation/Process	Outbound Logistics	Marketing, Sales and Services
(1) Farmers			
Silk maintenance equipment	Hatching / incubating silkworm eggs	Weighing together silk thread	In general, farmers do not carry out marketing activities
Acceptance of silkworm eggs	Maintenance of silkworms	Payment of the agreed price	
Taking mulberry leaves	Cocoon	Coarse yarn collection in the head of the farmer group	
	Pressing silk thread	Coarse thread storage	
	Yarn twisting		
	Release		
	Bleaching and coloring of silk thread		
(2) Local Traders			
Weighing silk thread	Silk thread bleaching	Loading into motorized vehicles	Sales of loose threads in the Salojampu market
Payment for the purchase of silk thread		Bringing to the sale location (salojampu market)	
		Weighing at trade locations	
		Payment of the agreed price	
(3) Weaver			
Purchase of yarn in the local market	Silk weaving	Bring to a silk sales shop	There is no marketing information between weavers and buyers. The weaving will be done when there is an order from a silk sales shop
Weaving tools		Use of motorized vehicles	
(4) Silk Clothing Store			
Purchase products from weavers	Sorting out products	Delivering goods through the sales shop	Loose sales at the store
Payment according to price agreement	Pricing on products	Through shipping to outside Wajo district	Product information through the exhibition of regional superior products
		Payment according to price agreement between seller and consumer	

3.2.2 Supporting Activities

Supporting activities help the company as a whole by providing infrastructure or inputs that enable key activities to be carried out on an ongoing basis. Supporting activities of silk commodity business actors begin with farmers planting mulberry plants that feed silkworms. Supporting activities are then carried out by local traders with their activities supporting infrastructure to facilitate the distribution of silk thread from farmers to weavers / craftsmen.

Supporting activities carried out by weavers / craftsmen possess knowledge in weaving, they own vehicles, and establish relationships with silk clothing stores, as well as supporting activities at the final actor level, namely the silk clothing store. This includes involvement at local product exhibitions in the city of Makassar and in other cities. Supporting activities of silk entrepreneurs are presented in Tables 2, 3, 4, 5.

Table 2. Farmer Support Activities

Inbound Logistics	Operation/Process	Outbond Logistics	Marketing, Sales and Services
(1) Infrastructure			
Picking up mulberry leaves on the home page	Maintenance of silkworms is done at home	Coarse silk threads are collected in a plastic bag	In general, farmers do not carry out marketing activities
(2) Human Resources			
Farmer ability to cultivate mulberry is gained from experience	Expertise is obtained from learning outcomes from parents The result of training in silkworm cultivation	In general, there is no special ability or expertise in removing goods	In general, farmers do not understand aspects related to marketing
(3) Technological Development			
Generally, there is no technology used in this activity	The technology used is still quite simple, equipment made of wood	Generally, there is no technology used in this activity	Generally, there is no technology used in this activity
(4) Procurement			
Taking mulberry leaves using household appliances Acceptance of silkworm eggs from the local district agency	Procurement of silkworm maintenance equipment Procurement of coloring material	Weighing equipment purchased by farmers	In general, farmers do not carry out marketing activities

For supporting activities, in terms of infrastructure for logistics, farmers require a field that is large enough to plant mulberry. Furthermore, the distance from the location to the maintenance area must not be too far. In addition, the process or maintenance of silkworms is done at home. For outbound logistics, first, coarse threads are collected in a container before being marketed. Meanwhile, in terms of marketing, farmers do not carry out these activities because farmers only wait for local traders to collect the yarn.

As with human resources for internal logistics, farmers only have the ability to cut mulberry leaves using knives when harvesting mulberry. In addition, in terms of maintenance, silkworm cultivation skills are derived from the experience of more experienced farmers and from trainings that have been conducted. In terms of outbound logistics, farmers do not have special abilities. This is because the expenditure of goods is only in the form of collection of silk thread so that it does not require specialized knowledge. This is similar in terms of marketing that there is no special ability in marketing goods, because farmers only wait for local traders to pick up the goods.

Whereas in terms of technology, farmers only use equipment that is quite simple, in the form of self-maintenance shelves made from wood and bamboo, cocoon making made with bamboo-based material, and spinning equipment purchased from equipment stores. Although farmers conduct supporting activities in terms of procurement, they only need equipment such as knives or machetes, maintenance tools, yarn coloring materials and scales to weigh the yarn.

Table 3. Supporting activities of local traders

Inbound Logistics	Operation/Process	Outbound Logistics	Marketing, Sales and Services
(1) Infrastructure			
The facilities passed or accessibility to the location of the farmers are quite good	Processing of coarse yarn into fine yarn is done at the merchant's house Location of selling yarn in the form of buildings in the Salojampu market	Transport of finished yarn to market locations using motorized vehicles	Infrastructure in terms of marketing products only through storefronts in yarn sales shops in the Salojampu market
(2) Human Resources			
In general, there is no specific qualification in the collection of yarn to farmers because the trust between traders and farmers is intertwined	Knowledge gained by traders in yarn bleaching and coloring is obtained from parents or passed down through generations	In general, there is no specific qualification in issuing merchant products	In general, there are no specific qualifications and activities in marketing merchant products because they are only in the form of loose sales on the market
(3) Technological Development			
Use of motorized vehicles in the transportation and collection of goods for farmers	The technology used is quite simple in the form of pots, stoves, and natural ingredients in the form of leaves used in coloring	In general, local traders do not use technology in removing goods	In general, traders do not use certain technologies in terms of marketing
(4) Procurement			
Provision of merchant operations is carried out individually	Provision of raw materials for bleaching and coloring	Operational provision is carried out individually and based on the trader's capital	There is no special activity in marketing because it is only limited to free sales in the Salojampu market

In terms of infrastructure of local traders, they are supported by facilities or accessibility in the form of a fairly decent road to the location of farmers. In addition, local traders are also supported by rather vibrant sales locations in the form of buildings in the Salojampu market; whereas in marketing goods, local traders have a storefront at the sales location. In supporting activities in terms of human resources, local traders are supported by knowledge gained from parents and neighboring friends in terms of yarn bleaching whereas in terms of marketing, local traders obtain knowledge from the experiences gained.

In addition, in terms of technology development, local traders only use motorized vehicles to support the transportation and retrieval of goods from farmers. For the process of yarn bleaching, local traders use technology that is quite simple in the form of pots, stoves and natural local ingredients. For supporting activities in terms of procurement, local traders conduct operations that are carried out independently based on their capital. Moreover, the procurement of coloring is provided from locally harvested plants.

Table 4. Weaving Support Activities

Inbound Logistics	Operation/Process	Outbound Logistics	Marketing, Sales and Services
(1) Infrastructure			
The facilities passed or accessibility to the location of the farmers are quite good	The weaving location is done at home	The facilities passed or accessibility to the location of the farmers are quite good	In general, there is no supporting activity
(2) Human Resources			
Expertise in driving a motorized vehicle	The ability of human resources to weave using looms that affect product quality	Expertise in driving a motorized vehicle	There are no specific specifications in marketing
(3) Technological Development			
The technology used is in the form of motorized vehicles	The technology used is a loom instead of a machine	The technology used is in the form of motorized vehicles	There is no technology used
(4) Procurement			
Yarn purchases are done individually	Procurement of weaving equipment from own purchases and government assistance	In general, operations are carried out at their own expense	In general, there are no specific activities to support the main activities
Procurement of yarn is also sometimes done			

Weavers or craftsmen are supported by fairly decent road access in the form of paved roads to buy goods from local traders. In addition, weavers are also supported by their own homes as a place to weave threads purchased from traders to be used as products in the form of clothing or sarongs. In terms of human resources for internal logistics, weavers are supported by their expertise in using motorized vehicles so as to facilitate the collection of threads from local traders while the farmers operations are supported by knowledge gained from the experiences of their parents or they are simply self-taught.

For technological development, weavers are supported by ownership of motorized vehicles in the collection of goods from local traders. In addition, the weaving process is supported by simple technologies in the form of unmechanized looms purchased from loom makers and also supported by local government assistance. Moreover, in terms of procurement for logistics, weavers are supported by silk clothing stores. In a sense, the yarn obtained by weavers is not only from purchases from local merchants, but also from silk clothing stores. This is due to the relationships between the silk clothing stores and the weavers. The bond is in the form of cooperation between the two entities. In addition, the technology used is also obtained from government assistance and buying from loom makers.

Table 5. Supporting activities of silk clothing stores

Inbound Logistics	Operation/Process	Outbond Logistics	Marketing, Sales and Services
(1) Infrastructure			
Sales location in the building itself	Product sorting container	Use of storefront as storage area	Loose sales in your own shop Follow the product at the exhibition
(2) Human Resources			
The ability of entrepreneurs to manage products from weavers	The ability of entrepreneurs to sort products and price valuations	The ability of entrepreneurs in communication	The ability of entrepreneurs to include products at the exhibition location
(3) Technological Development			
The technology used does not affect product quality	Technology in the form of labeling tools	The technology used in transportation is not very influential, because it is carried out by courier services	Communication and information technology plays a role in marketing
(4) Procurement			
Retrieval of products in weavers using motorized vehicles	Plastic packaging	Performed by transportation courier services	Exhibition materials or equipment

For supporting activities in terms of infrastructure for internal logistics, silk clothing stores are supported by sales locations in their own buildings being used as a place to store goods that have been purchased from weavers. In addition, the operations carried out by the silk clothing stores is supported by cross-section containers in the form of storefronts to sort out items that have been purchased from the weavers, and also for the logistics. In terms of marketing, silk clothing stores conduct loose sales supported by their own buildings and also include products at local product exhibitions in the city of Makassar and in other cities.

In the human resource category, silk clothing stores have expertise in managing incoming products from weavers to classify products. In addition, silk clothing stores have the ability to judge items for price. In terms of outbound logistics and marketing, owners of silk clothing stores are supported by communication and information seeking capabilities such as information about exhibitions. While in the technology development category, silk clothing stores have the technology in the form of labeling tools to label prices for each product so that it has a clear price. In addition, in terms of logistics, shop owners do not use technology because they only use freight forwarding services or couriers to send their products to consumers. Whereas in terms of marketing, owners of silk clothing stores are supported by communication and information technology in the form of smartphones to communicate with consumers and search for information about exhibitions and silk market prices.

Procurement at the silk clothing store is supported by considerable capital to buy goods from weavers. In addition, the silk clothing store also provides plastic packaging to support the packaging process for its products. Meanwhile, in terms of marketing, shop owners only provide silk materials or equipment to be included in exhibitions in the form of clothes hangers and others.

3.4 Marketing margin

In this study, marketing margins are calculated as the difference between the selling price of silk at the farm level and the selling price of silk at the merchant level. To find out the marketing margin value of each marketing agent, testing is carried out using a cost margin analysis tool, namely by calculating the amount of marketing margins, costs and marketing benefits and the share obtained by farmers. Marketing margin is the price difference between two or more levels of the marketing chain, or between the price at the producer level and the retail price at the consumer level. Marketing margins only represent differences in prices paid by consumers at prices received by producers, but do not indicate the quantity of product marketing.

The analysis of marketing margins would reveal how efficient the pricing is in the local market, and gives an indication of the importance of transaction costs facing traders, farmers and middleman, helping to identify and solve gaps in reducing marketing costs (SIFSIA, 2011). Marketing margins are presented in Table 6.

Table 6. Marketing Margin

Actors	Purchase Price (IDR)	Selling Price (IDR)	Marketing Margin (IDR)
Farmer	200000	400000	200000
Local Trader	400000	500000	100000
Weaver	500000	600000	100000
Silk Clothing Store	600000	900000	300000

From the results obtained in this study, the largest marketing margin is found in the silk clothing store, which is IDR. 300,000, while the smallest marketing margin is found in local traders and weavers, namely IDR. 100,000. The marketing margin of farmers is able to obtain an increased value due to the purchase price of silkworm eggs obtained free of charge from the assistance of the UTPD for Natural Silk. Meanwhile, the local traders and weavers experienced a lower marketing margin due to the higher purchase price at the previous level, and furthermore, production costs are also low enough to affect the selling price of the product.

3.5 Profit Margin

Profit margins, or net margins, are gross margins or marketing margins subtracted by marketing costs. Marketing margins are calculated to determine the actor who gets the most profit. Profit margins are presented in Table 7.

Table 7. Profit Margin

Actors	Purchase Price (IDR)	Selling Price (IDR)	Marketing Cost (IDR)	Profit Margin (IDR)
Farmer	400000	200000	4000	196000
Local Trader	500000	400000	32000	68000
Weaver	600000	500000	14000	86000
Silk Clothing Store	900000	600000	112000	188000

From the results of these calculations it can be seen that the larger profit margin is found among the farmers, at approximately IDR. 196,000, while the smallest profit margin of IDR. 68,000.00 is found among local traders. The profit margin obtained by farmers is greater due to the absence of marketing activities by farmers, whereby farmers just wait for local traders to come to pick up goods that have been produced by the farmer. The profit margins obtained by local traders are smaller due to the low selling prices coupled with labor costs when trading in the local market. The factor causing lesser profitability among local traders is due to the increasing component of the cost of product sold, and other costs (Hafid, 2016). The existence of community forestry (social forestry) programs can help to improve community welfare, especially improving community livelihoods of farmers. However, the government needs to regulate the system of trade in social forestry products, particularly in silk commodity, therefore the efficient trading and fair distribution of benefits can be achieved for each marketing channel.

4. Conclusions

Based on the research that has been conducted, it can be concluded as follows:

(i) From the results of the study, it can be seen that the main actors in the silk business are farmers, local traders, weavers, and silk clothing stores. While the actors who support the main actors of the silk businesses are the local district agency (UTPD Natural Silk) and the Office of Industry.

(ii) The main activities in the silk business are found among farmers, local traders, weavers and silk clothing stores. From these results it can be seen that farmers act as producers of coarse silk thread, local traders act as processors of coarse threads into fine threads, weavers act as processors of fine yarns into clothing products, and silk clothing stores act as shops selling silk clothing products.

The widest marketing margin is found in the silk clothing stores, which is IDR. 300,000, while the smallest marketing margin is found in local traders and weavers in the amount of IDR 100,000. The largest profit margin is found in farmers, which is IDR.196,000, while the smallest profit margin is found in local traders, which is IDR.68,000. Overall, this highlights that there are opportunities to enhance value and improve livelihoods in among forest farmers involved in social forestry schemes. This provides a unique opportunities to ensure that local farmers bargaining positions can continue to be further strengthened to create more vibrant rural livelihoods.

References

ACIAR. (2012). Membuat Rantai Nilai Lebih Berpihak Pada Kaum Miskin. *Australian Centre for International Agricultural Research* (hal. 100). Australian: Australian Goverment.

Hafid, I. (2016). The Effect of Margin Profit and Total Assets Towards Sustainable Growth Rate of the Distributor and trade Company. International Journal of Business Management 10 (4): 423-428.

Harbi, J., Nurrochmat, D. R., & Kusharto, C. M. (2015). *Analisis Kelayakan Usaha Persuteraan Alam, Diversifikasi dan Inovasi Produk di Kabupaten Wajo, Sulawesi Selatan.* Bogor: Institut Pertanian Bogor.

Hakim I., Setiasih, I., Murniatih, Asmanah W., Rachman E., dan Sri R. (2010). Social Forestry. Menuju Restorai Pembangunan Kehutanan Berkelanjutan. Pusat Penelitian dan Pengembangan Perubahan Iklim dan Kebijakan.

Nuraeni, S. (2017). Gaps in the thread: Disease, production, and opportunity in the failing silk industry of South Sulawesi. *Forest and Society, 1*(2), 110-120. doi:http://dx.doi.org/10.24259/fs.v1i2.1861

Popoko, S. (2013). Pengaruh Biaya Pemasaran Terhadap Tingkat Pendapatan Petani Kopra di Kecamatan Tobelo Selatan Kabupaten Halmahera Utara. *Jurnal UNIERA*, 2.

Porter, M. E. (1983). *Competitive Advantage.* New York: Free Press

SIFSIA. (2011). Price and Market-Structure Analysis for Some Agricultural Commodities: Marketing Costs and Margins. Agricultural Economics Directorate of the Ministry of Agriculture, Sudan. http://www.fao.org/fileadmin/user_upload/sifsia/docs/Marketing%20Cost%20Margin%20Final%20May%202011%20(2).pdf

Sihar, N. F. (2008). *Kriya Tekstil Tenunan Sutera Tradisional Gedongan Sulawesi Selatan, Studi Kasus: Kabupaten Wajo, Sulawesi Selatan.* Bandung: Institut Teknologi Bandung.

Widarsono, A. (2005). *Analisis Stratejik Rantai Nilai: Suatu Pendekatan Manajemen Biaya.* Bandung: BHMN

16

Making the case for institutional support on designing agroforestry technology models for rehabilitating critical lands

Andi Nuddin[1*], Muhammad Arsyad[2], Muhammad Ikbal Putera[1], Nuringsih Nuringsih[3], Temesgen Tilahun Teshome[4]

[1] Faculty of Agriculture, Muhammadiyah University of Parepare, Indonesia; andinuddin1956@gmail.com
[2] Faculty of Agriculture, Hasanuddin University, Indonesia
[3] Faculty of Economic, Muhammadiyah University of Parepare, Indonesia
[4] College of Development Studies, Addis Ababa University, Ethiopia
* Corresponding author

Abstract: Land and forest management practices in developing countries have resulted in millions of hectares of degraded lands. This is caused by policy implementation unable to synergize between conservation-ecological goals, and the economic needs of farmer households. This study aims to showcase a model for bringing together economic and ecological interests more closely in line with one another. Furthermore, the study also presents an institutional structure of a program that could help to establish agroforestry-based land rehabilitation policies. The research employed includes a combination of Farming Income Analysis and Interpretative Structural Modeling Analysis. The results show that farming income, when employing agroforestry technology is higher than non-agroforestry approaches. Furthermore, agroforestry technology supports critical land rehabilitation and provides conditions for longer term sustainability. Therefore, a programmatic institutional approach is needed to support these dual goals. We identify that a programmatic approach would include: (1) applying conditions of an agroforestry system as a holistic structured unit, (2) improvement of farmer knowledge and skills, (3) increasing the role and capacity of relevant institutions, (4) improving coordination between sectors, (5) developing conservation agriculture systems, (6) improving bureaucratic support systems, and (7) strengthening control and supervision functions. These elements imply that implementation of agroforestry technology requires institutional support in designing policy for critical land rehabilitation, of which would have significant economic and ecological outcomes on critical lands.

Keywords: Agroforestry; governance institutions; Interpretative Structural Modeling; monoculture; critical lands

---

1. Introduction

Forest management practices, particularly in extractive natural resource economies in developing countries, have created conditions whereby millions of hectares of land remain degraded and underutilized. Therefore, the outcomes result not only in conservation and ecological challenges, but also in the form of limited economic opportunity for local farmers in the long term. Knowledge management is critical for identifying potential innovations for these dual challenges, such as supporting farmer innovation practices to increase production volume and quality (Duangta, Borisutdhi, & Simaraks, 2018) and supported by adequate extension services (Gedgaew, Simaraks, & Rambo, 2018). This paper identifies a case study for engaging on the dual goals of conservation and economic improvement on critical lands in Sulawesi, particularly in degraded watershed areas that have become the source of flash floods in recent years. If continued excessive management and exploitation of these lands persist, so will the expansion of critical land/forest areas that are in poor conditions. Among the critical nutrient-poor lands, several are characterized by steep slopes. These topographic conditions require careful management so as not to deviate from the principles of soil and water conservation. Nevertheless, the reality throughout the upper watershed indicates that there are still significant land exploitation practices taking place that exacerbate conditions of critical lands. Worse still, because the condition of the land is poor in nutrients, farmers that cultivate these lands also experience conditions of low productivity, keeping rural populations poor. Therefore, these overall symptoms indicate that there is an intertwined vicious cycle that is reinforced by land conversion and farmer income interests.

Land management in rough topographical conditions like those in the upper watersheds of Sulawesi therefore require an appropriate agrotechnology that can reduce surface flow and ensure the reduction of erosion potential. One approach that we have considered is a model of agroforestry technology, which is appropriate for contexts of land rehabilitation. Agroforestry technology includes business activities that can integrate agricultural, forestry and ecological development activities. As stated by Rahu, et al. (2015) in the case of a Dayak community in Central Kalimantan, agroforestry not only contributes to the preservation of the balance of forest ecosystems, but also supports socio-economic interests and cultural aspects. The Dayak community is an ethnic group from Central Kalimantan, which have similar inland cultivation approaches as the *Bugis* people of South Sulawesi. They also both practice land cultivation systems by applying agroforestry technology systems. Therefore, based on an economic perspective, the application of agroforestry systems can help to integrate various types of plants in one area to meet economic interests and welfare concerns of farmers (Cordeiro, et al., 2018). However, for various reasons there are many cases that show farmers switching from agroforestry to "conventional" systems, which are indicative of single-crop agriculture. This raises new problems in relation to land rehabilitation and soil conservation, which also disrupt and affect local economic systems.

The failure of land rehabilitation and soil conservation is caused by various factors, including: (1) the weakness and/or inaccuracy of the technology applied, (2) limited funding, and/or (3) institutions that do not play an optimal role (Nuddin et al., 2007). All three interact with, and affect one another, so that even though the technology that is applied may be fairly precise, it can still fail to achieve its objectives. This is foremost due to lack of institutional support. Furthermore, a land and forest management technology will unlikely be adopted by farmers if there is weak coordination by the institutional structures designed to support local farmers in land rehabilitation.

Agroforestry technology adheres to a multi-cropping system with a variety of plants, from seasonal crops (foodstuffs) to long-term perennials. This cropping pattern produces a layered canopy structure that is able to reduce the kinetic energy of raindrops, and reduce erosion. Therefore, the agroforestry system carries two main functions, namely: (1) production/economic functions, and (2) conservation/ecological functions (Hakim, et al., 2018). In terms of production, farming of agroforestry systems is a source of production of food, animal feed, building materials (wood), rubber, and medicines. Whereas from a conservation point of view, the role of agroforestry includes erosion prevention, control of soil fertility, preservation of water systems, and protection of biodiversity (Moreno, et al., 2018).

Therefore, the purpose of this study is twofold. First, we describe farming technologies that are more economically and ecologically advantageous for the given study locations related to critical lands in the upper watersheds of sites in Sulawesi. In particular, we compare agroforestry models with non-agroforestry/monoculture models. Second, through an interpretive structural modeling analysis, we show whether institutional support is needed in designing policies for the application of agroforestry technology to rehabilitate critical land. In order to determine a model farming technology (agroforestry and non-agroforestry/monoculture) that is more economically profitable, an income analysis is carried out. Meanwhile, to measure the importance of institutional support in designing policies for implementing agroforestry technology in rehabilitating critical land, an Interpretative Structural Modeling (ISM) analysis was carried out.

2. Materials and method

2.1 Research site

The research was done in the Bila Watershed in South Sulawesi, Indonesia (Figure 1). The watershed covers three jurisdictional districts, namely Enrekang, Sidenreng Rappang, and Wajo. The location of the research was conducted on dry lands marked as farming land. If it is generally a landscape with a critical gradient, it has an impact on the magnitude of the fluctuations in the river flow between two different seasons (rainy and dry seasons).

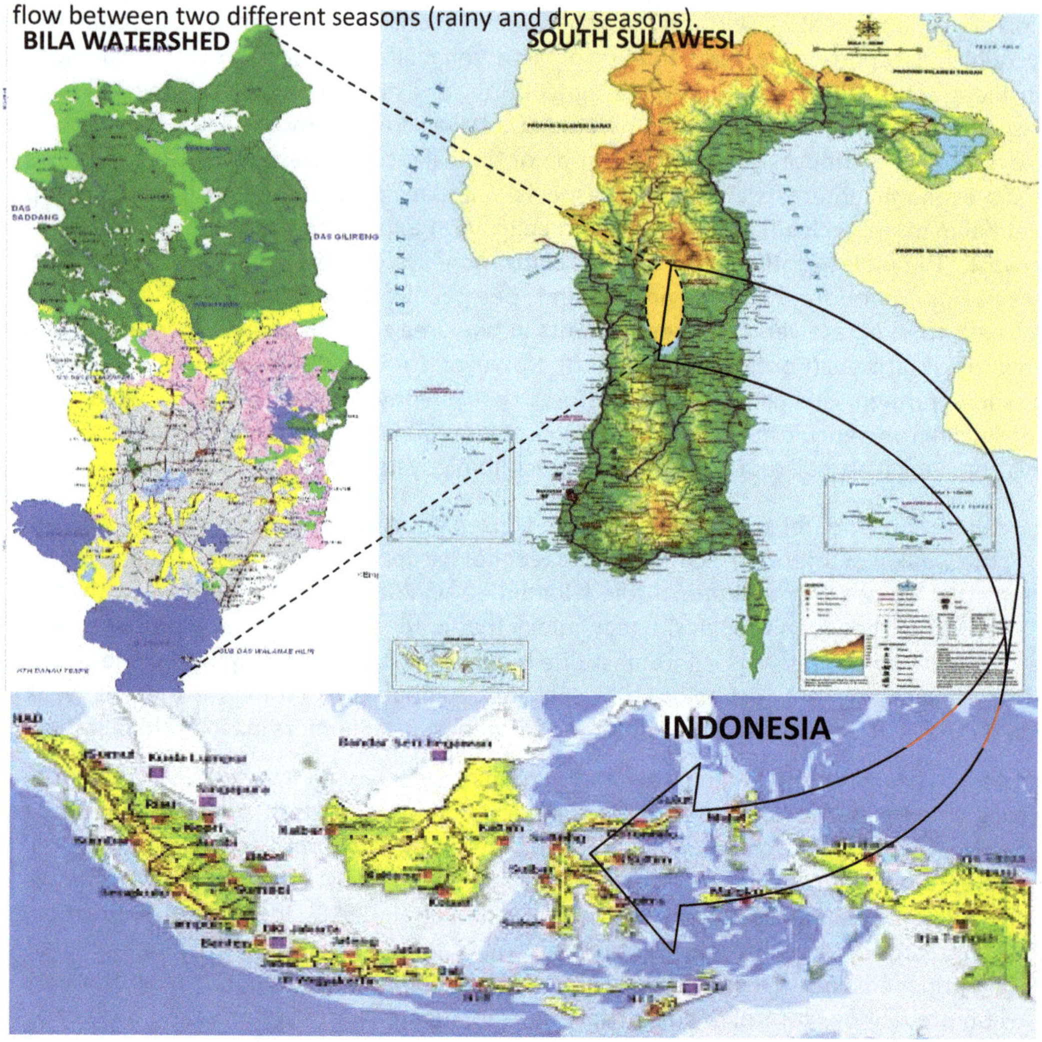

Figure 1. Watershed area as a research location

2.2 Research Design and Analysis

For this study, we selected 24 dryland farmers for each of the two categories. One group consisted of farmers applying agroforestry and the other non-agroforestry/monoculture technology. The location of this study consisted of two groups of very different biophysical characteristics, namely the biophysical structure of agroforestry plots and those with non-agroforestry/monoculture cultivation. This study was conducted in a descriptive manner with survey methods designed using two analytical methods as shown in Table 1.

The first methodological approach includes an income analysis to determine differences in farm income that applies for agroforestry and non-agroforestry systems. Second, the Interpretative Structural Modeling (ISM) method measured how much institutional support is present for determining critical land management policies (Didu, 2001). The ISM method begins with the determination of sub-elements (variables), which were identified through group discussions with a number of experts/practitioners; 13 sub-elements are alternative variables for determining priority land management programs. These alternative variables were chosen for consideration as a variable that can be implemented in critical land program if supported by institutions related to critical land management.

2.2.1 Farm income analysis

Farm income analysis was employed to develop an overall understanding of farmer income differentiation between those who apply agroforestry and non-agroforestry technology models. The following equations were applied:

P = TR - TC
Where:
P = Income
TR = Total agroforestry/non-agroforestry model farm revenue
= Total product x price
TC = Total expenditure = Fixed costs + Variable costs.

Table 1. Study design of the agroforestry technology model

Problem	Object	Analysis	Technological model applied by farmers	Data collection technique
1. Problems with management & utilization of critical land 2. Farmer activities to increase income.	1. Showing the most appropriate model can increase farmer income 2. Estimating and calculating the contribution of farming income	Technology based on local knowledge to improve land conditions, so that farmer income increases	Agroforestry technology model: 1. Model I: Combination of Pine Merkusii + Arabica Coffee + Cloves + Corn 2. Model II: Combination of Pine Merkusii + Arabica Coffee + Soybeans 3. Model III: Combination of Pine Merkusii + Coffee + Cloves + Soybeans Non-agroforestry model 4. Model IV: Shallot monoculture system 5. Model V: Vegetable monoculture system 6. Model V: Corn monoculture system	Observation, field and interview with farmers.
3. The importance of agroforestry technology as a priority program in forest / land rehabilitation institutions	3. Demonstrate the position of agroforestry technology as a priority program in institutional rehabilitation of forests / land	*Interpretative Structural Modeling*	Structural level and synergy of agroforestry technology with other programs in forest / land rehabilitation institutions	Observation, field interviews and interviews with stakeholders

2.2.2 Interpretative structural modeling

To design the structure and position model of the agroforestry technology model in the forest /land rehabilitation institution, Interpretative Structural Modeling (ISM) analysis was carried out. ISM is an assessment conducted by identifying groups to produce structural models of systems designed through images and sentences (Saxena 1999 and Didu., 2001). Fundamental reasons for using this tool is that ISM can prove the importance of institutions related to critical land management. The number of experts / practitioners in the required sample are not large (20 people), which were determined on a purposively based method to develop a level of understanding and/or relevance of their respective fields of work in the rehabilitation of critical lands. Data obtained in the field was obtained as a result of comparisons across the 13 sub-elements, which were analyzed through stages so as to produce a model of the interconnection structures of sub elements. Data obtained in the field were analyzed by developing a structural Self-Interaction Matrix (SSIM) using the results of the questionnaire tabulation; compiling the reachability matrix table by replacing the symbols V, A, X, O with numbers 1 and 0; compiling the Power-Dependent Matrix Driver (DP-D) consisting of four quadrants; and, developing a structural model (level) for each element.

3. Results and discussion

3.1 Technology model of critical land farming

This description compares two farming system models, which are described as distinct farming technologies, namely agroforestry and non-agroforestry/monoculture models. The agroforestry model is a form of intercropping farming that combines forestry and agricultural crops, such as a combination of forestry crops such as pine bran, plantation crops (coffee, cacao, and/or clove). Sometimes in certain seasons, plantation crops are interspersed with food crops (corn, cassava and/or soybeans). Furthermore, farming that has not applied agroforestry/monoculture technology include plots that cultivate corn, shallots, and/or other types of vegetables. The results of the field research show that the realization of the two farming technology models applied by farmers can further be broken down into several farming sub-models as presented in Table 2.

Table 2. The technology/farming system model applied by farmers

Farming Technology Model	Sub Model	Plant type
Agroforestry	I	*Pinus merkusii* + Arabica + Clove + Corn
	II	*Pinus merkusii* + Arabica coffee + Soybeans
	III	*Pinus merkusii* + Clove + Arabica Coffee + Soybeans
Non-agroforestry/ monoculture	IV	Red onion
	V	Cabbage
	VI	Corn

Source: field research

Both of these farming technology models are applied to critical categories of land uses with slope classes ranging from mild, moderate to steep.

3.2 Ecological impacts of applying the agroforestry technology model

The agroforestry technology model implemented by farmers (sub-models I, II, and III)

presented in Table 2 provides many ecological and environmental benefits. This is a result of differences in farming management in the two models as shown in Table 3.

Table 3. Differences in farming management according to the technology model

Technology Model / farming system	Farming Management			
	Land processing	Use of fertilizers (Urea, NPK, Phonska, and ZA)	Use of Pesticides (Insecticides, Fungicides, and Herbicides)	Use of labor
Agroforestry	Not intensive	Only at the beginning of the planting season	Light	Low
Non-agroforestry/ monoculture	Intensive	At the beginning, middle and end of the crop cultivation cycle	Without pesticides, production is not optimal	High

Source: field research

Differences in the application of agroforestry and non-agroforestry/monoculture technology models have a significant impact on ecological systems. First, agroforestry technology enables the creation of multi-strata canopy. Long-term perennials in the form of dominant trees occupy the topmost canopy, while food crops occupy the lowest strata. Such canopy structures have a positive effect on soil protection as part of interactions with rainfall. Likewise, the presence of leaves and twigs form a layer of litter that can increase water infiltration and reduce surface runoff. Increased infiltration and decreased surface flow have a further impact on erosion. Therefore, agroforestry technology strongly supports soil and water conservation efforts.

Second, non-intensive agroforestry land management will maintain the stability of the drainage system and maintain soil permeability so as to ensure the suitability of the soil structure. This factor is extremely important for plant growth. In addition to litterfall that covers the land surface in the agroforestry farming system, it can also help guarantee the balance of the content of organic elements needed by plants. Tongkaemkaew et al. (2018) also found that the application of rubber-based agroforestry technology increased litter volume so that the carbon cycle and nutrients were maintained at optimal levels, which resulted in a decrease of fertilizer used by farmers. In contrast, non-agroforestry/monoculture land management is very intensive. The land is always stretched to its maximum production capacity, and this is especially true at the beginning of the planting season. Such factors can increase surface erosion. Therefore, the level of land washing is very high, which depletes nutrient content.

Third, the agroforestry technology model is also able to minimize the use of fertilizers (urea, NPK, Phonska, and ZA). Table 3 shows the difference in fertilizer requirements, whereby the application of fertilizer to the agroforestry technology is carried out only at the beginning of planting. On the other hand, in the non-agroforestry/monoculture model, fertilizer application is carried out throughout the planting period. As farmers explained "my experience as a farmer using agroforestry technology has never used fertilizers, except for organic fertilizer that I obtained from litter. Therefore, the costs that I use in farming are very small because I do not use chemical fertilizers". Thus, there are differences in land characteristics between the two models but also other factors and dependencies at play. The land conditions in the agroforestry technology model are maintained without putting heavy pressure on the land, as a result of farmers not clearing all forest plants, and also affects the production costs. In contrast, the most intensive non-agroforestry/monoculture land processing system is onion farming. Land management is very intensive in this cultivation system, requiring first the clearing of forest plants resulting in the greater likelihood for erosion, soil washing, and loss of nutrients. This kind of tillage condition has an impact on the high use of fertilizers every planting season.

Fourth, farming in the agroforestry technology model does not require much use of pesticides, if any. In contrast there is a much larger dependency for pesticides in the monoculture farming systems as shown in Table 3. Because the need for pesticides in agroforestry technology models is low, maximizing production is not determined by pesticide use. This is very different from the monoculture farming system, where maximizing production is conducted through the use of pesticides, which sometimes exceed tolerated limits.

Fifth, the agroforestry technology model can also minimize the use of labor when compared to monoculture farming. The low number of workers needed is inseparable from: (1) land management that is not intensive, (2) fertilization carried out only at the beginning of planting, and (3) use of pesticides that are not continuous. This is the opposite of the labor requirements in the monoculture farming system, which includes: (1) intensive land management, (2) fertilization carried out continuously throughout cultivation from the beginning, middle, and end of the planting, and (3) continuous use of pesticides to maximize production.

3.3 Economic Impacts of the Application of the Agroforestry Technology Model

In general, farming financing can be divided into two parts, namely: fixed costs, which are costs incurred by farming that are not directly related to total production; and non-fixed costs, which incur a number of expenses for farming needs which directly affect production. Based on the results of observations and interviews with respondents in the field and subsequent calculation of production costs based on market price standards, we show a significant difference between the use of farming costs for agroforestry and non-agroforestry/monoculture technology models. The details are provided in Table 4, shown as differences in production costs, beginning with the fixed costs (purchase, equipment, depreciation of equipment), up to the variable costs (seeds, fertilizers, pesticides, and labor) as in Table 4.

Table 4. Differences in production costs based on farming technology model

Cost Type	Description	Production cost	
		Agroforestry (IDR/ha/year)	Non-agroforestry/ monoculture (IDR/ha/year)
Fixed cost:	Equipment Purchases	805, 000,00	1,282,000,00
	Land tax	15, 000.00	15,000,00
	Depreciation	22, 500. 00	38,055.00
		842,500.00	1,729,388.00
	Seed		
Variable cost:	Fertilizer:	1,166,667.00	2,708,333.00
	Urea		
	NPK	265,26.00	708,400.00
	Phonska	2,163,151.00	2,397,005.00
	ZA	320,000.00	426,667.00
		208,542.00	137,500.00
	Pesticides:		
	Insecticide	181,133.00	3,325,000.00
	Fungicide	480,000.00	2,913,000.00
	Herbicide	76,125.00	241,500.00
	Labor	455,000.00	1,165,000.00
		4,926,315.00	10,083,361,00
	Total	5,752,027.00	12,803,759.00

Source: calculated by authors

Differences in production costs are indicated as fixed costs (purchases, equipment, depreciation of equipment), to variable costs (seeds, fertilizers, pesticides, and labor). The application of the non-agroforestry/monoculture technology model uses the cost of purchasing smaller equipment (IDR 805,000/ha) compared to non-agroforestry technology (IDR 1,282,000/ha). This difference is caused by the farming of agroforestry technology models that do not use much production equipment.

The use of water machines in monoculture farming (shallots, cauliflower, and corn) is required, while in the application of agroforestry technology models, such equipment is not needed. Plants on agroforestry farms have sufficient water supply, both from rainwater and moisture by microclimate effects. Furthermore, the root system and litter layer on the floor of the land helps to retain rainwater. All of this is possible due to the interaction between forestry plants and agricultural crops (agroforestry). Furthermore, the cost of procurement of seeds also presented an additional difference. Specifically, in the application of the agroforestry technology model expenditures reached IDR 1,166,667/ha, while the non-agroforestry/monoculture technology model was up to IDR 2,708,333/ha. The low cost of procuring seedlings in agroforestry farming is an indication of the need for seeds in agroforestry farming activities, almost all of which can be prepared or can be seeded by each farmer. In addition, there are differences in the cost of pesticides between agroforestry technology models with the total cost of pesticides used reaching IDR 737,258/ha, while non-agroforestry/monoculture farming (shallots, cabbage and corn) amounted to a cost of procurement of costly pesticides at IDR 6,479,500 / ha.

Varied production inputs not only caused differences in total production costs, but also differences in labor costs between farming agroforestry and non-agroforestry/monoculture technology models, as shown in Table 4. Labor costs of non-agroforestry/monoculture farming are almost three times greater (IDR 1,165,000/ha) compared to agroforestry technology farming (IDR 455,000/ha). Thus, the total cost of production of non-agroforestry/monoculture farming is very large (IDR 12,803,759/ha), which is twice the total production cost of agroforestry technology (IDR 5,752,027/ha).

Table 5. Total farmer income based on farming technology model/sub model

Technology Model	Sub model	Combined / cultivated plants	Revenue (IDR/ha/year)
Agroforestry	I	*Pinus merkusii* + Arabica + Clove + Corn	15,364,621
	II	*Pinus merkusii* + Arabica coffee + Soybeans	2,337,138
	III	*Pinus merkusii* + Clove + Arabica coffee + Soybeans	18,535,542
			36,237,301
Non agroforestry	IV	Red onion	15,313,649
	V	Cabbage	4,015,338
	VI	Corn	3,611,354
			22,940,541

Source: calculated by authors

From an economic perspective, land use for farming by applying agroforestry technology models is more profitable compared to non-agroforestry/monoculture farming. This is due to the combination of forestry plants and agricultural crops that can produce various types of commodities, with higher economic value. For example, *Pinus merkusii* forestry plants can produce pine resin, in addition to agricultural crops that produce commodities such as arabica coffee, cloves, corn, and soybeans. Table 5 shows the differences in revenue of farmers according to combinations of cultivated crops (Table 5).

Table 5 clearly shows that the average revenue for farmers that have applied the agroforestry technology model is greater than that of farmers who apply the non-agroforestry/ monoculture technology model. Farmer selection to apply the agroforestry technology model is determined by commodity variations produced by the three sub-models of agroforestry technology, namely four commodities in sub-model I, three commodities in sub-model II, and four commodities in sub-model III. The remaining sub-models in the non-agroforestry/monoculture technology model also consists of three types with each producing one commodity, namely shallots in sub-model IV, cabbage in sub-model V, and corn in sub-model VI.

The amount of production costs affects the level of income obtained by farmers. There are differences in the income of farmers who apply the agroforestry technology model with farmers applying the non-agroforestry technology model as shown in Table 6. Farmers who apply the agroforestry model earn four times more income than farmers who apply the non-agroforestry/monoculture model. Thus, the application of agroforestry technology farming models is more economically profitable, compared to non-agroforestry/monoculture farming. This is similar to the findings of Jalon, et.al. (2018), that show the agroforestry model can increase community welfare to levels higher than non-agroforestry/monoculture systems.

Table 6. Farmer income by technology model applied

Technology model	*Benefit* (IDR/ha/year)	Total Biaya (IDR/ha/year)	*Revenue* (IDR/ha/year)
Agroforestry	36,237,301	5,752,027	30,485,274
Non-agroforestry/monoculture	22,940,541	12,803,759	7,646,822

Source: calculated by authors

3.3.1. The agroforestry technology model in critical land management

Based on the results of the analysis, both ecologically and economically, agroforestry technology is a viable alternative and substitution to be applied to improve the management of critical land. This technology can synergize agricultural production, forestry and improve land quality (Alao, J.S., & R.B. Shuaibu. 2013). However, further analysis is still needed, especially on how the strength of institutional support in implementing the agroforestry technology model as a strategic program in forest and land rehabilitation. Without such pre-requisites a transition to agroforestry to improve critical lands are unlikely.

In the study of watershed critical land management institutions, the results of the Interpretative Structural Modeling (ISM) analysis show that of the 13 sub-elements, nine of them are priority programs, while four are in the independent position, and five are in the linkage position. These are schematically listed in Table 7. One of the programs recommended immediate implementation of the agroforestry technology model. This position (linkage) indicates that agroforestry technology is a program with a large contribution to the successful management of critical land (DP = 0.62), but dependence on other sub-elements is also large (D = 0.69).

The continuity of the agroforestry technology program relies heavily on how government policies, including the support of the Ministry of Agriculture and Rural Development (Hoang. et al., 2017), and other programs, such as support from the Conservation Agriculture System (SPK) development program (DP = 0.77). The application of the SPK in the concept of critical land management is synonymous with the agroforestry technology model. In addition, the agroforestry technology program also depends on field extension program effectiveness (each DP = 0.92) and the program to increase knowledge and skills among administrators (DP = 0.62).

In addition to the programs stated above, the adoption of agroforestry technology also relies heavily on fellow programs in the linkage position. One of them is the enormous dependence on programs to improve farmer knowledge and skills. In this analysis, increasing farmer knowledge and skills (DP = 1.00) is also the same as Hoang et al.'s findings (2017). They also show that it is a key priority program in the management of critical land. The results of this study are relevant to the findings of Duangta et al. (2018) that for the sake of sustainability, the involvement of local communities is very important in supporting agroforestry. The role of local communities is not just physical participation, but must also be motivated through education and training, economic incentives to transition to multi-cropping systems, knowledge about the timing of increased productivity of agroforestry, and potential additional incentives emergent through the development of ecotourism and agro-tourism.

Table 7. Position and weight of priority programs in the management of critical watersheds

Position	Activity / Activity Program	score	
		DP	*D*
Independent (Influence on the program is strong, but its connection with other activities is weak).	1. Development of a conservation farming system	0.77	0.38
	2. Development of a social control function	0.67	0.46
	3. Effective field counseling	0.92	0.54
	4. Increased knowledge and skill apparatus	0,62	0.54
	Average	0.74	0.48
Linkage (Its influence on the program is strong and the level of its association with other activities is also strong).	5. Increased knowledge and farmer's skills	1.00	0.69
	6. Effectiveness of the role of relevant	0.92	0.69
	institutions	0.85	0.69
	7. Providing incentives	0.62	0.69
	8. Application of agroforestry technology	0.85	0.69
	9. Effective coordination between sectors		
	Average	0.82	0.74
Dependent (Its influence on the program is weak but its connection with other activities is strong).	10. Application of conservation technology *)	0.15	0,62
	11. Increased community participation *)	0.38	0.62
	12. Program socialization *)	0.31	0.77
	13. Development of community cultural wisdom *)	0.08	0.62
	Average	0.23	ı.66

Note: *) Programs are not a priority
Source: Analyzed by authors

Furthermore, the application of the agroforestry technology model relies heavily on the effectiveness of the role of the relevant institutions (DP = 0.92) both government and private, as well as coordination effectiveness between sectors (DP = 0.85). In the context of developing critical land management programs, support is needed in the form of incentives, especially for farmers who have and apply skills that support the Conservation Agriculture System program and agroforestry technology models.

Synergy between programs as illustrated in Figure 2, indicates the importance of institutional support in the application of agroforestry technology models, as indicated by the weight of power program driver in the role of the relevant institutions (DP = 0.92) and effective coordination between sectors (DP = 0.85). Both of these programs are very important because conflicts often occur between institutions as a cause of weak coordination functions in institutional agricultural production (Nuddin, et al., 2015). This institutional support for the application of agroforestry technology is urgently needed because agroforestry technology is not an independent conservation action, but rather fits in as an integral part of the strategy for land rehabilitation and soil conservation (Murthy.et al., 2016). Figure 2 shows that the 13 sub-elements as a program in managing critical land are distributed in three quadrants, namely four in independent positions, and five in linkage positions. Independent quadrants show all the programs in them have a large influence on the success of management of critical land (average DP = 0.74), in addition to dependence on other programs, which is small (average D = 0.48). The linkage quadrant shows that the program, in addition to having a large influence on the success of managing critical land, also relies on other programs. Therefore, all programs in this quadrant must be managed more effectively, because their dependence on other programs can create new problems if they are not managed carefully.

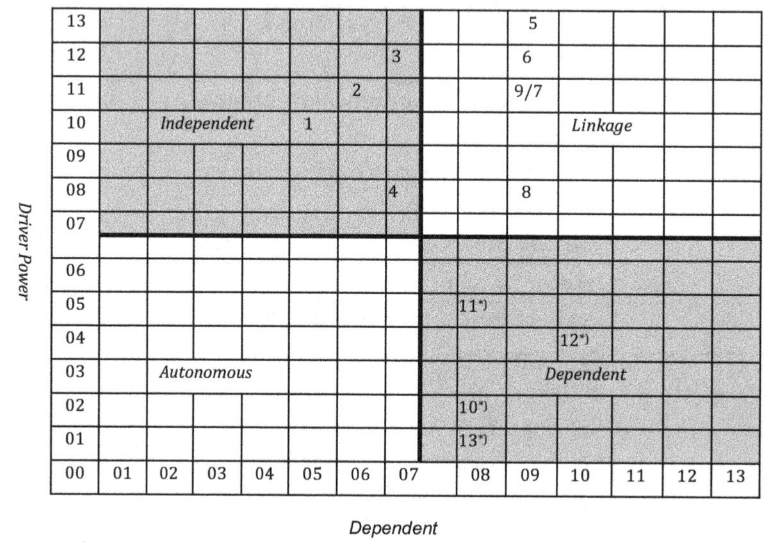

Information:

*) Programs that are not priorities
1 = Development of a conservation farming system
2 = Development of a social control function
3 = Effective field counseling
4 = Increased knowledge and skills of the apparatus
5 = Increased knowledge and skills of farmers
6 = Effectiveness of the role of the relevant institution
7 = Incentives
8 = Application of agroforestry technology
9 = Effective coordination between sectors
10 = Application of conservation technology
11 = Increased community participation
12 = Program socialization
13 = Development of the cultural wisdom of the community

Figure 2. Power-dependence driver position and frequency as an indicator priority program in the management of Watershed critical land.

Furthermore, awareness is dependent, which means that loading programs is not a priority. Its influence on critical land management programs is very weak, in addition to its dependence on other large programs. Therefore, the programs in it can run on their own, if other programs are in an independent quadrant and the linkage runs optimally. Based on the description above (Figure 2), it can be stated that the structural model of program linkages in agroforestry-based critical land management institutions in watersheds (Figure 3).

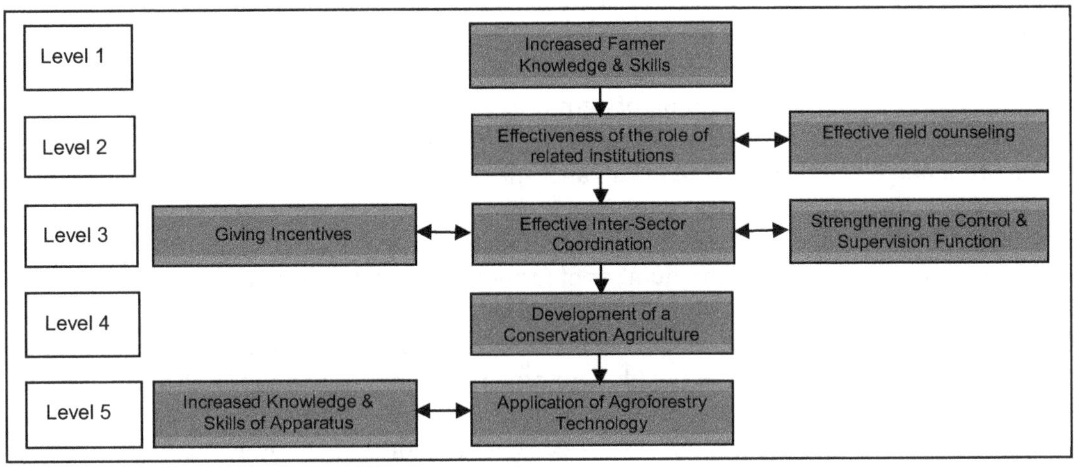

Figure 3. Program linkage structure model in agroforestry-based critical land management institutions.

Figure 3 shows that the increase in farmer knowledge and skills is a critical land management program at level 1. This indicates that the program is a key element of success (DP = 1.00) in managing critical lands. The importance of this program can be identified where farmers are the spearhead in the application of farming technology, including in terms of the application of agroforestry technology. The importance of increasing farmer knowledge has also been revealed by Nuddin et al. (2018), as one of the strategies in developing Robusta coffee through an agroforestry technology model in South Sulawesi.

At level 2 there is a priority program for the effectiveness of the roles of relevant institutions and the effectiveness of the role of extension agents. Related institutions include: Office of Agriculture, Forestry and Plantation, Field Extension Agency (Agriculture, Plantation, Forestry), Farmers' Groups. These institutions must collaborate together under one commitment so that the role of the field instructor is more effective in accordance with their duties and functions. At level 3 there are three strategic programs, namely: (1) effectiveness of coordination between sectors, (2) provision of incentives, and (3) strengthening of the control and supervision functions. Figure 3 shows that effective coordination between sectors is a program that is in a central position in strengthening critical land management institutions. This program will strengthen the functions of control / supervision and provide incentives for forest and critical land management. Control and supervision in forest and land management have been very weak so far, while the provision of incentives including payments for ecosystem services (Villamor et al., 2014) has not been implemented. One problem that has even become a national problem is that forest fires every dry season are never extinguished as a result of weak control and supervision. Control and supervision will be more effective if sectors coordinate better, and in turn, the role of related institutions will become more effective.

At level 4 there is only one development program for conservation agriculture systems (SPK). Through this system, a crop structure is created that guarantees the achievement of economic goals and the goal of conservation of natural resources and conservation of land and water. The conservation farming system is synonymous with agroforestry systems in which there is a combination of farming and agriculture approaches such as fruits, nuts, bananas, and several other products that can be sold in markets (Mercer et al., 2014). Therefore, SPK can help to realize agroforestry patterns as a buffer against protected areas and cultivation areas (Asase & D.A. Tetteh. 2010). In level 5, there is a program to implement agroforestry technology and a program to increase the knowledge and skills of the supporting agencies. Appropriate knowledge and skills will be improved if the role of relevant institutions can be made effective. This has been revealed by Knapp, and Sadorsky, (2010), especially in terms of developing marketing programs and increasing overall knowledge of officials. The implementation of agroforestry will be more efficient, therefore, if it is supported by the institutional structure of forest and land management programs, especially the relevant institutional effectiveness programs, and the effectiveness of coordination between sectors.

Even so, until now institutional support has not been in line with expectations. The facts in the field show that in terms of quantity, the number of farmers who switch from agroforestry systems to monoculture is still high. This is true even though it is widely accepted that an agroforestry system is strategic both from an environmental and economic point of view. This finding is similar to that revealed by Camilli et al. (2018) that although agroforestry is very positive in terms of production and environment, if it has not been supported by a good management structure, it will not come to fruition. Therefore, capacity building for institutions is an effective strategy to promote the development of agroforestry.

4. Conclusion

This paper has provided a strong case for agroforestry in the rehabilitation of degraded lands both economically and ecologically. Some principal findings are: (1) farming income with agroforestry technology is higher than non-agroforestry (or the economic purpose), (2) to enhance benefits and sustainability of agroforestry technology in critical land rehabilitation, an institutional and programmatic approach is required. These include: (1) the application of an agroforestry system that is a holistic and structured unit, (2) improvement of farmer knowledge and skills, (3) enhancement of the role of relevant institutions, (4) effective coordination between sectors, (5) development of conservation agriculture systems (the ecological purpose), (6) improvement in government institutions knowledge and skills, and (7) strengthening control and supervision functions. The presence of leaves and twigs on the surface of the land managed through agroforestry technology, produces litter layers which function to increase water infiltration /minimize surface runoff, and which neutralize the organic material needed by plants. In addition, the need for fertilizer and pesticide use in agroforestry systems tends to be relatively small, or as the case study sites have shown amount to half of the needs for fertilizer and pesticides for non-agroforestry /monoculture farming. Therefore, farming that applies the agroforestry technology model is an environmentally friendly farming technology that is able to neutralize and better maintain ecosystem balance avoiding degradation and encouraging restoration. These facts imply that implementation of agroforestry technology really needs institutional support in designing policy for critical land rehabilitation.

Conflicts of Interest

The authors declare that we have no conflict of interest.

Acknowledgments

The authors would like to thank the Provincial Government of South Sulawesi for their recommendation to do this research. Special thanks also goes to the Watershed Office of South Sulawesi for providing all data that enabled us to conduct the initial mapping of all stakeholders (especially experts) and for their valuable participation during discussions that supported earlier formulations of this paper.

References

Alao, J.S., & R.B. Shuaibu. (2013). Agroforestry Practices and Concepts in Sustainable Land Use Systems in Nigeria. *Journal of Horticulture and Forestry.* *5*(10): 156-159. doi: https://doi.org/10.5897/JHF11.055

Asase, A., & D.A. Tetteh. (2010). The Role of Complex Agroforestry systems in the Conservation of Forest Tree Diversity and Structure in Southeastern Ghana. *Agroforestry Systems. 79*(3): 355-368. doi: https://doi.org/10.1007/s10457-010-9311-1

Camilli, F., A. Pisanelli., G. Seddaiu., A. Franca., V. Bondesan., A. Rosati., G.M. Moreno., A. Pantera., J.E. Hermansen, & P.J. Burgess. (2018). How local Stakeholders Perceive Agroforestry Systems: *an Italian Perspective, 92*(4): 849–862. doi:https://doi.org/10.1007/s10457-017-0127-0

Cordeiro, A.A.C., S.D. Coelho., N.C. Ramos. & J.A.A.M, Neto. (2018). Agroforestry Systems Reduce Invasive Species Richness and Diversity in the Surroundings of Protected Areas. Agroforestry Systems 92(6): 1495-1505. doi: https://doi.org/10.1007/s10457-017-0095-4

Didu, M.S., (2001). Analisis Posisi dan Peran Lembaga serta Kebijakan dalam Proses Pembentukan Lahan Kritis. Jurnal Teknologi Lingkungan. 2(1): 93-105.doi: http://dx.doi.org/10.29122/jtl.v2i1.202

Duangta, K., Borisutdhi, Y., & Simaraks, S. (2018). Knowledge management of pomelo production system in Northeast Thailand: A case study of the pomelo farmer group in Ban Thaen district. *Forest and Society, 2*(2), 162-172. doi:http://dx.doi.org/10.24259/fs.v2i2.5122

Gedgaew, C., Simaraks, S., & Rambo, A. (2018). Factors influencing long term tomato seed production under contract farming. *Forest and Society, 2*(2), 150-161. doi:http://dx.doi.org/10.24259/fs.v2i2.4340

Moreno, G., S, Aviron., S, Berg., J.C. Duran., A. Franca., S. G, de Jalón., T. Hartel., J. Mirck., A. Pantera., J. H. N. Palma., J. A. Paulo., G. A. Re., F. Sanna., C. Thenail., A. Varga., V. Viaud., P. J. Burgess. (2018). Agroforestry Systems of High Nature and Cultural Value in Europe: Provision of Commercial Goods and Other Ecosystem Services. Agroforestry Systems 92(6): 877–891. https://doi.org/10.1007/s10457-017-0126-1

Hakim, L., N.R. Pamungkas., K.P. Wicaksono. & Soemarno. (2018). Conservation Osingnese Traditional Home Garden Agroforestry in Banyuwangi, East Java, Indonesia. AGRIVITA Journal of Agricultural Science. 40(3):506-514. doi: http://doi.org/10.17503/agrivita. v40i3.1605.

Hoang, L.T., J.M. Roshetko., T.P. Huu., T. Pagella. & P.N. Mai. (2017). Agroforestry - The Most Resilient Farming System for the Hilly Northwest of Vietnam. *International Journal of Agriculture System (5)*1: 1-23. doi: http://dx.doi.org/10.20956/ijas.v5i1.1166

Jalon, S.G., A. Graves.,J.H.N. Palma., A. Williams., M. Upson.P.J. Burgess. (2018). Modelling and Valuing the Environmental Impacts of Arable, Forestry and Agroforestry Systems: A Case Study. *Agroforestry Systems 92*(2): 1059–1073.doi:https://doi.org/10.1007/s10457-017-0128-z

Knapp, K.C. & P.A, Sadorsky. (2010). Economics of Agroforestry Production in Irrigated Agriculture. *Journal of Agricultural and Resource Economics 25*(1): 286-306.

Mercer, D.E., G.E. Frey. & F.W. Cubbage. (2014). *Economics of Agroforestry*. In: Kant S, and J.R.R. Alavalapati (eds.). Hanbook of Forest Economics. Earthcan from Routledge. New York. 188-209.

Murthy, I. K., Dutta, S., Varghese. V., Joshi. P.P., & Kumar, P. (2016). Impact of Agroforestry Systems on Ecological and Socio-Economic Systems: A Review. *Glob J Sci Front Res: H Environ Earth Sci, 16*(5), 15-27.

Nuddin, A., Sinukaban, N., Murtilaksono, K., Alikodra, H.S. (2007). *Analysis of institution system on planning and strategy of critical land on Bila Watershed management.* IPB (Bogor Agricultural University).

Nuddin, A., Reskiana., Arman., S. Yusuf., Nuringsih., M. Arsyad., A. Bustan. & A. Laapo. (2018), Designing New Strategy for Enhancing Robusta Coffee Production: An Application of Interpretative Structural Modeling. *International Journal Agriculture System. 6*(2): 156-163.

Nuddin, A., S. Yusuf. & N.M.V. Sulianderi. (2015). An Alternative Model of Cocoa Production Institution: A Solution in Facing Asean Economic Community. *International Journal Agriculture System. 3*(2): 143-156.

Rahu, A. A., K. Hidayat., M. Ariyadi. & L. Hakim. (2013). Ethnoecology of Kaleka: Dayak's agroforestry in Kapuas, Central Kalimantan Indonesia. *Research Journal of Agricultureand Forestry Sciences, 1*(8): 5–12.

Saxena, J.J.P. (1999). Hierarchy and Classification of Program Plan Elements Using Interpretative Structural Modeling. 5(6): 651–670.doi: https://doi.org/10.1007/BF01083616

Tongkaemkaew, U., Sukkul, J., Sumkhan, N., Panklang, P., Brauman, A., & Ismail, R. (2018). Litterfall, litter decomposition, soil macrofauna, and nutrient contents in rubber monoculture and rubber-based agroforestry plantations. *Forest and Society, 2*(2), 138-149. doi:http://dx.doi.org/10.24259/fs.v2i2.4431

Villamor, G. B., Le, Q. B., Djanibekov, U., van Noordwijk, M., & Vlek, P. L. (2014). Biodiversity in rubber agroforests, carbon emissions, and rural livelihoods: An agent-based model of land-use dynamics in lowland Sumatra. *Environmental Modelling & Software, 61,* 151-165.doi: https://doi.org/10.1016/j.envsoft.2014.07.013

Permissions

All chapters in this book were first published in F&S, by Universitas Hasanuddin; hereby published with permission under the Creative Commons Attribution License or equivalent. Every chapter published in this book has been scrutinized by our experts. Their significance has been extensively debated. The topics covered herein carry significant findings which will fuel the growth of the discipline. They may even be implemented as practical applications or may be referred to as a beginning point for another development.

The contributors of this book come from diverse backgrounds, making this book a truly international effort. This book will bring forth new frontiers with its revolutionizing research information and detailed analysis of the nascent developments around the world.

We would like to thank all the contributing authors for lending their expertise to make the book truly unique. They have played a crucial role in the development of this book. Without their invaluable contributions this book wouldn't have been possible. They have made vital efforts to compile up to date information on the varied aspects of this subject to make this book a valuable addition to the collection of many professionals and students.

This book was conceptualized with the vision of imparting up-to-date information and advanced data in this field. To ensure the same, a matchless editorial board was set up. Every individual on the board went through rigorous rounds of assessment to prove their worth. After which they invested a large part of their time researching and compiling the most relevant data for our readers.

The editorial board has been involved in producing this book since its inception. They have spent rigorous hours researching and exploring the diverse topics which have resulted in the successful publishing of this book. They have passed on their knowledge of decades through this book. To expedite this challenging task, the publisher supported the team at every step. A small team of assistant editors was also appointed to further simplify the editing procedure and attain best results for the readers.

Apart from the editorial board, the designing team has also invested a significant amount of their time in understanding the subject and creating the most relevant covers. They scrutinized every image to scout for the most suitable representation of the subject and create an appropriate cover for the book.

The publishing team has been an ardent support to the editorial, designing and production team. Their endless efforts to recruit the best for this project, has resulted in the accomplishment of this book. They are a veteran in the field of academics and their pool of knowledge is as vast as their experience in printing. Their expertise and guidance has proved useful at every step. Their uncompromising quality standards have made this book an exceptional effort. Their encouragement from time to time has been an inspiration for everyone.

The publisher and the editorial board hope that this book will prove to be a valuable piece of knowledge for researchers, students, practitioners and scholars across the globe.

List of Contributors

Messalina Lovenia Salampessy
Nusa Bangsa University Bogor

Indra Gumay Febryano
University of Lampung

Dini Zulfiani
University of Mulawarman

Sukanlaya Choenkwan
Department of Agricultural Extension, Faculty of Agriculture, Khon Kaen Univesity, Khon Kaen, Thailand

Sitti Nuraeni
Forestry Faculty, Universitas Hasanuddin, Tamalanrea, 90245, Indonesia

Bobby Anderson
School of Oriental and African Studies, University of London

Patamawadee Jongruck
Chiang Mai University's School of Political Science and Public Administration

Bobby Anderson
Research Associate, School of Oriental and African Studies, University of London

Uraiwan Tongkaemkaew, J. Sukkul and Narathorn Sumkhan
Faculty of Technology and Community Development, Thaksin University, Phatthalung Campus, 93120 Thailand

Phantip Panklang
Faculty of Natural resources, Songklanakarin University, Hatyai Campus, 90112, Thailand

Alain Brauman
Representative of Institut de recherche pour le dévelloppement (IRD), French

Roslan Ismail
Faculty of Agriculture, Universiti Putra Malaysia, 43400, Malaysia

Kanchana Wiset, Rawee Thaworn, Seth Kane and David Gritten
RECOFTC – The Center for People and Forests, Kasetsart University, Bangkok, 10903, Thailand

Ahmad Dhiaulhaq
RECOFTC – The Center for People and Forests, Kasetsart University, Bangkok, 10903, Thailand
Crawford School of Public Policy, Australian National University. Stanner Building, 37 Lennox Crossing, ANU, Canberra, ACT 2601

Oding Affandi, Anita Zaitunah and Ridwanti Batubara
Forestry Faculty, University of Sumatera Utara

Larry A. Fisher
School of Natural Resources and the Environment, University of Arizona, Tucson, AZ 85721, USA

Yeon-Su Kim
School of Forestry, Northern Arizona University, Flagstaff, AZ 86011, USA

Sitti Latifah
Foretsry Study Program, University of Mataram, Indonesia

Madani Makarom
West Rinjani Forest Management Unit, Lombok, Indonesia

Agung Wibowo
Department of Forestry, Faculty of Agriculture, University of Palangka Raya, Indonesia

Lukas Giessen
Chair of Forest and Nature Conservation Policy, University of Göttingen, Germany
European Forest Institute (EFI), Bonn Office

Moira Moeliono
Center for International Forestry Research (CIFOR), Jalan CIFOR, Situ Gede, Sindang Barang, Bogor (Barat) 16115, Indonesia

Pham Thu Thuy
CIFOR. 172E Quan Thanh Street, Hanoi, 10000, Vietnam

Indah Waty Bong
CIFOR. Jalan CIFOR, Situ Gede, Sindang Barang, Bogor (Barat) 16115, Indonesia

Grace Yee Wong
Stockholm Resilience Centre, Stockholm University, SE-10691 Stockholm

Maria Brockhaus
Department of Forest Sciences, University of Helsinki, Finland

Divine O. Appiah
Department of Geography and Rural Development, KNUST, Kumasi, Ghana

Stephanie E. A. Gbeddy
Department of Geography and Rural Development, KNUST, Kumasi, Ghana

Seth Kane, Ahmad Dhiaulhaq, Lok Mani Sapkota and David Gritten
RECOFTC - The Center for People and Forests, Kasetsart University, Bangkok, 10903, Thailand

Albert Empawi Tindit, Seca Gandaseca, Laurna Nyangon and Ahmad Mustapha Mohamad Pazi
Department of Forest Production, Faculty of Forestry, Universiti Putra Malaysia, 43400 Serdang, Selangor Darul Ehsan Malaysia

Ferdinal Asmin
West Sumatra Forestry Office, Indonesia

Dudung Darusman, Iin Ichwandi and Didik Suharjito
Bogor Agricultural University, Indonesia

Andi Gunawan Pratama, Supratman Supratman and Makkarennu Makkarennu
Forestry Faculty of Hasanuddin University, Jl. Perintis Kemerdekaan Km. 10, Makassar, Indonesia

Andi Nuddin and Muhammad Ikbal Putera
Faculty of Agriculture, Muhammadiyah University of Parepare, Indonesia

Muhammad Arsyad
Faculty of Agriculture, Hasanuddin University, Indonesia

Nuringsih Nuringsih
Faculty of Economic, Muhammadiyah University of Parepare, Indonesia

Temesgen Tilahun Teshome
College of Development Studies, Addis Ababa University, Ethiopia

Index

www.ingramcontent.com/pod-product-compliance
Lightning Source LLC
LaVergne TN
LVHW060547210926
848485LV00005B/1273